BORON PROXIES IN PALEOCEANOGRAPHY AND PALEOCLIMATOLOGY

New Analytical Methods in Earth and Environmental Science Series

Introducing New Analytical Methods in Earth and Environmental Science, a new series providing accessible introductions to important new techniques, lab and field protocols, suggestions for data handling and interpretation, and useful case studies.

New Analytical Methods in Earth and Environmental Science represents an invaluable and trusted source of information for researchers, advanced students, and applied earth scientists wishing to familiarize themselves with emerging techniques in their field.

All titles in this series are available in a variety of full-color, searchable e-book formats.

See below for the full list of books from the series.

BORON PROXIES IN PALEOCEANOGRAPHY AND PALEOCLIMATOLOGY

BÄRBEL HÖNISCH
Lamont-Doherty Earth Observatory and
Department of Earth and Environmental Sciences of Columbia University
Palisades
NY, USA

STEPHEN M. EGGINS
The Australian National University
Canberra
Australian Capital Territory, Australia

LAURA L. HAYNES
Lamont-Doherty Earth Observatory and
Department of Earth and Environmental Sciences of Columbia University
Palisades
NY, USA

KATHERINE A. ALLEN
University of Maine
Orono
ME, USA

KATHERINE D. HOLLAND
The Australian National University
Canberra
Australian Capital Territory, Australia

KATJA LORBACHER
The University of Melbourne
Melbourne, Victoria, Australia

WILEY Blackwell

This edition first published 2019
© 2019 John Wiley & Sons Ltd

The right of Bärbel Hönisch, Stephen M. Eggins, Laura L. Haynes, Katherine A. Allen, Katherine D. Holland, and Katja Lorbacher be identified as the authors of this work has been asserted in accordance with law.

Registered Offices
John Wiley & Sons, Inc., 111 River Street, Hoboken, NJ 07030, USA
John Wiley & Sons Ltd, The Atrium, Southern Gate, Chichester, West Sussex, PO19 8SQ, UK

Editorial Office
9600 Garsington Road, Oxford, OX4 2DQ, UK

For details of our global editorial offices, customer services, and more information about Wiley products visit us at www.wiley.com.

Wiley also publishes its books in a variety of electronic formats and by print-on-demand. Some content that appears in standard print versions of this book may not be available in other formats.

Library of Congress Cataloging-in-Publication Data

Names: Hönisch, Bärbel, 1974– author. | Eggins, Stephen Malcolm, author. | Haynes, Laura Louise, 1991– author. | Allen, Katherine Ann, author. | Holland, Katherine Davina, author. | Lorbacher, Katja, author.
Title: Boron proxies in paleoceanography and paleoclimatology / Bärbel Hönisch, Stephen Malcolm Eggins, Laura Louise Haynes, Katherine Ann Allen, Katherine Davina Holland, Katja Lorbacher.
Description: Hoboken, NJ : John Wiley & Sons, 2018. | Series: New analytical methods in earth and environmental science series | Includes bibliographical references and index. |
Identifiers: LCCN 2018032341 (print) | LCCN 2018057134 (ebook) | ISBN 9781119010647 (Adobe PDF) | ISBN 9781119010623 (ePub) | ISBN 9781119010630 (hardcover)
Subjects: LCSH: Paleoceanography. | Seawater–Carbon dioxide content. | Paleoclimatology. | Boron–Isotopes.
Classification: LCC QE39.5.P25 (ebook) | LCC QE39.5.P25 H66 2018 (print) | DDC 551.46–dc23
LC record available at https://lccn.loc.gov/2018032341

Cover Design: Wiley
Cover Image: Courtesy of Bärbel Hönisch

Set in 10/12.5pt Minion by SPi Global, Pondicherry, India

Printed and bound by CPI Group (UK) Ltd, Croydon, CR0 4YY

10 9 8 7 6 5 4 3 2 1

Contents

Preface

Atmospheric carbon dioxide levels are rising at a pace that may be unprecedented in Earth history, and it is unclear how much this will warm our planet and whether marine life can adapt to acidifying oceans. To understand where our climate and oceans are headed, we seek information from Earth history, for instance through the geochemical signals stored in the fossil remains of marine organisms. The boron isotope proxy for past seawater pH was first introduced two decades ago, but its application has only started to gain momentum over the past decade, when the biological and inorganic constraints on boron incorporation into marine carbonates became better understood, studies confirmed the potential for reconstructing atmospheric pCO_2 beyond ice cores, and new analytical techniques were developed. The related B/Ca proxy is based on the same principles as the boron isotope proxy, but B/Ca was traditionally considered a temperature proxy in corals, and its potential for reconstructing pH had not been explored until about a decade ago. Several complications have been encountered over the years, and selecting the best samples for answering a specific question, sample preparation and analysis are complexities that have restricted analyses to a handful of laboratories worldwide. Premature interpretation of unsuitable sample material has created confusion about whether the proxies are reliable, or which technique should be used. Therefore, as more scientists embark on characterizing past ocean acidity and atmospheric pCO_2, it is important to provide a resource that helps to educate and train geoscientists in the opportunities and complications of this method. We hope that this book will provide a useful guideline for the interested researcher.

Acknowledgments

We would like to thank the many people that helped us write this book – students, friends, and colleagues who provided data, discussed aspects of boron and carbonate chemistry with us, taught us how to read the chemical parlance of the past century, or how to implement up and coming methods of this century, who simply shared their enthusiasm, and encouragement for boron and the product in hand, and provided comments on earlier drafts. Too many to list, but we are deeply indebted to Michael Henehan and Claire Rollion-Bard for reviewing this book and providing many valuable comments and suggestions. We do not agree on all aspects discussed herein, but we all concur that there are many opportunities to strengthen boron proxies even further, and that there are many avenues to reach this goal. In addition, we would like to specifically thank the following friends and colleagues for their support (in alphabetical order): Jelle Bijma, Oscar Branson, Aaron Celestian, Rob DeConto, Jesse Farmer, Mathis Hain, Gil Hanson, Gary and Sidney Hemming, Damien Lemarchand, Chiara Lepore, Tim Lowenstein, Alberto Malinverno, Gianluca Marino, Miguel Martínez-Botí, Vasileios Mavromatis, Helen McGregor, Oded Nir, Mo Raymo, Andy Ridgwell, Dana Royer, Mats Rundgren, Abhijit Sanyal, Gavin Schmidt, Paolo Stocchi, Daniel Storbeck, Taro Takahashi, Joji Uchikawa, Avner Vengosh, Richard Zeebe. And finally, we are grateful to the research stations on Santa Catalina Island and One Tree Island, where this book took its first steps.

About the Companion Website

Don't forget to visit the companion website for this book:

www.wiley.com/go/Hönisch/Boron_Paleoceanography

There you will find valuable material designed to enhance your learning, including:

- Calculation sheets
- MATLAB Scripts

Scan this QR code to visit the companion website

1

Introduction and Concepts

Abstract

This chapter presents a brief introduction to marine carbonate chemistry systematics, including definitions of different pH scales. As a starting point, published estimates of Pleistocene and Cenozoic pCO_2 reconstructions from boron isotopes and B/Ca ratios in planktic foraminifera are shown in the context of ice core records and reconstructions from terrestrial leaf stomata and marine alkenones. These published boron proxy records form the foundation for discussing boron proxy systematics and sensitivity studies presented in the following chapters.

Keywords: atmospheric pCO_2; seawater carbonate chemistry; seawater pH; pH scales

1.1 Why Are we Interested in Reconstructing Marine Carbonate Chemistry?

It has been known since the early studies of Arrhenius (1896) that anthropogenic emissions of carbon dioxide from fossil fuel burning and land use changes will warm our planet, but direct evidence for increasing atmospheric pCO_2 levels emerged only in 1958, when Charles Keeling started continuous measurements at the Mauna Loa Observatory on Hawaii and initially observed an average annual value of 315 parts per million (ppm) (Keeling et al. 1976). These atmospheric pCO_2 levels varied seasonally, steadily increased year upon year and were finally put into perspective when Raynaud and Barnola (1985) presented the first pCO_2 measurements from Antarctic ice cores, which revealed pre-anthropogenic background levels as

Boron Proxies in Paleoceanography and Paleoclimatology, First Edition. Bärbel Hönisch, Stephen M. Eggins, Laura L. Haynes, Katherine A. Allen, Katherine D. Holland, and Katja Lorbacher. © 2019 John Wiley & Sons Ltd. Published 2019 by John Wiley & Sons Ltd. Companion website: www.wiley.com/go/Hönisch/Boron_Paleoceanography

low as 260 ppmv (parts per million by volume). Subsequent studies expanded the ice core records to 800 000 years ago and constrained the pre-industrial range of atmospheric pCO_2 to 172–300 ppmv, together with concomitant Antarctic temperature fluctuations of ~12 °C (Barnola et al. 1987; Jouzel et al. 1987; Lüthi et al. 2008; Petit et al. 1999; Siegenthaler et al. 2005). In 2014 atmospheric pCO_2 hit 400 ppm for the first time (Dlugokencky and Tans 2017) and levels are projected to climb to 420–940 ppm by the end of this century, depending on future emissions (Figure 1.1).

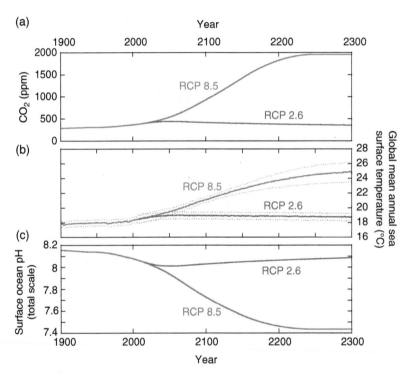

Figure 1.1 Historical observations and future trends in (a) atmospheric pCO_2 (Meinshausen et al. 2011), (b) global sea surface temperature, and (c) surface ocean pH as a function of two CO_2 emission scenarios – Representative Concentration Pathways RCP2.6 and RCP8.5. These scenarios represent the full range of possible future CO_2 emissions used by scientists to predict future climate trends (see also IPCC 2013). Future sea surface temperature trends are globally averaged multi-model estimates extracted from CMIP5 numerical experiments (Moss et al. 2010; Taylor et al. 2011), where temperature uncertainties are based on differences between individual model outputs. pH has been calculated in CO2SYS (Pierrot et al. 2006), with K_1 and K_2 according to Lueker et al. (2000), K_{SO4} according to Dickson (1990), total [B] after Lee et al. (2010). Calculations have been parameterized using pCO_2 and T as displayed in (a) and (b), S = 35, and adopting total alkalinity AT = 2300 µmol kg⁻¹ as the second parameter required for seawater carbonate chemistry calculations. The temperature uncertainties displayed in (b) exert negligible influence on the pH estimates and do not exceed the thickness of the lines displayed in (c). Depending on the actual extent of future emissions, ocean acidification may peak at pH ~ 8.0 (TS, RCP2.6) or fall to pH < 7.3 (TS, RCP8.5). Predicting ocean ecosystem responses to such acidification remains a challenge but may be improved by studying carbonate chemistry perturbations in Earth's geological past.

While discussion of the consequences of rising atmospheric pCO_2 initially concentrated on global warming, research over the past two decades has increasingly addressed the dissolution of CO_2 in seawater and its consequences for marine life. Briefly, as CO_2 dissolves in the ocean, it hydrates and reacts with water to form carbonic acid, which then dissociates into bicarbonate, carbonate, and hydrogen ions according to the following reactions:

$$CO_2 + H_2O \longleftrightarrow H_2CO_3 \longleftrightarrow HCO_3^- + H^+ \longleftrightarrow CO_3^{2-} + 2H^+ \quad (1.1)$$

The more CO_2 dissolves, the more hydrogen ions are created but these ions do not immediately accumulate, as they are buffered by the carbonate ions already in solution:

$$CO_3^{2-} + H^+ \longleftrightarrow HCO_3^- \quad (1.2)$$

However, a small fraction of the resulting bicarbonate ions will dissociate, ultimately increasing the hydrogen ion concentration and therefore the acidity of seawater (i.e. lowering pH):

$$HCO_3^- \longleftrightarrow CO_3^{2-} + H^+ \quad (1.3)$$

A detailed description of marine carbonate chemistry systematics and calculations can be found in Zeebe and Wolf-Gladrow (2001); here we will limit the discussion to a few basic details. The reactions between carbonate and hydrogen ions are governed by dissociation constants (K_1 and K_2), which depend on the thermodynamic seawater properties pressure (p), temperature (T) and salinity (S). The associated shift in carbonate ion speciation is shown in Figure 1.2, which displays the relative concentrations of $[CO_2]$, $[HCO_3^-]$ and $[CO_3^{2-}]$ versus seawater-pH at typical surface (T = 25 °C, S = 35, and p = 1 bar) and deep ocean conditions (T = 4 °C, S = 34.8, p = 401 bar). In contrast, the sum of all dissolved inorganic carbon (DIC) species and their alkalinity (i.e. the sum of their charges) are independent of T, S, and p when expressed in gravimetric units (i.e. $\mu mol\,kg^{-1}$, as opposed to the volumetric $\mu mol\,l^{-1}$). Because these six parameters are interrelated, the entire carbonate system can be determined if two of its components, in addition to temperature, salinity, and pressure, are known. Several programs facilitate computation of the carbonate system; see Further Reading for details.

One aspect that requires specific attention is the choice of pH scale. Four scales have been defined, the National Bureau of Standards (NBS), free hydrogen, seawater, and total scale; they differ in the chemical composition of their respective reference material and pH values determined for identical solutions differ by up to 0.15 units (Table 1.1). While this pH difference may appear small, it has significant consequences for carbon system calculations, as demonstrated in Table 1.1. For example, assuming the same T, S, p, pH, and DIC value to calculate pCO_2, but with pH defined on different scales, calculated pCO_2 differs by >150 µatm. Such large differences are inacceptable for carbon system determinations and must be avoided by all means. Fortuitously, pH scales are interrelated and values can be converted (see Zeebe and Wolf-Gladrow 2001), but this is only possible if studies cite the pH scale used.

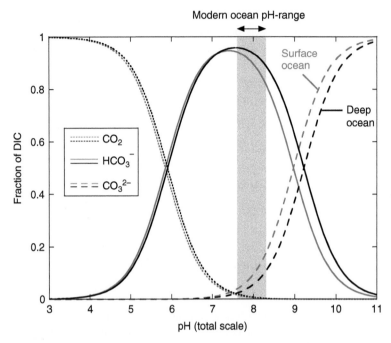

Figure 1.2 This Bjerrum plot displays relative concentrations of dissolved carbon species versus seawater-pH. Relative species concentrations were calculated using the CO2SYS program by Pierrot et al. (2006) with K_1 and K_2 according to Lueker et al. (2000), K_{SO4} according to Dickson (1990), total [B] after Lee et al. (2010), and using T = 25 °C, S = 35, p = 1 bar for the sea surface (red lines) and T = 4 °C, S = 34.8 p = 401 bar for the deep ocean (black lines). The modern range of seawater-pH is indicated by the gray bar.

Table 1.1 Definitions of pH scales, differences in scale-specific pH values in solutions of the same composition, and differences in pCO$_2$ calculated from solutions of similar composition but assuming pH = 8.10 for all four pH-scales.

Scale	Definition	pH value at TA = 2400 µmol kg⁻¹, DIC = 2100 µmol kg⁻¹, T = 25 °C, S = 35, p = 1 bar	pCO$_2$ at pH = 8.10, DIC = 2100 µmol kg⁻¹, T = 25 °C, S = 35, p = 1 bar
NBS (µmol kg⁻¹ H$_2$O)	$pH_{NBS} = -\log a^{H+}$	8.162	513
Free (µmol kg⁻¹ SW)	$pH_F = -\log [H+]_F$	8.133	477
Total (µmol kg⁻¹ SW)	$pH_T = -\log ([H^+]_F + [HSO_4^-])$	8.025	363
Seawater (µmol kg⁻¹ SW)	$pH_{SWS} = -\log ([H^+]_F + [HSO_4^-] + [HF])$	8.016	354

Calculations performed using the CO2SYS program (version 2.1) by Pierrot et al. (2006) with K_1 and K_2 according to Lueker et al. (2000), K_{SO4} according to Dickson (1990) and total [B] after Lee et al. (2010).

Because the boron equilibrium constants are reported for the total scale (Dickson 1990; Millero 1995), this book will present all data on the total scale.

Modern surface ocean pH is ~8.1 (total scale, TS), which is already ~0.1 pH units lower compared to the preindustrial, when atmospheric pCO_2 was ~120 ppm lower compared to today (Figure 1.1). Surface ocean pH continues to drop by ~0.002 units annually (Takahashi et al. 2014) and anthropogenic CO_2 slowly enters the intermediate and deep ocean via thermohaline circulation (Feely et al. 2004; Khatiwala et al. 2012; Sabine et al. 2004). Although the incremental accumulation of hydrogen ions resulting from dissolution of CO_2 will not actually turn seawater acidic (i.e. pH will not drop below 7), the trend towards decreasing pH has been termed *Ocean Acidification* (Caldeira and Wickett 2003). Depending on the source and extent of future anthropogenic carbon emissions, surface seawater pH is projected to decrease by an additional 0.1–0.7 pH units by the year 2200 (Figure 1.1). Laboratory experiments with various marine organisms and observations of naturally acidified ecosystems have highlighted the vulnerability of marine life to ocean acidification, but also the diversity of the biotic response (for a review see Doney et al. 2009).

Despite a wealth of experimental and observational work, projections of future ecosystem changes in the warming and acidifying ocean suffer from limited diversity and typically short duration of laboratory experiments, a shortcoming that can be compensated by the study of the geological record (e.g. Hönisch et al. 2012). Similarly, improving estimates of future warming requires better estimates of climate sensitivity, and the geological record offers a multitude of opportunities to study the interplay of CO_2 and temperature (Foster et al. 2017, PALEOSENS-project-members 2012). While polar ice provides the best archive for past CO_2 concentrations, continuous ice core records are currently limited to the past 800 000 years (Lüthi et al. 2008). Horizontal drilling into Antarctic blue ice has recovered isolated sections ~1 million years old (Higgins et al. 2015) and ~2.7 million years old (Yan et al. 2017), but the prospect of a continuous vertical record may not exceed 1.5 million years (Fischer et al. 2013). The study of geological archives therefore requires the use of proxies, i.e. measurable stand-ins for environmental parameters that can no longer be measured directly. CO_2- proxies have been developed for the terrestrial and the marine realm, and include the stomata density of fossil leaves, the carbon isotopic composition ($\delta^{13}C$) of marine biomarkers, and the boron isotopic composition and B/Ca ratios recorded in foraminifer shells, among others (e.g. Beerling and Royer 2011; Foster et al. 2017). Figures 1.3 and 1.4 display a selection of reconstructions over the past 800 000 and 65 million years, respectively. The functioning of the systematics, advantages, and shortcomings of the proxies displayed in these figures have been reviewed in Royer et al. (2001a) and Allen and Hönisch (2012). Because this book focuses on boron proxies, we will only mention the systematics of other proxies briefly.

Of the proxies shown, only the stomata (breathing cells) of vascular land plants are directly related to atmospheric pCO_2 – the stomatal index

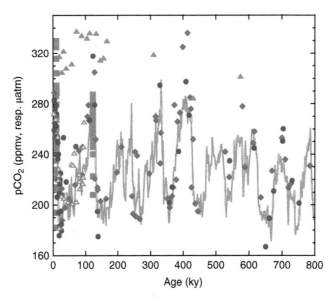

Figure 1.3 800 000 year record of atmospheric pCO_2 extracted from Antarctic ice cores (Lüthi et al. 2008; Petit et al. 1999; Siegenthaler et al. 2005) and reconstructed from planktic foraminiferal boron isotopes (dark blue circles, Henehan et al. 2013; Hönisch and Hemming 2005, Hönisch et al. 2009) and B/Ca ratios (light blue diamonds, Tripati et al. 2009; Yu et al. 2007), alkenones (orange triangles, Jasper and Hayes 1990; Zhang et al. 2013) and leaf stomata (green squares, Rundgren and Bennike 2002, Steinthorsdottir et al. 2013). Note that the calibration of Jasper and Hayes (1990) scaled the alkenone amplitude to the ice core pCO_2 amplitude. This is therefore not a completely independent reconstruction, but the shape of the proxy reconstruction matches the ice core data well. pCO_2 uncertainties are ~20 µatm for boron isotope estimates, ~30 µatm for B/Ca estimates, ~60 µatm for alkenone estimates and ~20 ppmv for leaf stomata estimates.

decreases as atmospheric pCO_2 increases, such that water loss via evaporation can be minimized when CO_2 is abundant (e.g. Royer et al. 2001b), but see also Franks et al. (2014) for additional environmental and stomatal anatomy controls on leaf gas exchange. Alkenone pCO_2 estimates are based on the carbon isotope fractionation that occurs during photosynthesis performed by marine haptophytes, where $\delta^{13}C_{alkenone}$ is inversely related to aqueous $[CO_2]$, but also depends on algal growth rate (i.e. nutrient supply) and cell geometry (e.g. Henderiks and Pagani 2007). As such, alkenone reconstructions require a few auxiliary data, including estimates of $\delta^{13}C$ of DIC, temperature, nutrients, and cell geometry (e.g. Zhang et al. 2013), all of which can be estimated from respective marine proxy records. Boron isotopes and B/Ca ratios in planktic foraminifer shells are not directly related to pCO_2 but rather to seawater acidity, and thus require a second parameter of the carbonate system to estimate pCO_2 via pH. The second parameter is often given by an assumption of total alkalinity, which changes little on Pleistocene time scales, but is more uncertain on multi-million year time scales (Caves et al. 2016; Ridgwell 2005; Tyrrell and Zeebe 2004). Boron proxy-to-pCO_2 translations also require estimates of temperature and salinity, in addition to knowledge of the boron isotopic composition ($\delta^{11}B_{sw}$),

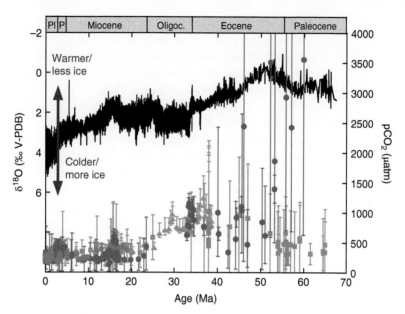

Figure 1.4 Proxy estimates of Cenozoic climate change (as inferred from benthic foraminiferal δ¹⁸O, Zachos et al. 2008) and atmospheric pCO₂ reconstructed from planktic foraminiferal boron isotopes (dark blue circles, Anagnostou et al. 2016; Badger et al. 2013; Bartoli et al. 2011; Foster et al. 2012; Greenop et al. 2014; Hönisch et al. 2009; Martínez-Botí et al. 2015; Pearson et al. 2009; Seki et al. 2010) and B/Ca ratios (light blue diamonds, Tripati et al. 2009), alkenones (orange triangles, Zhang et al. 2013) and leaf stomata (green squares, van der Burgh et al. 1993; Kürschner et al. 1996, 2001; McElwain 1998; Royer et al. 2001a; Beerling et al. 2002; Greenwood et al. 2003; Royer 2003; Kürschner et al. 2008; Retallack 2009; Smith et al. 2010; Doria et al. 2011). Geological epochs are indicated by the gray bar. Whereas Pleistocene (Pl) and Pliocene (P) pCO₂ estimates compare fairly well, Miocene records diverge by ~150 µatm and Paleocene/Eocene records by ~2000 µatm. Similarly, pCO₂ uncertainties vary by proxy and increase further back in time. Such large uncertainties compromise estimates of past climate sensitivity.

boron and calcium concentrations of seawater. The details of these parameters and translations will be explained later in this book, for now it suffices to say that pCO₂ reconstructions from proxies are more complicated than the extraction of actual CO₂ from air trapped in polar ice. However, despite the complexity of the respective translation process, validation of proxy estimates relative to ice core pCO₂ (Figure 1.3) shows convincing results. Going further back in time, pCO₂ estimates from different proxies show relatively consistent values until ~40 Ma, but diverge greatly during the early Eocene and Paleocene, with δ¹¹B estimates showing the highest pCO₂ values (Figure 1.4).

While these proxy estimates have greatly enhanced our understanding of Earth's climate system, the uncertainties associated with all of these pCO₂ estimates preclude accurate estimates of climate sensitivity (PALEOSENS-project-members 2012). Improvements have been made over the past few years but are still needed for all proxies. In particular, some of the boron proxy records shown in Figures 1.3 and 1.4 are no longer considered scientifically sound, and we will discuss individual boron proxy records in detail. However, this book will not only focus on atmospheric pCO_2. Estimates of seawater pH in coral reefs and carbon storage in the deep ocean are all aspects that contribute to our understanding of the marine carbon system, climate, and ecosystem dynamics. These properties can be reconstructed with boron proxy estimates in marine carbonates as different as shallow and deep-water coral skeletons, planktic, and benthic foraminifer shells, brachiopod shells, and inorganic precipitates. In addition to pCO_2 estimates beyond ice cores, boron proxies thereby provide a plethora of opportunities to decipher the causes of past carbon cycle variations and their effect on marine ecosystems.

Acknowledgments

We acknowledge the World Climate Research Programme's Working Group on Coupled Modeling, which is responsible for CMIP, and we thank the climate modeling groups (listed at http://cmip-pcmdi.llnl.gov/cmip5/docs/CMIP5_modeling_groups.pdf) for producing and making available their model output. For CMIP, the U.S. Department of Energy's Program for Climate Model Diagnosis and Intercomparison provides coordinating support and led development of software infrastructure in partnership with the Global Organization for Earth System Science Portals.

References

Allen, K.A. and Hönisch, B. (2012). The planktic foraminiferal B/Ca proxy for seawater carbonate chemistry: a critical evaluation. *Earth and Planetary Science Letters* 345–348: 203–211.

Anagnostou, E., John, E.H., Edgar, K.M. et al. (2016). Changing atmospheric CO_2 concentration was the primary driver of early Cenozoic climate. *Nature* 533 (7603): 380–384.

Arrhenius, S. (1896). On the influence of carbonic acid in the air upon the temperature on the ground. *Philosophical Magazine* 41: 237–276.

Badger, M.P.S., Lear, C.H., Pancost, R.D. et al. (2013). CO_2 drawdown following the middle Miocene expansion of the Antarctic Ice Sheet. *Paleoceanography* 28 (1): 42–53.

Barnola, J.M., Raynaud, D., Korotkevich, Y.S., and Lorius, C. (1987). Vostok ice core provides 160,000-year record of atmospheric CO_2. *Nature* 329 (6138): 408–414.

Bartoli, G., Hönisch, B., and Zeebe, R.E. (2011). Atmospheric CO_2 decline during the Pliocene intensification of Northern Hemisphere glaciations. *Paleoceanography* 26 (4): PA4213.

Beerling, D.J. and Royer, D.L. (2011). Convergent Cenozoic CO_2 history. *Nature Geoscience* 4 (7): 418–420.

Beerling, D.J., Lomax, B.H., Royer, D.L. et al. (2002). An atmospheric pCO($_2$) reconstruction across the cretaceous-tertiary boundary from leaf megafossils. *Proceedings of the National Academy of Sciences of the United States of America* 99 (12): 7836–7840.

van der Burgh, J., Visscher, H., Dilcher, D.L., and Kürschner, W.M. (1993). Paleoatmospheric signatures in Neogene fossil leaves. *Science* 260: 1788–1790.

Caldeira, K. and Wicket, M.E. (2003). Anthropogenic carbon and ocean pH. *Nature* 425: 365.

Caves, J.K., Jost, A.B., Lau, K.V., and Maher, K. (2016). Cenozoic carbon cycle imbalances and a variable weathering feedback. *Earth and Planetary Science Letters* 450: 152–163.

Dickson, A.G. (1990). Standard potential of the reaction: $AgCl(s)+1/2H_2(g)=Ag(s)+HCl(aq)$, and the standard acidity constant of the ion HSO_4^- in synthetic seawater from 273.15 to 318.15K. *The Journal of Chemical Thermodynamics* 22: 113–127.

Dlugokencky, E. and Tans, P. (2017) *Mauna Loa CO2 annual mean data*, NOAA/ESRL (www.esrl.noaa.gov/gmd/ccgg/trends/)

Doney, S.C., Fabry, V.J., Feely, R.A., and Kleypas, J.A. (2009). Ocean acidification: the other CO_2 problem. *Annual Review of Marine Science* 1: 169–192.

Doria, G., Royer, D.L., Wolfe, A.P. et al. (2011). Declining atmospheric CO_2 during the late middle Eocene climate transition. *American Journal of Science* 311: 63–75.

Feely, R.A., Sabine, C.L., Lee, K. et al. (2004). Impact of anthropogenic CO_2 on the $CaCO_3$ system in the oceans. *Science* 305: 362–366.

Fischer, H., Severinghaus, J., Brook, E. et al. (2013). Where to find 1.5 million yr old ice for the IPICS "oldest-ice" ice core. *Climate of the Past* 9 (6): 2489–2505.

Foster, G.L., Lear, C.H., and Rae, J.W.B. (2012). The evolution of pCO$_2$, ice volume and climate during the middle Miocene. *Earth and Planetary Science Letters* 341–344: 243–254.

Foster, G.L., Royer, D.L., and Lunt, D.J. (2017). Future climate forcing potentially without precedent in the last 420 million years. *Nature Communications* 8: 14845.

Franks, P.J., Royer, D.L., Beerling, D.J. et al. (2014). New constraints on atmospheric CO_2 concentration for the Phanerozoic. *Geophysical Research Letters* 41: 4685–4694.

Gattuso, J.-P., Epitalon, J.-M., Lavigne, H. et al. (2017) seacarb: Seawater Carbonate Chemistry, CRAN-R.project.org, https://cran.r-project.org/package=seacarb.

Greenop, R., Foster, G.L., Wilson, P.A., and Lear, C.H. (2014). Middle Miocene climate instability associated with high-amplitude CO_2 variability. *Paleoceanography* 29 (9): 845–853.

Greenwood, D.R., Scarr, M.J., and Christophel, D.C. (2003). Leaf stomatal frequency in the Australian tropical rainforest tree *Neolitsea dealbata* (Lauraceae) as a proxy measure of atmospheric *p*CO$_2$. *Palaeogeography, Palaeoclimatology, Palaeoecology* 196: 375–393.

Henderiks, J. and Pagani, M. (2007). Refining ancient carbon dioxide estimates: significance of coccolithophore cell size for alkenone-based pCO$_2$ records. *Paleoceanography* 22.

Henehan, M.J., Rae, J.W.B., Foster, G.L. et al. (2013). Calibration of the boron isotope proxy in the planktonic foraminifera *Globigerinoides ruber* for use in palaeo-CO$_2$ reconstruction. *Earth and Planetary Science Letters* 364: 111–122.

van Heuven, S., Pierrot, D., Lewis, E., and Wallace, D.W.R. (2009) MATLAB Program Developed for CO_2 System Calculations, ORNL/CDIAC-105b.

Higgins, J.A., Kurbatov, A.V., Spaulding, N.E. et al. (2015). Atmospheric composition 1 million years ago from blue ice in the Allan Hills, Antarctica. *Proceedings of the National Academy of Sciences* 112 (22): 6887–6891.

Hönisch, B. and Hemming, N.G. (2005). Surface ocean pH response to variations in pCO_2 through two full glacial cycles. *Earth and Planetary Science Letters* 236 (1–2): 305–314.

Hönisch, B., Hemming, N.G., Archer, D. et al. (2009). Atmospheric carbon dioxide concentration across the mid-Pleistocene transition. *Science* 324 (5934): 1551–1554.

Hönisch, B., Ridgwell, A., Schmidt, D.N. et al. (2012). The geological record of ocean acidification. *Science* 335 (6072): 1058–1063.

IPCC (2013) Climate Change 2013: The Physical Science Basis. Contribution of Working Group I to the Fifth Assessment Report of the Intergovernmental Panel on Climate Change Cambridge University Press, Cambridge. http://www.ipcc.ch/report/ar5/

Jasper, J.P. and Hayes, J.M. (1990). A carbon isotope record of CO_2 levels during the late quaternary. *Nature* 347 (6292): 462–464.

Jouzel, J., Lorius, C., Petit, J.R. et al. (1987). Vostok ice core: a continuous isotope temperature record over the last climatic cycle (160,000 years). *Nature* 329 (6138): 403–408.

Keeling, C.D., Bacastow, R.B., Bainbridge, A.E. et al. (1976). Atmospheric carbon dioxide variations at Mauna Loa Observatory, Hawaii. *Tellus* 28: 538–551.

Khatiwala, S., Primeau, F., and Holzer, M. (2012). Ventilation of the deep ocean constrained with tracer observations and implications for radiocarbon estimates of ideal mean age. *Earth and Planetary Science Letters* 325–326: 116–125.

Kürschner, W.M., Kvacek, Z., and Dilcher, D.L. (2008). The impact of Miocene atmospheric carbon dioxide fluctuations on climate and the evolution of terrestrial ecosystems. *Proceedings of the National Academy of Sciences of the United States of America* 105: 449–453.

Kürschner, W.M., van der Burgh, J., Visscher, H., and Dilcher, D.L. (1996). Oak leaves as biosensors of late Neogene and early Pleistocene paleoatmospheric CO_2 concentrations. *Marine Micropaleontology* 27: 299–312.

Kürschner, W.M., Wagner, F., Dilcher, D.L., and Visscher, H. (2001). Using fossil leaves for the reconstruction of Cenozoic paleoatmospheric CO_2 concentrations. In: *Geological Perspectives of Global Climate Change* (ed. L.C. Gerhard, W.E. Harrison and B.M. Hanson), 169–189. Tulsa: The American Association of Petroleum Geologists.

Lee, K., Kim, T.-W., Byrne, R.H. et al. (2010). The universal ratio of boron to chlorinity for the North Pacific and North Atlantic oceans. *Geochimica et Cosmochimica Acta* 74 (6): 1801–1811.

Lueker, T.J., Dickson, A.G., and Keeling, C.D. (2000). Ocean pCO_2 calculated from dissolved inorganic carbon, alkalinity, and equations for K1 and K2: validation based on laboratory measurements of CO_2 in gas and seawater at equilibrium. *Marine Chemistry* 70 (1–3): 105–119.

Lüthi, D., Le Floch, M., Bereiter, B. et al. (2008). High-resolution carbon dioxide concentration record 650,000–800,000 years before present. *Nature* 453 (7193): 379–382.

Martínez-Botí, M.A., Foster, G.L., Chalk, T.B. et al. (2015). Plio-Pleistocene climate sensitivity evaluated using high-resolution CO_2 records. *Nature* 518 (7537): 49–54.

McElwain, J.C. (1998). Do fossil plants signal palaeoatmospheric CO_2 concentration in the geological past? *Philosophical Transactions of the Royal Society London B* 353: 83–96.

Meinshausen, M., Smith, S.J., Calvin, K. et al. (2011). The RCP greenhouse gas concentrations and their extensions from 1765 to 2300. *Climatic Change* 109 (1–2): 213–241.

Millero, F.J. (1995). Thermodynamics of the carbon dioxide system in the oceans. *Geochimica et Cosmochimica Acta* 59 (4): 661–667.

Moss, R.H., Edmonds, J.A., Hibbard, K.A. et al. (2010). The next generation of scenarios for climate change research and assessment. *Nature* 463 (7282): 747–756.

PALEOSENS-project-members (2012). Making sense of paleoclimate sensitivity. *Nature* 419: 683–691.

Pearson, P.N., Foster, G.L., and Wade, B.S. (2009). Atmospheric carbon dioxide through the Eocene-Oligocene climate transition. *Nature* 461 (7267): 1110–U204.

Petit, J.R., Jouzel, J., Raynaud, D. et al. (1999). Climate and atmospheric history of the past 420,000 years from the Vostok ice core, Antarctica. *Nature* 399: 429–436.

Pierrot, D., Lewis, E., and Wallace, D.W.R. (2006) MS Excel Program Developed for CO_2 System Calculations, ORNL/CDIAC-105a.

Raynaud, D. and Barnola, J.M. (1985). An Antarctic ice core reveals atmospheric CO_2 variations over the past few centuries. *Nature* 315 (6017): 309–311.

Retallack, G.J. (2009). Greenhouse crises of the past 300 million years. *Geological Society of America Bulletin* 121: 1441–1455.

Ridgwell, A. (2005). A mid Mesozoic revolution in the regulation of ocean chemistry. *Marine Geology* 217 (3–4): 339–357.

Royer, D.L. (2003). Estimating latest cretaceous and tertiary atmospheric CO_2 concentration from stomatal indices. In: *Causes and Consequences of Globally Warm Climates in the Early Paleogene*, Geological Society of America Special Paper 369 (ed. S.L. Wing, P.D. Gingerich, B. Schmitz and E. Thomas), 79–93. Boulder: Geological Society of America.

Royer, D.L., Berner, R.A., and Beerling, D.J. (2001a). Phanerozoic atmospheric CO_2 change: evaluating geochemical and paleobiological approaches. *Earth-Science Reviews* 54: 349–392.

Royer, D.L., Wing, S.L., Beerling, D.J. et al. (2001b). Paleobotanical evidence for near present-day levels of atmospheric CO_2 during part of the tertiary. *Science* 292 (5525): 2310–2313.

Rundgren, M. and Bennike, O. (2002). Century-scale changes of atmospheric CO_2 during the last interglacial. *Geology* 30 (2): 187–189.

Sabine, C.L., Feely, R.A., Gruber, N. et al. (2004). The oceanic sink for anthropogenic CO_2. *Science* 305: 367–371.

Seki, O., Foster, G.L., Schmidt, D.N. et al. (2010). Alkenone and boron-based Pliocene pCO_2 records. *Earth and Planetary Science Letters* 292 (1–2): 201–211.

Siegenthaler, U., Stocker, T.F., Monnin, E. et al. (2005). Stable carbon cycle-climate relationship during the Late Pleistocene. *Science* 310 (5752): 1313–1317.

Smith, R.Y., Greenwood, D.R., and Basinger, J.F. (2010). Estimating paleoatmospheric pCO_2 during the early Eocene climatic optimum from stomatal frequency of *Ginkgo*, Okanagan Highlands, British Columbia, Canada. *Palaeogeography, Palaeoclimatology, Palaeoecology* 293: 120–131.

Steinthorsdottir, M., Wohlfarth, B., Kylander, M.E. et al. (2013). Stomatal proxy record of CO2 concentrations from the last termination suggests an important role for CO2 at climate change transitions. *Quaternary Science Reviews* 68: 43–58.

Takahashi, T., Sutherland, S.C., Chipman, D.W. et al. (2014). Climatological distributions of pH, pCO_2, total CO_2, alkalinity, and $CaCO_3$ saturation in the global surface ocean, and temporal changes at selected locations. *Marine Chemistry* 164: 95–125.

Taylor, K.E., Stouffer, R.J., and Meehl, G.A. (2011). An overview of CMIP5 and the experiment design. *Bulletin of the American Meteorological Society* 93 (4): 485–498.

Tripati, A.K., Roberts, C.D., and Eagle, R.A. (2009). Coupling of CO_2 and ice sheet stability over major climate transitions of the last 20 million years. *Science* doi: 10.1126/science.1178296.

Tyrrell, T. and Zeebe, R.E. (2004). History of carbonate ion concentration over the last 100 million years. *Geochimica et Cosmochimica Acta* 68 (17): 3521–3530.

Yan, Y., J. Ng, J. Higgins et al. (2017) 2.7-Million-Year-Old Ice from Allan Hills Blue Ice Areas, East Antarctica Reveals Climate Snapshots Since Early Pleistocene, Goldschmidt Conference, Paris.

Yu, J., Elderfield, H., and Hönisch, B. (2007). B/Ca in planktonic foraminifera as a proxy for surface seawater pH. *Paleoceanography* 22: doi: 10.1029/2006PA001347.

Zachos, J.C., Dickens, G.R., and Zeebe, R.E. (2008). An early Cenozoic perspective on greenhouse warming and carbon-cycle dynamics. *Nature* 451 (7176): 279–283.

Zeebe, R.E. and Wolf-Gladrow, D.A. (2001). *CO_2 in seawater: Equilibrium, kinetics, isotopes*, vol. 65, 346. Elsevier.

Zhang, Y.G., Pagani, M., Liu, Z. et al. (2013). A 40-million-year history of atmospheric CO_2. *Philosophical Transactions of the Royal Society A: Mathematical, Physical and Engineering Sciences* 371 (2001).

Further Reading/Resources

CO_2 in seawater: Equilibrium, Kinetics, Isotopes, by Zeebe and Wolf-Gladrow, Elsevier Oceanography Series, Volume 65, 360 pp., eBook ISBN: **9780080529226, 2001** – the resource for all questions on ocean carbonate chemistry

Ocean carbonate chemistry calculation programs can be found at:

CO2SYS for EXCEL (Pierrot et al. 2006) and Matlab (van Heuven et al. 2009): http://cdiac.ornl.gov/oceans/co2rprt.html

csys.m, a Matlab program that accompanies *CO_2 in seawater* by Zeebe and Wolf-Gladrow (2001): http://www.soest.hawaii.edu/oceanography/faculty/zeebe_files/CO_2_System_in_Seawater/csys.html

seacarb (Gattuso et al. 2017): https://CRAN.R-project.org/package=seacarb

2

Boron Systematics

Abstract

This chapter presents the theoretical background of the boron isotope and B/Ca proxies, starting with the dissociation of dissolved boron in aqueous solution, and constraints on the boron isotope fractionation between the dominant dissolved boron species in seawater. Using laboratory culture experiments with foraminifers, corals and inorganically precipitated calcium carbonate, as well as observations of naturally grown samples collected from the ocean, a strong empirical framework has been established for the boron isotope proxy. However, it is also clear that an organism's calcifying fluid is often chemically distinct from ambient seawater, and partial dissolution of skeletal remains at the seafloor can further modify original proxy records. These aspects need to be taken into account when selecting sample material and interpreting proxy records. Deep-time paleoreconstructions further need to consider secular variations in the seawater boron concentration and isotopic composition; currently available estimates are summarized.

In comparison to the boron isotope proxy, the relationship between B/Ca and marine carbonate chemistry has only been studied over the past decade, and it is clear that B incorporation in different calcifiers responds to different environmental controls. While B/Ca in corals may be more sensitive to temperature than carbonate chemistry, planktic foraminiferal B/Ca increases with experimental seawater-pH in laboratory culture, but sediment observations suggest that additional environmental parameters complicate the proxy. In contrast, strong and reproducible relationships have been established for B/Ca in benthic foraminifers and ocean bottom water carbonate saturation, albeit with significant species effects and without a mechanistic explanation as to why carbonate saturation should be the controlling parameter.

Keywords: Aqueous speciation; boron isotope partitioning; B/Ca, $\delta^{11}B$; empirical calibrations; vital effects; pH elevation; dissolution; temperature effect; $\delta^{11}B_{sw}$

Boron Proxies in Paleoceanography and Paleoclimatology, First Edition. Bärbel Hönisch, Stephen M. Eggins, Laura L. Haynes, Katherine A. Allen, Katherine D. Holland, and Katja Lorbacher.
© 2019 John Wiley & Sons Ltd. Published 2019 by John Wiley & Sons Ltd.
Companion website: www.wiley.com/go/Hönisch/Boron_Paleoceanography

2.1 Introduction

The geochemistry of boron remained largely unstudied until (1932), when Goldschmidt and Peters replaced boron-containing graphite electrodes with copper electrodes to spectrographically measure the boron concentration of various substrates more accurately. Their observation of "surprisingly high boron concentrations in clay sediments" fueled several decades of boron studies in clay deposits, where boron was first considered a proxy for silicon (Goldberg and Arrhenius 1958) or aluminium (Harder 1961) concentrations in the clay mineral illite, leading to further intensive investigation of the relationship between boron content in clays and the salinity of the depositional environment (Landergren 1945; Walker and Price 1963). Although the boron concentration in marine clays is generally greater than in lacustrine clays, the relationship with salinity was eventually found to be weak due to a multitude of competing controls – temperature, boron concentration and pH of the parent solution, grain size, mineral composition and sedimentation rate (e.g. Fleet 1965; Harder 1970).

In contrast, the study of boron in carbonates had largely been neglected due to the fact that the boron concentration in carbonates is generally low (<100 ppm), and that most of the boron in limestones is associated with the adsorbed clay fraction (Harder 1959). An early study of boron in travertines by Ichikuni and Kikuchi (1972) concluded that boron does in fact reside in the calcite lattice and that the incorporation starts with the adsorption at the mineral surface. The authors suggested that borate $(B(OH)_4^-)$, as the charged species, is the most likely boron species to be adsorbed, and that its abundance ultimately depends on the temperature and pH of the thermal waters that travertines precipitate from. The first study to make a direct connection between the boron isotopic composition $(\delta^{11}B)$ of inorganic boron minerals and pH was made by Oi et al. (1989), who compared minerals of similar geologic origin and therefore precipitated from solutions of supposedly similar aqueous $\delta^{11}B$. The authors interpreted the differences in mineral-$\delta^{11}B$ as being indicative of the pH of the parent solution, and the associated abundance ratio of $^{10}B(OH)_4^-$ and $^{11}B(OH)_3$. Two years later the studies of Vengosh et al. (1991) and Hemming and Hansson (1992) finally set the stage for detailed investigations of boron and its isotopes in biogenic marine carbonates. Both studies measured the boron concentration and $\delta^{11}B$ in the shells and skeletons of various marine organisms, including foraminifers, corals, mollusks, ostracods, coralline algae, and sea urchins, as well as some carbonate sediments, limestones, dolomites, and ooids. Whereas the boron concentration of these carbonates varies greatly, from 2 to 75 ppm (Hemming and Hanson 1992; Vengosh et al. 1991), their isotopic composition is relatively low and consistent, with $\delta^{11}B$ values ranging 14–32 ‰ (Hemming and Hanson 1992; Vengosh et al. 1991). These values are lower than the mean $\delta^{11}B$ of seawater $(\delta^{11}B_{sw})$ of ~40 ‰ (Foster et al. 2010; Hemming and Hanson 1992; Spivack and Edmond 1987; Vengosh et al. 1991), suggesting that marine carbonates must select only a portion of the total dissolved boron in seawater, and the proportion incorporated must be characterized by low $\delta^{11}B$. Vengosh et al. (1991) and Hemming and Hanson (1992) identified $B(OH)_4^-$ as the most likely species incorporated and suggested that the $\delta^{11}B$ of fossil carbonates may be used to infer the chemistry of the seawater from which these carbonates precipitated. These studies laid the foundation for $\delta^{11}B$ and boron-to-calcium ratios (B/Ca) in marine carbonates as recorders of seawater pH and pCO_2. This chapter will explain the current mechanistic understanding of the boron proxies and how it has evolved over the past two decades.

2.2   What Determines the Sensitivity of $\delta^{11}B$ and B/Ca to Marine Carbonate Chemistry?

2.2.1   Dissolved Boron and Equilibrium Reactions in Seawater

In aqueous solution boron occurs as the weak monoprotonic boric acid, which reacts with water and dissociates into borate and polyborate species via reactions that all depend on pH (Figure 2.1a and b). While formation of polyborate species is insignificant in solutions with a total boron concentration $[B_T]$ <0.025 mol kg^{-1} (Ingri et al. 1957), and polyborates therefore do not need to be considered in seawater (with [BT] = 0.0004 mol kg^{-1}, Lee et al. 2010), some ion pairing occurs in solutions containing Ca^{2+}, Mg^{2+} and Na^+, and such pairing affects about 44% of dissolved borate in modern seawater (Byrne and Kester 1974). The dissociation of boric acid into borates can be described by the following equilibrium reactions:

$$B(OH)_3 + H_2O \longleftrightarrow B(OH)_4^- + H^+ \tag{2.1}$$

$$3B(OH)_3 + OH^- \longleftrightarrow B_3O_3(OH)_4^- + 3H_2O \tag{2.2}$$

$$3B(OH)_3 + 2OH^- \longleftrightarrow B_3O_3(OH)_5^{2-} + 3H_2O \tag{2.3}$$

The *stoichiometric dissociation constant* for the boric acid/borate equilibrium is defined as

$$K^*_B = \left[H^+\right]\left[B(OH)_4^-\right]/\left[B(OH)_3\right] \tag{2.4}$$

and the negative decadic logarithm of K^*_B is termed pK^*_B. The value of pK^*_B is equal to the pH at which boric acid and borate concentration occur in equivalent concentrations.

$$pK^*_B = -\log K^*_B \tag{2.5}$$

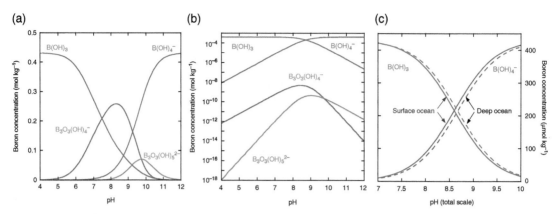

Figure 2.1 Relative abundance of dissolved boron species in aqueous solutions. (a) represents boron speciation in 3 M Na(ClO)$_4$ solution, with [B$_T$] = 0.432 mol kg^{-1}. With such high boron concentration polyborate formation is significant. (b) In a similar solution containing [B$_T$] = 432.6 µmol kg^{-1} (i.e. as in modern seawater with S = 35, Lee et al. 2011) polyborate formation becomes negligible and boric acid and borate ions dominate aqueous boron. (a) and (b) use the equilibrium constants of Ingri (1963), and it should be noted that (b) is shown on a log scale, so that the vanishing concentrations of B$_3$O$_3$(OH)$_4^-$ and B$_3$O$_3$(OH)$_5^{2-}$ can be visualized. (c) is shown on a linear scale again and uses [B$_T$] = 432.6 µmol kg^{-1} but pK$^*_B$ = 8.60 for the surface ocean (with S = 35, T = 25 °C and p = 1 bar, solid lines) and pK$^*_B$ = 8.67 for the deep ocean (with S = 34.8, T = 2 °C and p = 401 bar, dashed lines); all calculations done after Dickson (1990a) and Millero (1995). The shift in pK$^*_B$ affects boron proxy studies, because at the same pH boric acid is relatively more abundant in the deep compared to the surface ocean.

Dissociation constants are also termed *stability or equilibrium* constants in the literature, or *K-values*. Because their experimental determination is based on ion concentrations and not on ion activities, they are termed *stoichiometric* dissociation constants (as opposed to *thermodynamic* dissociation constants), and stoichiometric dissociation constants are denoted with an asterisk. Stoichiometric dissociation constants disregard any complex formation or ion pairs (Millero 1995), such that complexes and ion pairs are included in the concentration of B(OH)$_4^-$. Figure 2.1a shows the relative distributions of dissolved boron in water with [B$_T$] = 0.432 mol kg^{-1} (i.e. 1000 times more concentrated than in seawater). To display the abundance of polyborates at different boron concentrations, these distributions were calculated using the dissociation constants of Ingri (1963), which were determined in boron doped 3 M Na(ClO)$_4$ solutions. Under Ingri's experimental conditions B(OH)$_3$ and B(OH)$_4^-$ are the most abundant species, but the polyborates B$_3$O$_3$(OH)$_4^-$ and B$_3$O$_3$(OH)$_5^{2-}$ occur in significant quantities. This speciation changes at lower boron concentration, as in modern seawater (i.e. [B$_T$] =432.6 µmol kg^{-1} at S = 35, Lee et al. 2010), where boric acid and borate become by far the most abundant species in solution, and the concentration of polyborate species is negligible (Figure 2.1b). While the boron system in seawater is thus relatively simple, the reader should note that polyborate speciation has not yet been studied in seawater. Ingri (1963) merely provided evidence that the equilibrium constants change with the elemental composition of the experimental solution. The boron speciation estimated for Figure 2.1b therefore only approximates the equilibria in actual seawater.

Figure 2.1c presents the boron speciation currently accepted to best represent modern seawater conditions. The pK_B^* used for Figure 2.1c has been determined in artificial seawater on the total pH scale, and under variable temperature and salinity conditions (Dickson 1990a). In addition, the effect of pressure can be estimated from changes in partial molal volume and molal compressibility (Millero 1995; Millero et al. 2012). Following these studies, surface ocean pK_B^* equals 8.60 at $T = 25\,°C$, $S = 35$ and $p = 1\,bar$, and this value decreases at higher T, S, and p. For a given deep-water condition with $T = 2\,°C$, $S = 34.8$, $p = 401\,bar$, pK_B^* is therefore 8.67, so that at a given pH the relative abundance of dissolved boron species shifts in favor of boric acid in the deep compared to the surface ocean. Boron proxy reconstructions therefore need to determine pK_B^* for the specific growth conditions (T, S, p) of the proxy carrier. An EXCEL spreadsheet to make this calculation for the modern ocean can be found in the online supplement (online Table A2.1).

In the distant past, when the seawater elemental composition differed from today (e.g. Lemarchand et al. 2000; Lowenstein et al. 2014), pK_B^* will have varied with changes in anion-cation pairing. The implications of geologically variable ion pairing effects for pH reconstructions have recently been discussed by Nir et al. (2015) and Hain et al. (2015), and will be elaborated in Chapter 3.

2.2.2 Aqueous Boron Isotope Fractionation

In addition to pH-dependent relative abundance changes, aqueous boric acid, and borate also differ in their isotopic composition. Boron has two stable isotopes, ^{10}B and ^{11}B, and they contribute approximately 19.9 and 80.1% to the total boron on Earth, respectively. The isotope exchange reaction between $B(OH)_3$ and $B(OH)_4^-$ in seawater is given by the following equation:

$$^{10}B(OH)_3 + {}^{11}B(OH)_4^- \longleftrightarrow {}^{11}B(OH)_3 + {}^{10}B(OH)_4^- \qquad (2.6)$$

Because coordination of the boric acid molecule is trigonal-planar, and that of borate is tetrahedral, the difference in molecular coordination and vibrational frequencies creates an energy difference that leads to isotope fractionation between the two molecules in solution. The exact value of this fractionation is fundamental to using the boron isotope proxy and the process of its determination deserves some detail.

The determination of the boron isotope fractionation factor between dissolved boron species, $\alpha_{B(OH)3 - B(OH)4-}$ (hereafter α_{B3-B4}), in thermodynamic equilibrium has a long history, which gained momentum with Kakihana et al. (1977) and Kakihana and Kotaka (1977) at the height of the nuclear power era. Using spectroscopically determined vibrational frequencies, Kakihana and Kotaka (1977) calculated the molecular energy difference and predicted $\alpha_{B3-B4} = 1.0194$ at 25 °C, which means that ^{11}B is enriched in boric acid by $\varepsilon_{B3-B4} = 19.4$ ‰ relative to borate (Figure 2.2). The boron isotope fractionation ε_{B3-B4} is defined as

$$\varepsilon_{B3-B4}\left(‰\right)=\left(\alpha_{B3-B4}-1\right)*1000 \qquad (2.7)$$

Due to the lack of an experimentally determined value, Kakihana and Kotaka's (1977) α_{B3-B4} remained unquestioned until Palmer et al. (1987) studied B adsorption and isotope fractionation on marine clays and determined a larger $\alpha_{B3-B4} = 1.023$. Continued interest in the use of boron for nuclear reactor control, neutron tumor therapy and the rising interest in reconstructing seawater-pH from marine carbonates triggered renewed efforts to improve the theoretical approach for estimating α_{B3-B4} (Liu and Tossell 2005; Oi 2000; Rustad and Bylaska 2007; Rustad et al. 2010; Sanchez-Valle et al. 2005; Zeebe 2005). Although most of these studies estimated $\alpha_{B3-B4} \sim 1.03$, values ranged from as low as 1.0176 (Sanchez-Valle et al. 2005) to as high as ~1.05 (Zeebe 2005), depending on the different theoretical approaches and data parameterizations used to calculate the molecular

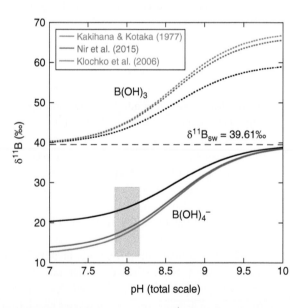

Figure 2.2 Boron isotope partitioning between dissolved boric acid and borate at T = 25 °C, S = 35, p = 1 bar, $\delta^{11}B_{sw}$ = 39.61 ‰ (Foster et al. 2010) and [B_T] = 432.6 µmol kg^{-1} (Lee et al. 2011), but using three different α_{B3-B4} values. Kakihana and Kotaka (1977) predicted α_{B3-B4} = 1.0194 theoretically, Klochko et al. (2006) determined α_{B3-B4} = 1.0272 from isotopically induced differences in aqueous boron speciation, and Nir et al. (2015) measured α_{B3-B4} = 1.026 across a semipermeable membrane. Note that the fractionation ε_{B3-B4} is constant with pH. The gray box indicates the boron isotopic composition of modern marine carbonates, which suggests predominant incorporation of borate ion.

forces. Rustad and Bylaska (2007) discovered that a major fractionating vibrational mode had been improperly assigned in calculations that yield α_{B3-B4} at the lower end of the spectrum (i.e. ~1.02), which makes those results incorrect. Using density functional and correlated molecular orbital theory (MP2), Rustad et al. (2010) finally determined $\alpha_{B3-B4} = 1.026–1.028$ as the currently best possible computationally predicted estimates.

Direct experimental determination of α_{B3-B4} was finally made possible by significant analytical advances in spectrophotometric pH measurements, which allowed determination of α_{B3-B4} from differences in the dissociation constants of $^{11}B(OH)_3$ and $^{10}B(OH)_3$ (Byrne et al. 2006). Practically, this analysis was achieved by measuring small pH-differences between alkaline solutions containing only ^{10}B or only ^{11}B, in addition to the pH-sensitive dye thymol blue. In conjunction with precise constraints on the buffering characteristics of these two solutions, their isotopically induced pH-difference could then be translated to α_{B3-B4}. Following this experimental design, Klochko et al. (2006) determined $\alpha_{B3-B4} = 1.0272 \pm 0.0006$ in synthetic seawater solutions with ionic strength similar to $S = 35$, $[B_T] = 0.01–0.05 \, mol \, kg^{-1}$ and at $T = 25\,°C$ (Figure 2.2) and $\alpha_{B3-B4} = 1.0308 \pm 0.0023$ in pure water with $[B_T] = 0.05 \, mol \, kg^{-1}$ and at $T = 25\,°C$. Because analyses of KCl and synthetic seawater solutions produced similar results with $\alpha_{B3-B4} = 1.0250$ and $\alpha_{B3-B4} = 1.0272$, respectively, the authors concluded that ion pairing with Na^+, Ca^{2+} and Mg^{2+} exerts only a negligible influence on α_{B3-B4}. Similarly, the identical result of $\alpha_{B3-B4} = 1.0272$ at $[B_T] = 0.01 \, mol \, kg^{-1}$ and $[B_T] = 0.05 \, mol \, kg^{-1}$ led them to argue against any significant influence of polyborate formation on α_{B3-B4}. Consequently, the experimentally derived value of $\alpha_{B3-B4} = 1.0272 \pm 0.0006$ should be applicable over a wide range of ocean water masses.

Recently, Nir et al. (2015) developed an alternative approach to estimate α_{B3-B4}, where boric acid and borate ions of a known solution were spatially separated by the preferential passage of the uncharged boric acid through a reverse osmosis membrane. These experiments were done at constant pH, so that the relative proportion of the two boron species could be calculated from pK^*_B, and the spatial isolation and subsequent isotopic analysis of the boric acid fraction allowed calculation of $\alpha_{B3-B4} = 1.026 \pm 0.001$. Within analytical uncertainty this value is in excellent agreement with $\alpha_{B3-B4} = 1.0272 \pm 0.0006$ determined by Klochko et al. (2006) and thus confirms the greater boron isotope fractionation in aqueous solutions compared to the original theoretical estimate by Kakihana and Kotaka (1977). It is worth noting that $\alpha_{B3-B4} \sim 1.027$ is larger than estimated for boron adsorption onto clays (i.e. $\alpha_{B3-B4} = 1.023$, Palmer et al. 1987), suggesting that surface effects may play a role in boron adsorption onto solid minerals.

In summary, observations on dissolved boron in aqueous solutions suggest that boron dissociation depends on pH, with the concentration of borate increasing at higher pH. The boron dissociation constant pK^*_B varies with T, S, p, and varying seawater ionic composition (cf. ion pairs). This value needs to be determined for the specific growth environment of marine carbonates, so that the relative abundance of $B(OH)_3$ and $B(OH)_4^-$ can be determined accurately. Modern seawater has a $\delta^{11}B_{sw} = 39.61 \pm 0.04$ ‰ (Foster et al. 2010), and of the two stable isotopes of boron, ^{10}B is preferentially fractionated into borate, whereas ^{11}B resides preferentially in boric acid. This fractionation is described by the fractionation factor α_{B3-B4}, which is a constant. Because isotope measurements find the $\delta^{11}B$ of marine carbonates to fall close to the $\delta^{11}B$ of borate in seawater, it has been suggested that borate is the dominant species incorporated into marine carbonates. Consequently, the boron concentration and isotopic composition in marine carbonates should increase at higher pH. This hypothesis forms the basis of the B/Ca and $\delta^{11}B$ proxies, which had to be tested over a wide range of seawater-pH. Because seawater-pH is relatively uniform in the ocean (i.e. $pH_{TS} \sim 8.1 \pm 0.1$ in the surface ocean and $pH_{TS} \sim 7.85 \pm 0.1$ in the deep ocean, Figure 2.3), boron proxies have often been calibrated in the laboratory, where marine organisms and synthetic $CaCO_3$ can be grown over a wide range of experimental seawater pH.

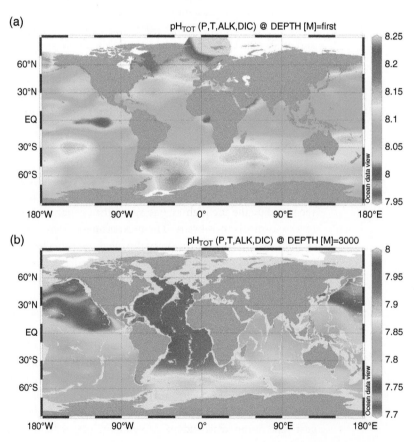

Figure 2.3 Seawater-pH_{TS} maps projected from local concentrations of alkalinity, DIC and salinity for (a) the surface ocean, and (b) at 3000 m water depth. Maps produced in Ocean Data View (Schlitzer 2012), using the GLODAP database (Key et al. 2004). Note that pH is relatively uniform within the surface ocean and within the deep ocean environment, but a ~0.3-unit offset exists between the two depth horizons.

Boron proxy validation in inorganic marine carbonates has made major advances over the last couple of years and because proxy relationships in living organisms often deviate from inorganic theory, we will start our survey of boron proxy calibrations by laying out the inorganic framework.

2.3 Boron Proxy Systematics in Synthetic Carbonates

2.3.1 Boron Isotope Partitioning into Synthetic Marine Carbonates – Evidence from Laboratory Precipitation Experiments

Synthetic $CaCO_3$ precipitation experiments can help to decipher inorganic systematics, and several studies have been published describing such precipitates from B-containing solutions. Before we describe the results in detail, the reader should note that pH values in these next two sections are given on the NBS scale, which is the pH scale used by all available inorganic precipitation studies to date. Unfortunately, most inorganic studies do not provide sufficient information to allow conversion between pH scales; presenting all data on the reported NBS scale therefore allows greater accuracy than attempting conversions with estimated carbonate chemistry values.

The first study aiming to determine the boron isotope partitioning into inorganic $CaCO_3$ was performed by Hemming et al. (1995), who grew calcite, aragonite and high magnesium calcite at pH ~8 (presumably on the NBS scale, i.e. <pH 7.9 on the TS) from solutions containing $CaCl_2$, NH_4Cl, $(NH_4)_2CO_3$, and different concentrations of $B(OH)_3$ (0.45–498.7 ppm). $MgCl_2$ was only added to aragonite and high-magnesium calcite experiments, as the presence of Mg^{2+} promotes the precipitation of aragonite and inhibits formation of calcite. While boron uptake in these three $CaCO_3$ polymorphs decreased from aragonite to high-magnesium calcite to low-magnesium calcite, their $\delta^{11}B = -16.5 \pm 0.7$ ‰ was the same within error for all three polymorphs and at all B concentrations. Assuming that the $\delta^{11}B$ of the solution was zero (Hemming used NBS SRM 951 as the boron source), the $\delta^{11}B$ of these carbonates would approximate to 22.4 ‰ if they had been grown in natural seawater (Figure 2.4, see also Box 2.1 for estimating $\delta^{11}B$ of experimental carbonates grown under $\delta^{11}B_{fluid} \neq \delta^{11}B_{sw}$). Estimating the exact $\delta^{11}B$ of $CaCO_3$ grown in seawater and precipitation pH from these data is somewhat limited by the limited available information on the elemental composition of the experimental solution, however, it is clear that the molecular source of boron incorporated into all three polymorphs is the same, and the $\delta^{11}B$ of this source falls close to $\delta^{11}B$ of borate.

Hemming et al. (1995) had performed their experiments only at a single pH condition, follow-up experiments were designed by Sanyal et al. (2000) and Noireaux et al. (2015) to span a wide pH-range (Figure 2.4). Both studies grew calcite in Mg-free solutions, with Sanyal et al. (2000) using an artificial seawater

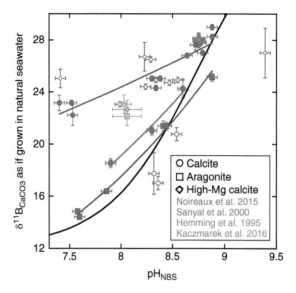

Figure 2.4 $\delta^{11}B_{CaCO3}$ sensitivity versus seawater pH in synthetic carbonates. Calcite data are shown as circles, aragonite as squares and high-Mg calcite as diamonds. Data are from Hemming et al. (green symbols, 1995), Sanyal et al. (red symbols, 2000), Noireaux et al. (blue symbols, 2015), and Kaczmarek et al. (orange symbols, data from their 22 °C experiments only, 2016). Open symbols reflect data that were precipitated at $[B_T] < 14\,mmol\,kg^{-1}$ and $[B_T] > 17\,mmol\,kg^{-1}$; these data were excluded from trend lines. Note that $\delta^{11}B_{aragonite}$ is generally smaller than $\delta^{11}B_{calcite}$, but the pH-sensitivity in $\delta^{11}B_{calcite}$ is smaller than in $\delta^{11}B_{aragonite}$. The $\delta^{11}B_{borate}$ curve is shown for T = 25 °C, S = 35, p = 1 and using $\alpha_{B3-B4} = 1.0272$ (Klochko et al. 2006) and $pK^*_B = 8.60 + 0.14$ to approximate the difference between the total and NBS pH scale.

medium, whereas Noireaux et al. (2015) performed their experiments in 0.2 M NaCl solutions. Noireaux et al. (2015) also performed a second set of experiments in which aragonite was precipitated by doping their experimental solution with 0.025 M $MgCl_2$. For the sake of completeness, we also include the 22 °C experimental data of Kaczmarek et al. (2016), who precipitated calcite in a 0.7 M NaCl solution (Figure 2.4). All data from synthetic $CaCO_3$ experiments are presented as if precipitated in seawater (see Box 2.1) and we display all data relative to the NBS pH-scale reported in each study.

Consistent with Hemming et al. (1995), all synthetic carbonates incorporate $\delta^{11}B$ similar to aqueous $\delta^{11}B_{borate}$. Importantly, all studies performed over a wide pH range show $\delta^{11}B$ increases with seawater pH, generally consistent with the theoretical background that the abundance and isotopic composition of aqueous borate in seawater increase at higher pH (Figures 2.1 and 2.2). However, in terms of absolute values different studies yield different $\delta^{11}B$ values for the same $CaCO_3$ polymorph, and the sensitivity of $\delta^{11}B$ in calcite and aragonite versus experimental pH differs as well (Figure 2.4).

In terms of proxy sensitivity to pH, the aragonite calibration (Noireaux et al. 2015) is broadly consistent with $\delta^{11}B_{borate}$ (Klochko et al. 2006; Nir et al. 2015), although significant data variability was introduced by precipitating $CaCO_3$ across a range of solution compositions (e.g. with $[B_T]$ ranging from

Box 2.1 Calibration Experiments Using $\delta^{11}B_{\text{experimental water}} \neq \delta^{11}B_{\text{natural seawater}}$

Inorganic precipitation experiments using artificial seawater solutions and some laboratory culture experiments enriching their natural seawater with laboratory boric acid expose their proxy carrier to $\delta^{11}B_{\text{experimental seawater}}$ that differs from $\delta^{11}B_{\text{natural seawater}}$, and comparability between studies and to natural samples therefore requires estimation of $\delta^{11}B_{\text{CaCO3}}$ as if it had grown in natural seawater. In most cases the isotopic composition of laboratory boric acid is significantly lower than that of natural seawater (i.e. $\delta^{11}B_{\text{B(OH)3}} \leq 0$, versus $\delta^{11}B_{\text{sw}} = 39.61$ ‰, Foster et al. 2010), and experimental $[B_T]$ enrichment to ~10x the natural seawater concentration has been applied in a range of studies to reduce the amount of material necessary for analysis (e.g. Hönisch et al. 2003; Howes et al. 2017; Kaczmarek et al. 2015; Noireaux et al. 2015; Sanyal et al. 2000, 2001). Importantly, a tenfold increase in $[B_T]$ to ~0.005 mol kg^{-1} falls significantly below $[B_T] = 0.025$ mol kg^{-1}, such that the abundance of polyborate species remains insignificant in such experimental seawater (Ingri et al. 1957).

Estimating the $\delta^{11}B$ of a sample as if it had grown in seawater requires the determination of $\delta^{11}B_{\text{exp. sw}}$ in addition to $\delta^{11}B_{\text{CaCO3}}$. For their inorganic precipitation study Sanyal et al. (2000) calculated the boron isotope fractionation between experimental seawater and calcite precipitates as the difference between these two substrates:

$$\Delta\delta^{11}B = \delta^{11}B_{\text{exp.sw}} - \delta^{11}B_{\text{CaCO3}} \tag{2.8}$$

This value was then compared to the equivalent $\Delta\delta^{11}B$ of other experimental calibrations (Sanyal et al. 2000). In a follow up study with cultured planktic foraminifers grown under 10x $[B_T]$ (Sanyal et al. 2001), the $\delta^{11}B$ value of a *Globigerinoides sacculifer* shell grown in natural seawater was then calculated as

$$\delta^{11}B_{\text{CaCO3}} = \delta^{11}B_{\text{nat.sw}} - \Delta\delta^{11}B \tag{2.9}$$

In contrast, Noireaux et al. 2015) aimed to follow a similar approach but subtracted $\delta^{11}B_{\text{exp.sw}}$ from $\delta^{11}B_{\text{CaCO3}}$, which merely shifts all $\delta^{11}B_{\text{CaCO3}}$ data by a constant value but does not allow for the estimation of a natural equivalent.

While Sanyal's approach produces a reasonable approximation to $\delta^{11}B$ of a carbonate grown in natural seawater, accurate expression of the difference between natural and experimental seawater requires calculation of the fractionation factor

$$\alpha_{\text{nat.sw/exp.sw}} = \left(\delta^{11}B_{\text{nat.sw}} + 1000\right) / \left(\delta^{11}B_{\text{exp.sw}} + 1000\right)$$
$$\left(\text{Zeebe and Wolf} - \text{Gladrow} 2001\right), \tag{2.10}$$

The $\delta^{11}B$ of the "natural" carbonate can then be calculated as

$$\delta^{11}B_{\text{CaCO3}} = \alpha_{\text{nat.sw/exp.sw}} * \delta^{11}B_{\text{exp.CaCO3}}$$
$$+ \left(\alpha_{\text{nat.sw/exp.sw}} - 1\right) * 1000 \tag{2.11}$$

How significant is the difference between these calculations? For example, in the synthetic calcite precipitation experiments of Sanyal et al. (2000), the $\delta^{11}B_{\text{CaCO3}}$ according to Eq. 2.11 is 0.78 ‰ higher at $pH_{TS} = 7.77$ than results from Eq. 2.9, but only 0.57 ‰ higher at $pH_{TS} = 8.47$. This indicates that Eq. 2.9 not only overestimates the true $\delta^{11}B$ value of an individual sample, but also slightly underestimates the overall sensitivity of the $\delta^{11}B_{\text{CaCO3}}$ vs. pH calibration.

Calibration data for all different marine carbonates studied to date are presented in Table A2.2 in the online Appendix, where $\delta^{11}B_{\text{CaCO3}}$ of any samples collected in solutions with $\delta^{11}B_{\text{exp. sw.}} \neq 39.61$ ‰ have been converted using Eq. 2.11.

0.4 to 20.2 mmol kg^{-1} between individual precipitates). For Figure 2.4, we have restricted all regression fits for Noireaux's data to experiments conducted within narrow range of [B$_T$] = 14–17 mmol kg^{-1}. This does not change the slopes of the regression fits, but highlights the much-improved data coherency when carbonates are precipitated under similar conditions.

In comparison to aragonite, δ^{11}B in calcite (Noireaux et al. 2015; Sanyal et al. 2000) is enriched in ^{11}B compared to δ^{11}B$_{borate}$ (i.e. δ^{11}B$_{calcite}$ > δ^{11}B$_{borate}$), and the pH sensitivity appears to be lower than predicted from exclusive borate ion incorporation. The slopes of the regression fits shown in Figure 2.4 are significantly smaller than calculated for δ^{11}B$_{borate}$, and also differ between the two studies, thus complicating identification of the inorganic baseline. While the general pattern of δ^{11}B$_{calcite}$ > δ^{11}B$_{aragonite}$ holds true across all studies and likely indicates a crystallographic control on boron incorporation (Noireaux et al. 2015), significant offsets occur between calcites and aragonites precipitated in different studies (Figure 2.4). For instance, δ^{11}B$_{calcite}$ data in the experiments of Noireaux et al. (2015) are 2.6–4.5 ‰ higher than Sanyal et al. (2000), and data of Hemming et al. (1995) fall in between these two studies. Unfortunately, due to differences in experimental techniques applied by all inorganic precipitation studies performed to date, it is not yet possible to identify the reason for these offsets. Because of these uncertainties, we recommend that comparisons between the absolute values of specific calibrations and natural carbonates should not be made until experiments have been performed in solutions that are truly comparable to natural seawater. In contrast, several hypotheses have been discussed to explain the lesser pH sensitivity of empirical carbonate calibrations compared to δ^{11}B$_{borate}$. We will combine the review of these discussions with additional evidence from B/Ca in synthetic carbonates.

2.3.2 *B/Ca Evidence from Synthetic Calcite Precipitation Experiments*

Based on the observation that the δ^{11}B of marine carbonates falls close to the δ^{11}B of aqueous borate in seawater (Figure 2.2), Hemming and Hanson (1992) suggested the tetrahedral, charged B(OH)$_4^-$ species is attracted to mineral surfaces and subsequently incorporated into the CaCO$_3$ lattice. Because HBO$_3^{2-}$ (or BO$_2$(OH)$^{2-}$, Balan et al. 2016) and CO$_3^{2-}$ are similar in ionic size and charge, Hemming and Hanson (1992) suggested that aqueous carbon species must also play a role in the incorporation process and therefore proposed the following substitution reaction:

$$CaCO_{3,solid} + B(OH)_{4,aq}^- \longleftrightarrow CaHBO_{3,solid} + HCO_{3,aq}^- + H_2O \quad (2.12)$$

The partition coefficient for this reaction is:

$$K_D = \frac{\left[HBO_3^{2-}/CO_3^{2-}\right]_{CaCO3}}{\left[B(OH)_4^-\right]/\left[HCO_3^-\right]_{seawater}} \quad (2.13)$$

Yu et al. (2007) expressed this equation in terms of measurable geochemical quantities:

$$K_D = \frac{(B/Ca)_{CaCO3}}{\left(\left[B(OH)_4^-\right]/\left[HCO_3^-\right]\right)_{seawater}} \qquad (2.14)$$

The B concentration in the numerator of this equation is directly related to $[B(OH)_4^-]$ in the denominator as the aqueous source concentration, but the relation of $[Ca^{2+}]$ to $[HCO_3^-]$ may seem odd at first sight. The reasoning behind this construct is that the Ca^{2+} concentration in $CaCO_3$ can be considered equal to its CO_3^{2-} concentration (the small contribution of other trace cations is negligible in this regard), but charge balance requires HCO_3^- to be the aqueous source species in Eq. 2.12. Consequently, the Ca^{2+} concentration in $CaCO_3$ can be considered proportional to aqueous $[HCO_3^-]$. These equations form the foundation of the B/Ca proxy and the following presentation of synthetic precipitation experiments builds on this foundation.

The first studies investigating B incorporation into synthetic carbonates applied parent solutions that were compositionally quite different from seawater. Kitano et al. (1979) studied $Ca(HCO_3)_2$ –NaCl - $MgCl_2$ solutions and observed B incorporation into synthetic $CaCO_3$ is proportional to the boron concentration of the parent fluid; aragonite generally incorporates more B than calcite. Unfortunately, the authors did not control pH in their experiments, which hampers insightful comparison with other studies. Hemming et al. (1995) precipitated low-Mg calcite, high-Mg calcite and aragonite from supersaturated solutions containing $CaCl_2$, NH_4Cl, and $(NH_4)_2CO_3$ at constant pH_{NBS} ~8.0. Similar to Kitano et al. (1979), the authors found increasing B uptake with $[B_T]$, and overall B incorporation ranks aragonite > high-Mg calcite > low-Mg calcite (Figure 2.5a), i.e. opposite to the $\delta^{11}B$ pattern (low-Mg calcite > high-Mg calcite > aragonite) presented in Figure 2.4. Although overall B uptake increased with higher $[B_T]$, Hemming et al. (1995) observed that K_D is smaller at high $[B_T]$ compared to low $[B_T]$ (<5 ppm), and K_D is constant in solutions containing >5 ppm boron.

Hobbs and Reardon (1999) expanded the observations of Hemming et al. (1995) by focusing specifically on precipitation rates of calcites formed from solutions containing vaterite and aragonite polymorphs. In their experiments boron uptake increased with pH and $[B_T]$, and was 5x greater in calcites precipitated from vaterite compared to aragonite solutions, an observation that is consistent with greater supersaturation of the vaterite-doped parent fluid (and consequently greater precipitation rate), and/or different crystal morphologies and therefore face dependent incorporation of boron into calcite. While generally consistent with other studies, these experiments were performed at an ionic strength of I = 0.01 and pH_{NBS} ~9, so results may not be directly comparable to calcite grown in seawater (with a typical ionic strength of I = 0.7).

The first synthetic boron calcification experiment in artificial seawater was done by Sanyal et al. (2000), who used calcite seeds as a nucleation

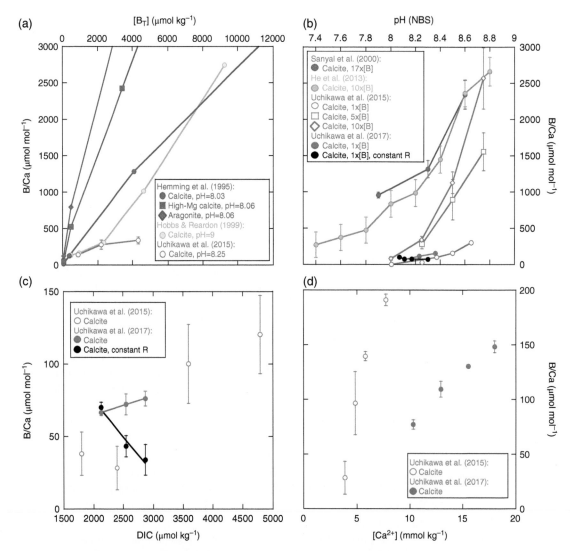

Figure 2.5 B/Ca systematics in synthetic CaCO₃ precipitated in laboratory experiments. Individual studies and specific experimental conditions are presented by different symbols as defined in the respective legends. (a) B/Ca generally increases with increasing [B$_T$] and B uptake is more sensitive in aragonite and high-Mg calcite compared to low-Mg calcite, but recent data of Uchikawa et al. (2015) show relatively decreasing uptake sensitivity at greater [B$_T$]. (b) B/Ca generally increases with experimental water pH, and the pH sensitivity appears to increase with [B$_T$]. However, precipitation rates typically increase at higher pH. To distinguish between [B(OH)$_4^-$] and precipitation rate control on B uptake in synthetic calcite Uchikawa et al. (2017) varied [Ca²⁺] and pH simultaneously and found B/Ca does not increase with [B(OH)$_4^-$] when precipitation rate is constant (black symbols). (c) B/Ca in synthetic calcite increases with DIC at constant pH (red symbols), but decreases with DIC when precipitation rate is maintained constant in response to lower [Ca²⁺] at higher DIC (black symbols). (d) At constant pH boron uptake increases with [Ca²⁺]. Note that the experiments of Uchikawa et al. (2015) were performed in reduced elemental complexity saline solutions, whereas those of Uchikawa et al. (2017) were performed in Mg-free artificial seawater. Error bars are shown whenever provided in the publications.

template. The parent solution used was Mg-free to precipitate calcite instead of aragonite, contained ~17x more boron than modern seawater, and the ionic strength was similar to natural seawater (i.e. I = 0.7). Although this experiment was designed to study the boron isotopic composition of synthetic calcite across a range of pH, the boron concentration was also determined by isotope dilution, corrected for the amount and chemical composition of the seeds. Estimated B/Ca ratios for these experiments are shown in Figure 2.5b and show a clear increase in B incorporation with experimental seawater pH, similar to the results of Hobbs and Reardon (1999). This observation is consistent with the increase of aqueous borate in solution (Figure 2.1), but a growth rate effect could not be excluded with the limited experimental parameterization (pH was the only variable; neither dissolved inorganic carbon (DIC) nor $[Ca^{2+}]$ experiments were performed).

A similar experiment by He et al. (2013) precipitated calcite by dissolving Li_2CO_3 as the carbonate source in Mg-free artificial seawater; $[B_T]$ and $[Ca^{2+}]$ of their experimental solution were 10x and 5x natural seawater, respectively. This experiment supports the growing number of studies that find increased B incorporation at greater pH (Figure 2.5b). However, the study also observed that K_D decreases at higher pH (i.e. $K_D = 0.00213$ at pH_{NBS} 7.4 and $K_D = 0.00127$ at pH_{NBS} 8.6), suggesting $[B(OH)_4^-]$ and $[HCO_3^-]$ are unlikely the only species involved in boron partitioning into synthetic and biogenic calcite.

The most comprehensive experimental matrices were presented by Uchikawa et al. (2015, 2017), Mavromatis et al. (2015), Kaczmarek et al. (2016) and Holcomb et al. (2016). While all of these studies suggest the precipitation rate plays a major role in B incorporation into synthetic carbonates (see also Figure 2.6 and Gabitov et al. 2014), individual studies vary in the precipitated $CaCO_3$ polymorph (i.e. calcite and aragonite), and in particular the experiments of Mavromatis et al. (2015) and Holcomb et al. (2016) varied several environmental parameters simultaneously, which complicates targeted analysis of individual parameter effects (i.e. pH, T, DIC, etc.) on B/Ca. The remainder of this discussion will therefore focus predominantly on the experiments of Uchikawa et al. (2015, 2017), who varied individual environmental parameters from the same starting solution while keeping other potential controls constant. Importantly, Uchikawa et al. (2015) performed their experiments in simple $NaCl$-$CaCl_2$-$B(OH)_3$-Na_2CO_3 solutions with an ionic strength of I = 0.47 (i.e. S = 23), while Uchikawa et al. (2017) used Mg-free artificial seawater with S~40. As displayed in Figures 2.5 and 2.6, the elemental composition of the experimental solution causes significant shifts in B uptake.

Figure 2.5 shows increasing B uptake at higher pH, $[B_T]$, DIC, and $[Ca^{2+}]$. In the experiments of Uchikawa et al. (2015, 2017) saturation state ($\Omega_{calcite}$) and precipitation rates (R) were tightly monitored and increase with pH, DIC and $[Ca^{2+}]$.

$$\Omega_{calcite} = \left(\left[Ca^{2+} \right]_{sw} \times \left[CO_3^{2-} \right]_{sw} \right) / K_{sp} \qquad (2.15)$$

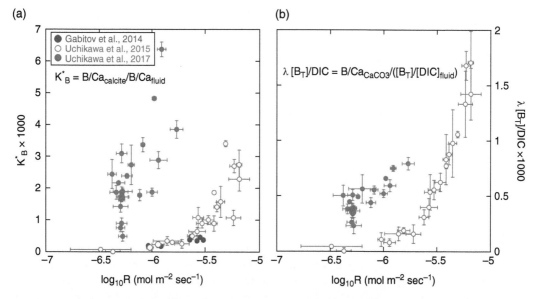

(a)

(b)

Figure 2.6 Several studies have found that B uptake correlates with precipitation rate (R) in synthetic CaCO₃ (Gabitov et al. 2014; Kaczmarek et al. 2016; Mavromatis et al. 2015; Uchikawa et al. 2015, 2017), but only a small subset of studies is shown here. (a) displays the relationship between boron partitioning (shown as the exchange coefficient K^*_B) in multiple experimental conditions versus $\log_{10}R$. Note that the experiments of Uchikawa et al. (2015) and Gabitov et al. (2014) were performed in reduced elemental complexity saline solutions, whereas those of Uchikawa et al. (2017) were performed in Mg-free artificial seawater. (b) Uchikawa et al. (2015, 2017) assessed a number of apparent partition coefficients (λ), which do not represent stoichiometric reactions (i.e. K_D) but simple theoretical considerations. By relating B/Ca to fluid ratios of dissolved boron and carbon species, they found the strongest correlation with $[B_T]/DIC$ (shown here) or $[B(OH)_4^-]/(CO_3^{2-})$ (similar to Holcomb et al. 2016).

where $[Ca^{2+}]_{sw}$ and $[CO_3^{2-}]_{sw}$ are the concentrations of Ca^{2+} and CO_3^{2-} in the calcifying fluid and K_{sp} is the solubility product at local T, S, and p. $\Omega > 1$ thereby indicates the calcifying fluid is oversaturated with respect to calcite and $\Omega < 1$ indicates the calcifying fluid is corrosive to calcite.

Because the DIC and $[Ca^{2+}]$ experiments were performed at constant pH and thus constant $[B(OH)_4^-]$, the increase in B/Ca appears controlled by precipitation rate rather than $B(OH)_4^-/HCO_3^-$ (see Eq. 2.14), but because both $[B(OH)_4^-]$ and precipitation rate increase in the pH experiments, Uchikawa et al. (2015, 2017) achieved the concluding test of the controlling parameter on B/Ca by performing experiments at constant R but variable pH, $[Ca^{2+}]$ and DIC. The results of these experiments are shown in Figure 2.5b (black symbols), where B/Ca remains constant when pH increases but R is constant. This suggests that the pH effect observed by those studies that only varied pH but did not control R (Figure 2.5b) is ultimately a precipitation rate effect, i.e. higher B/Ca is driven by the greater saturation state at higher pH, and not the greater abundance of aqueous borate ion.

In contrast, Uchikawa et al. (2017) observed decreasing B/Ca with increasing DIC at constant R (black symbols in Figure 2.5c), which is consistent with observations made with planktic foraminifera (Allen et al. 2012; Haynes et al. 2017). This agreement led Uchikawa et al. (2017) to suggest that B/Ca in planktic foraminifera is at least partly controlled by inorganic processes. B/Ca systematics in foraminifera will be discussed in Section 2.6.

Uchikawa et al. (2015) explored the possible kinetic effect on boron incorporation by determining the boron distribution coefficient of their synthetic calcites (i.e. $K_B^* = (B/Ca)_{CaCO_3}/([B_T]/[Ca^{2+}])_{aq}$) and relating it to precipitation rate. They found a tight correlation that fits not only the B/Ca data of their own experiments but is also in agreement with the data of Gabitov et al. (2014), Uchikawa et al. (2017) (Figure 2.6a), Mavromatis et al. (not shown, 2015) and Holcomb et al. (not shown, 2016). These experiments thus confirm that precipitation rate exerts a unifying control on B incorporation into synthetic $CaCO_3$.

Given scatter in the relationship between K_B^* and precipitation rate (Figure 2.6a), Uchikawa et al. (2015) evaluated a number of alternative boron partition coefficients, denoted λ. Specifically, they evaluated the relationship between B/Ca in calcite versus the ratios of $[B(OH)_4^-]/DIC$, $[B_T]/[HCO_3^-]$ and $[B_T]/DIC$ in the parent solution, and found the best correlation with $[B_T]/DIC$ (Figure 2.6b). Importantly, the scatter in the relationship was significantly reduced when $[B(OH)_4^-]$ was replaced by $[B_T]$, which Uchikawa et al. (2015) interpreted as indirect evidence for boric acid incorporation into synthetic calcite, particularly at high precipitation rates. In contrast, Holcomb et al. (2016) evaluated B/Ca in their aragonite precipitates and suggested $[B(OH)_4^-]/[CO_3^{2-}]^{0.5}$ described their data best, proposing that borate and carbonate ion are the most likely controlling quantities for B/Ca in synthetic aragonite. Uchikawa et al. (2017) evaluated B partitioning in their synthetic calcites with respect to $[B_T]/DIC$ and $[B(OH)_4^-]/[CO_3^{2-}]$ and found both ratios describe their data equally well. However, Uchikawa et al. (2017) did not consider $[B(OH)_4^-]/[CO_3^{2-}]^{0.5}$ as in Holcomb et al. (2016). We calculated $\lambda_{[B(OH)4-]/[CO32-]^{0.5}}$ for the data of Uchikawa et al. (2017) (not shown) but it appears that the correlation is actually weaker ($R^2 = 0.42$ for a simple exponential regression) than for $\lambda_{[B(OH)4-]/[CO32-]}$ ($R^2 = 0.58$). This comparison does not verify or question the suggested controlling aqueous source species (in particular as boron adsorption at the aragonite surface may differ from calcite, see e.g. Balan et al. 2016; Noireaux et al. 2015), but it is clear that boron and carbon species compete for inclusion into synthetic carbonates, and growth rate plays a significant role in the incorporation process (Gabitov et al. 2014; Holcomb et al. 2016; Kaczmarek et al. 2016; Mavromatis et al. 2015; Uchikawa et al. 2015, 2017).

2.3.3   Temperature Control on B/Ca in Synthetic Calcite

The effect of temperature on B/Ca in synthetic $CaCO_3$ has been tested in two studies. Mavromatis et al. (2015) precipitated both calcite and aragonite at two temperatures and across a range of elemental compositions, and

Kaczmarek et al. (2016) performed a series of synthetic calcite precipitation experiments across a range of temperatures at constant pH (Figure 2.7). Because precipitation rate in Kaczmarek's experiments tends to increase with temperature (Figure 2.7c), the authors added a second experiment, in which they increased the precipitation rate by increasing $[Ca^{2+}]$ at the same temperature and otherwise constant conditions. While Mavromatis et al. (2015) could not identify any significant temperature control, Kaczmarek et al. (2016) concluded that boron partitioning increases with precipitation rate and decreases with temperature (Figure 2.7b). While this result disagrees with the unifying precipitation control suggested by Uchikawa et al. (2015, 2017) and Mavromatis et al. (2015) (Figure 2.6), it is important to realize that the interpretation of the temperature effect depends on how the partitioning coefficient K_D is defined (Figure 2.7).

Figure 2.7a compares B/Ca ratios in temperature experiments from Mavromatis et al. (2015) and Kaczmarek et al. (2016). To reduce the influence of other controls, we have reduced the presentation of Mavromatis' data to a narrow $[B_T]$ range (5.0–5.8 mM); the experiments of Kaczmarek et al. (2016) were performed at $[B_T]\sim3$ mM. While Kaczmarek's B/Ca data show no significant change with temperature, their experiment

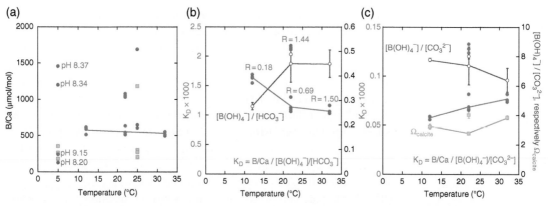

Figure 2.7 Boron uptake in response to temperature variations has been studied by Mavromatis et al. (aragonite in green squares and calcite in blue circles in (a), 2015) and Kaczmarek et al. (red symbols in (a)–(c), 2016). Whereas B/Ca shows no change with temperature in either study (a), the partitioning coefficient K_D in Kaczmarek's experiments decreases with temperature when K_D is defined via $[B(OH)_4^-]/[HCO_3^-]$ (b), and increases with temperature when K_D is defined via $[B(OH)_4^-]/[CO_3^{2-}]$ (c). The temperature-dependent shift in the $[B(OH)_4^-]/[HCO_3^-]$ and $[B(OH)_4^-]/[CO_3^{2-}]$ ratios is shown in (b) and (c) as gray symbols. The scatter in Mavromatis' data cannot be explained by differences in pH, as B/Ca should increase with pH (see example pH conditions for calcite data in (a)). Temperature experiments in (b) and (c) are connected by lines, the precipitation rate R (in mg h^{-1}) is indicated next to the respective K_D values. $\Omega_{calcite}$ increases from 22 to 32 °C in Kaczmarek's temperature experiments, and is further elevated in the high $[Ca^{2+}]$ experiment (not connected by lines). Kaczmarek et al. (2016) interpret the higher K_D in the $[Ca^{2+}]$ experiment as an indication of impurity entrapment at high precipitation rate, whereas the lower K_D at high temperature but the same R may be due to higher diffusivity of boron and consequent detachment from the crystal surface. Note that the latter interpretation only holds for K_D as defined in (b) but not in (c).

at elevated $[Ca^{2+}]$ has B/Ca ratios nearly twice as high as their other experiments. In comparison, the calcite and aragonite precipitates of Mavromatis et al. (2015) display a slight positive B/Ca response to temperature, but the overall data spread obscures whether this is due to temperature or other experimental variables. To determine whether precipitation rate might mask the temperature response in the study of Mavromatis et al. (2015), we indicate the experimental pH conditions to their calcite data at 5 °C (Figure 2.7a), and find that B/Ca at this single temperature does not consistently increase with pH and thus precipitation rate. The same observation is true for their aragonite precipitates and for their experiments performed at 25 °C (not plotted herein). While precipitation rate changes associated with pH do not appear to explain the lack of a temperature effect, the complexity of the experimental conditions investigated by Mavromatis et al. (2015) precludes identifying individual environmental controls on B incorporation.

Kaczmarek et al. (2016) interpreted their data in the form of boron partitioning, i.e. the B uptake relative to the controlling seawater elemental composition. As indicated in Eqs. 2.13 and 2.14, Hemming and Hanson (1992) and Yu et al. (2007) hypothesized that the borate to bicarbonate ratio may control B uptake and using this equation, boron partitioning in Kaczmarek's calcites decreases at higher temperatures (Figure 2.7b). However, the fact that B/Ca shows no trend with temperature (Figure 2.7a) but K_D does, indicates that this trend is determined by the denominator. The temperature effect on boron and carbon speciation in seawater produces an increase of the $[B(OH)_4^-]/[HCO_3^-]$ ratio with temperature, and therefore a decrease in K_D (Figure 2.7b). Because $[B(OH)_4^-]/[HCO_3^-]$-control is a hypothesis, Kaczmarek et al. (2016) also tested whether boron partitioning may be driven by the $[B(OH)_4^-]/[CO_3^{2-}]$ ratio (see also Holcomb et al. 2016) and found a positive trend with temperature (Figure 2.7c).

$$K_D = \frac{B/Ca_{CaCO3}}{\left[B(OH)_4^-\right]/\left[CO_3^{2-}\right]_{seawater}} \tag{2.16}$$

Although the $[B(OH)_4^-]/[CO_3^{2-}]$ ratio also varies with temperature, this comparison highlights that identifying the true controls on B incorporation into marine carbonates requires sophisticated experiments that test one parameter at a time, while maintaining all other parameters constant. Identifying these controls is complicated in the experiments of Kaczmarek et al. (2016) because both $[Ca^{2+}]$ and DIC were varied in order to precipitate calcite at constant pH, and the individual effects of $[Ca^{2+}]$ and DIC were not tested over a range of specific experimental conditions. In the experiments of Uchikawa et al. (2015, 2017) B uptake increases at higher DIC, $[Ca^{2+}]$ and $\Omega_{calcite}$ (Figure 2.5c and d), but in the experiments of Kaczmarek et al. (2016) DIC and $[Ca^{2+}]$ were highest in the 12 °C temperature experiment, and $\Omega_{calcite}$ of that experiment was intermediate compared to the other experimental conditions. Because the greater abundance of $[B(OH)_4^-]$ at higher pH predicts greater B uptake, and the $[B(OH)_4^-]$ abundance is greatest in their high temperature experiments, Kaczmarek et al. (2016) interpreted their observed decrease in $K_D = B/Ca/([B(OH)_4^-]/[CO_3^{2-}])$ with temperature as a temperature dependent decrease in boron partitioning.

In contrast to the interpretation made by Kaczmarek et al. (2016), we caution that it is unclear from Figure 2.7b and c and Eqs. 2.13, 2.14 and 2.16 whether $[HCO_3^-]$ or $[CO_3^{2-}]$ participate in boron partitioning. Based on existing experimental evidence we also cannot discern whether and how B partitioning is affected by temperature. The effect of the saturation state on B/Ca in the NaCl solutions used by Kaczmarek et al. (2016) would need to be tested in similar detail to Uchikawa et al. (2015, 2017). While multiple experiments have shown that $\Omega_{calcite}$ exerts a role on B incorporation into synthetic calcites (Gabitov et al. 2014; Hobbs and Reardon 1999; Kaczmarek et al. 2016; Mavromatis et al. 2015; Uchikawa et al. 2015, 2017), it is puzzling that Kaczmarek's high temperature and high $\Omega_{calcite}$ experiments display different B/Ca at the same precipitation rate (Figure 2.7). Clearly, targeted experiments are required to verify whether temperature counteracts the precipitation rate effect under constant $[B(OH)_4^-]/[HCO_3^-]$ and $[B(OH)_4^-]/[CO_3^{2-}]$.

2.3.4 Mechanisms of Boron Incorporation into Synthetic CaCO$_3$

The synthetic precipitation experiments described in the previous section are generally coherent in displaying increasing B incorporation with pH, $[B_T]$, DIC and precipitation rate, greater B incorporation into aragonite than calcite (Figure 2.5), and higher $\delta^{11}B$ in calcite compared to aragonite (Figure 2.4). However, few studies have tested all of these experimental conditions, and results vary both in absolute values of B incorporation, $\delta^{11}B$, sensitivity to a single experimental parameter, and sometimes even in trend (Figures 2.4–2.7). Most of this is likely due to the different experimental approaches and conditions in each study (e.g. elemental composition and ionic strength of the parent solution, presence or absence of seed crystals, precipitation rates, free drift or static conditions), which makes it difficult to discern which behaviors represent the correct inorganic baseline against which to compare biogenic CaCO$_3$ precipitation.

The similarity between $\delta^{11}B$ of marine carbonates and $\delta^{11}B_{borate}$ (Figure 2.2) led Hemming and Hanson (1992) to suggest that borate ion is the predominant species adsorbed and incorporated into marine carbonates, and the synthetic CaCO$_3$ data shown in Figure 2.4 are consistent with this initial suggestion. Support for the borate uptake hypothesis comes from crystal habit observations by Differential Interference Contrast Microscopy (Hemming et al. 1995) and Atomic Force Microscopy (Ruiz-Agudo et al. 2012), which clearly show strong interaction between the adsorbed molecule and the CaCO$_3$ surface, including modified crystal morphology and rounding of growth steps, pitting of the crystal surface, and roughening and thickening of growth steps at higher $[B_T]$. Given that Hemming et al. (1995) and Ruiz-Agudo et al. (2012) observed more sensitive boron incorporation at low $[B_T]$ and reduced partitioning at greater $[B_T]$ (Figure 2.5a), Hemming et al. (1995) hypothesized that this observation may reflect a difference in adsorption efficiency at crystal defect (i.e. non-lattice) sites over structural sites. If B were adsorbed more readily at defect sites, and the number of defect sites is limited, B adsorption and incorporation may be more sensitive to variations in $[B_T]$ at low $[B_T]$, whereas defect sites would be quickly occupied at higher $[B_T]$, and B incorporation would depend predominantly on the B affinity of structural sites. Using material from the same experiments, Hemming et al. (1998a) observed greater B incorporation at acute growth steps relative to obtuse steps, but their experiments were performed exclusively at $pH_{NBS} = 8.0$. In contrast, Ruiz-Agudo et al. (2012) observed that the spreading rate of obtuse growth steps is retarded in $CaCO_3$ precipitated from boron-bearing solutions at $pH_{NBS} > 9$. The difference in experimental setup between the two studies ($pH_{NBS} = 8.0$, respectively variable pH_{NBS}) complicates a direct comparison, as Ruiz-Agudo et al. (2012) did not observe any strong effects on calcite growth at $pH_{NBS} < 8.5$. However, both studies agree that the tetra-hedrally-coordinated borate ion is the main boron species incorporated into $CaCO_3$. This is because $B(OH)_4^-$ predominates over $B(OH)_3$ at $pH_{NBS} > 8.6$ (as in the experiments of Ruiz-Agudo et al. 2012), its charge interacts more firmly with the crystal surface, and the isotopic composition of carbonates more closely agrees with $B(OH)_4^-$ incorporation. Ruiz-Agudo et al. (2012) concluded that boron is preferentially incorporated via obtuse growth steps, and growth rate differences may be key controlling factors on B incorporation. Notably, calcite precipitation rate also decreases with increasing B concentration in the experiments of Uchikawa et al. (2015), supporting evidence from Hemming et al. (1995) and Ruiz-Agudo et al. (2012) that B adsorption may have an inhibitory effect on $CaCO_3$ growth.

In contrast, Gabitov et al. (2014) and Kaczmarek et al. (2016) observed increased boron partitioning at higher growth rates and decreasing partitioning at higher temperatures, respectively (Figures 2.6 and 2.7). Both studies applied a growth entrapment model that predicts the capture of impurities at higher growth rates (Gabitov et al. 2014; Kaczmarek et al. 2016), respectively greater diffusivity of B and detachment from the crystal surface occurring at higher temperatures (Kaczmarek et al. 2016). If correct, the disequilibrium between the solid and solution (i.e. K_D) should be greatest at high growth rates and low temperatures. Unfortunately, the experimental conditions explored by Gabitov et al. (2014) do not allow for testing of the model for predicted decrease in boron incorporation at low precipitation rates, and the lack of crystal surface area constraints in the study of Kaczmarek et al. (2016) (they present their growth rates in mg growth per mg seed crystals and hours of precipitation, $mg\,mg^{-1}\,h^{-1}$) does not allow comparison of their data to other studies in Figure 2.6. Existing data from Gabitov et al. (2014), Hemming et al. (1995), and Ruiz-Agudo et al. (2012)

are therefore not necessarily in conflict because the different reporting units do not allow direct comparison. Similarly, impurity entrapment at high growth rates cannot be ruled out by the studies of Hemming et al. (1995) and Ruiz-Agudo et al. (2012). For example, Ruiz-Agudo et al. (2012) observed thickening of growth edges in some experiments, which may indicate B trapping in non-lattice sites in synthetic calcite.

2.3.5   Is $B(OH)_4^-$ the Only Species Incorporated? Structural $CaCO_3$-Lattice Observations from NMR and Synchrotron Studies

One pattern that seems consistent across studies is the greater B uptake in aragonite compared to calcite (Figure 2.5, Hemming et al. 1995; Mavromatis et al. 2015). Hemming et al. (1995) aimed to explain this observation by discussing molecular size and coordination of the substituting boron species with regard to the anion site in $CaCO_3$. Using nearest neighbor ion volume calculations, Hemming et al. (1995) estimated the size of the anion site to be smaller in aragonite than calcite, and the B-O bond to be approximately 7% longer than the C-O bond. Based on these estimates, one would expect greater B uptake into calcite compared to aragonite, opposite to what is observed. To explain the greater B incorporation into aragonite, Hemming et al. (1995) therefore considered the coordination of boron in the $CaCO_3$ lattice. As already mentioned in Section 2.2.1, coordination of the boric acid molecule is trigonal-planar, and that of borate is tetrahedral. Boron speciation at the $CaCO_3$ surface has not yet been observed, but if boron coordination does not change from the aqueous B species adsorbed at the crystal surface to the B species incorporated in the $CaCO_3$, boron coordination in the crystal lattice would constrain the adsorbed species. This hypothesis has been explored by several studies, using [11]B Magic Angle Spinning Nuclear Magnetic Resonance (MAS NMR) and synchrotron X-Ray spectromicroscopy.

Sen et al. (1994) used ^{11}B MAS NMR to analyze the same synthetic calcite and aragonite samples precipitated by Hemming et al. (1995), in addition to a natural aragonitic *Montastrea* sp. coral and the high-Mg calcitic algae *Goniolithon* sp. Their NMR results suggested that boron coordination in aragonite was 100% tetrahedral, whereas 90% of boron in calcite occurred in trigonal coordination. Similarly, B coordination in the Mg-calcite sample was ~80% trigonal and 20% tetrahedral. In addition to these original biogenic carbonates, Sen et al. (1994) also transformed a portion of their aragonitic coral to calcite at 450–500 °C and found that B coordination changed to mostly or entirely trigonal coordination. Based on this change in B coordination with transformation between the different polymorphs, Sen et al. (1994) concluded that B resides in the $CaCO_3$ lattice, and is not simply trapped in interstitial fluid inclusions. This provides an important first constraint on the reliability of boron proxies in marine carbonates, as fluid inclusions may be subject to control by an infinite number of variables.

Tetrahedral B coordination in aragonite and predominantly trigonal coordination in calcite was a surprising result because as indicated above, Hemming et al. (1995) estimated the CO_3-site in aragonite to be smaller than in calcite. The larger tetrahedral molecule should therefore fit better into calcite and the smaller trigonal molecule better into aragonite. If, as boron isotope analyses of marine carbonates suggest (Figure 2.2), boron incorporation into $CaCO_3$ starts with adsorption of the tetrahedral, charged $B(OH)_4^-$ ion at the crystal surface, the presence of trigonally coordinated B in $CaCO_3$ therefore requires a coordination change. Consequently, Sen et al. (1994) and Hemming et al. (1995) suggested that the similarity of the B coordination state in aragonite and aqueous borate supports greater B incorporation into aragonite (Figure 2.5a), whereas the required coordination change into calcite and its associated energy demand may explain comparably limited B uptake into calcite.

While this interpretation seems reasonable, the question remains why the coordination change would take place in calcite and not in aragonite, where the smaller size of the anion site should favor the coordination change even more than in calcite. However, because the nearest neighbor calculations of Hemming et al. (1995) do not take into account the compressibility of the electron cloud in minerals (Lindsay and Jackson 1994; Skinner et al. 1994), the resulting volume estimates are not necessarily accurate. In particular, the CO_3^{2-} group in aragonite is somewhat non-planar (de Villiers 1971), which results in an asymmetry that suggests shorter C—O bonds on one side of the CO_3^{2-} group, but a longer C—O bond on the other side. This asymmetry may allow aragonite to better accommodate the tetrahedral B ion, despite the overall smaller size of the CO_3^{2-} group compared to calcite (Oscar Branson, personal communication).

To make things more complicated, more recent evidence from ^{11}B MAS NMR analyses indicates the presence of roughly equal trigonal and tetrahedral coordination of boron in both aragonite and calcite samples (e.g. Klochko et al. 2009). Similar observations of trigonal B in biogenic carbonates have subsequently been made by Rollion-Bard et al. (2011) on the deep-sea coral *Lophelia pertusa* and by Cusack et al. (2015) on the coralline algae *Lithothamnion glaciale*. While Rollion-Bard et al. (2011) found ~48% trigonal B in the calcification centre and ~18% trigonal B in the fibers of the *L. pertusa* skeleton, Cusack et al. (2015) found ~30% trigonal B in the coralline algae. The remainder was attributed to tetrahedral B, which possibly exists as two different species, although one of them makes up only ~2% of total incorporated boron (Cusack et al. 2015; Rollion-Bard et al. 2011). Finally, a recent study using synchrotron X-Ray microscopy to investigate B coordination in the benthic foraminifer *Amphistegina lessonii* found B predominantly, if not completely, present in trigonal coordination (Branson et al. 2015), similar to the original results of Sen et al. (1994), Branson et al. (2015) argued that NMR results showing variable proportions of trigonal and tetrahedral B may actually have analyzed interstitial impurities and misinterpreted them within their analytical constraint of having to analyze bulk carbonate samples. While this may seem reasonable, Branson et al. (2015) did not observe any evidence for such impurities in their analyses. This leaves questions as to which method is correct, and would synchrotron analyses find variable coordination in carbonates other than the single foraminifer species analyzed to date? While we cannot answer these questions at this time, we can examine the arguments made by the different studies to date.

In general, the greater abundance of the tetrahedral species in aragonite appears solid (Table 2.1), and B coordination in a Lost City hydrothermal Mg-calcite was found to be 100% tetrahedral, which is consistent with almost exclusive presence of aqueous borate at the assumed precipitation pH > 10 (Klochko et al. 2009). However, the presence of trigonally coordinated boron in $CaCO_3$, in addition to $\delta^{11}B_{CaCO3}$ displaying an increasingly positive deviation from $\delta^{11}B_{borate}$ at lower pH (Figure 2.4), have led some researchers to suggest that boric acid must also play a role in boron uptake by marine carbonates (Cusack et al. 2015; Klochko et al. 2009; Rollion-Bard et al. 2011). Furthermore, the greater coherency of data in Figure 2.6b compared to a denominator defined by $[B(OH)_4^-]/DIC$ (see original study) has also led Uchikawa et al. (2015) to suggest that boric acid may be involved in the incorporation stoichiometry. Some studies have therefore tried to reconcile the abundance of trigonally coordinated boron with evidence from $\delta^{11}B$ analyses in the same carbonates. For instance, Blamart et al. (2007) measured boron isotopes on the same *Lophelia* specimen later studied by NMR (Rollion-Bard et al. 2011), which allowed Rollion-Bard et al. (2011) to combine the isotopic and molecular coordination evidence and establish a new boron isotope incorporation equation for *L. pertusa*. Assuming that seawater-pH is the controlling factor on $\delta^{11}B$ in marine carbonates, Rollion-Bard et al. (2011) obtained reasonable $\delta^{11}B_{Lophelia}$ vs. pH-agreement for the centres of calcification, but the fibers reflect a much higher pH of ~9.15. While such a high pH-value is consistent with microelectrode observations of calcifying fluid-pH elevated up to $pH_{TS} = 9.14$ in shallow water corals (see also Sections 2.4.2 and 2.4.6, Al-Horani et al. 2003; Cai et al. 2016; Venn et al. 2011), these pH estimates may be questioned due to conflicting evidence for the accuracy of $\delta^{11}B$ values measured for *L. pertusa*. Whereas SIMS analyses by Blamart et al. (2007) yield $\delta^{11}B_{Lophelia} = 27.9-38.5$ ‰, McCulloch et al. (2012b) measured $\delta^{11}B_{Lophelia} = 26.6-28.7$ ‰ by P-TIMS, and recent SIMS analyses by Wall et al. (2015) yield $\delta^{11}B_{Lophelia} \sim 20-32$ ‰. Although P-TIMS analyses are based on bulk material, the range of $\delta^{11}B_{Lophelia}$ data obtained in different laboratories and by different techniques needs to be resolved, and suggests that a technique-specific $\delta^{11}B$ elevation may be involved. While such a technique-specific elevation does not diminish the value of SIMS $\delta^{11}B$-analyses, significant offsets have been observed upon bulk analysis of the same standard in different laboratories (C. Rollion-Bard, personal communication) and because the reported $\delta^{11}B$ of a carbonate

Table 2.1 ^{11}B MAS NMR observations of trigonal and tetrahedral B coordination in synthetic and biogenic marine carbonates, corresponding prediction of bulk $\delta^{11}B_{carbonate}$, and comparison to actual $\delta^{11}B$ measurements on the same or similar marine carbonates.

	% BO$_3$	% BO$_4$	$\delta^{11}B_{boric\ acid}$ (‰)	$\delta^{11}B_{borate}$ (‰)	B coordination - predicted bulk $\delta^{11}B_{carbonate}$ (‰)	Observed bulk $\delta^{11}B_{carbonate}$ (‰)
Sen et al. (1994)						
Synthetic calcite	90	10	46.4	18.7	43.6	22.4
Goniolithon sp. (Mg calcite)	80	20	46.4	18.7	40.8	22.4
Synthetic aragonite	0	100	46.4	18.7	18.7	22.4
Montastrea sp. (aragonite)	0	100	46.4	18.7	18.7	24.7
Montastrea calcite (phase transformed from original aragonite)	100	0	46.4	18.7	N/A	N/A
Klochko et al. (2009)						
Assilina ammonoides (calcite)	46	54	46.4	18.7	31.3	23.9[a]
Diploria strigosa (aragonite)	36	64	46.4	18.7	28.5	24.0[b]
Porites sp. (aragonitic coral)	36	64	46.4	18.7	28.5	24.0[b]
Lost City hydrothermal Mg-calcite	0	100	66.8	38.6	38.6	35.5[c]
Rollion-Bard et al. (2011)						
Lophelia pertusa (aragonite, center of calcification)	48	52	46.4	18.7	31.8	31.1[d], 27.6[e]
L. pertusa (aragonite, fibers)	18	82	46.4	18.7	23.6	35.7[d], 27.6[e]
Cusack et al. (2015)						
Lithothamnion glacialis (Mg calcite)	30	70	46.4	18.7	26.9	24.7[f]
(Mavromatis et al. 2015; Noireaux et al. 2015)						
Synthetic aragonite	13	87	56.0	28.0	31.6	21.4
Synthetic calcite	15	85	43.9	16.2	20.3	24.9

Note that $\delta^{11}B_{carbonate}$ predictions are calculated for pH$_{TS}$ = 8.1, T = 25 °C, S = 35, p = 1 bar for all carbonates except the Lost City hydrothermal calcite, which is assumed to have precipitated at pH$_{TS}$ = 10, and the synthetic aragonite and calcite data of Noireaux et al. (2015) and Mavromatis et al. (2015), for which we selected two representative examples where the experiments were performed at pH$_{TS}$ ~ 8.74 and 7.85, respectively. Lost City calcite, *L. pertusa* and *L. glacialis* were precipitated at lower T and higher p in the deep ocean but the exact values are not known for all carbonates. However, adjustment to deep ocean conditions would increase the $\delta^{11}B_{borate}$ fraction, thus elevating the bulk $\delta^{11}B_{carbonate}$ estimate for these carbonates even further. References for actual $\delta^{11}B$ observations are as follows:
[a] value representative for shallow benthic foraminifer *Amphistegina lobifera* (Rollion-Bard and Erez 2010).
[b] representative value for multiple species of shallow water corals, see Figure 2.7c.
[c] estimated from extension of inorganic calibration after Sanyal et al. (2000).
[d] SIMS data measured on the same samples by Blamart et al. (2007).
[e] bulk P-TIMS analysis by McCulloch et al. (2012b).
[f] Fietzke et al. (2015).

sample shifts with the standard reference value, it seems likely that taking the SIMS data at face value and comparing them directly to seawater-pH (Rollion-Bard et al. 2011) is premature. We recommend that the accuracy of a given boron isotope method will first have to be demonstrated before interpretations of biological pH-deviations from seawater can be drawn from $\delta^{11}B$ values.

While most NMR studies have concluded the boron isotope proxy in marine carbonates may be complicated by boric acid incorporation in addition to the presumed dominant borate incorporation, most studies realized that any boric acid adsorption cannot result in quantitative incorporation (see also Balan et al. 2016). This can be demonstrated by a simple mass balance calculation to predict bulk carbonate $\delta^{11}B$ from $\delta^{11}B_{boric\ acid}/\delta^{11}B_{borate}$ with the fractions of trigonal/tetrahedral coordinated B in the crystal lattice (Table 2.1). Comparison of mass balance predictions with actual $\delta^{11}B$ measurements of natural and synthetic $CaCO_3$ indicates the $\delta^{11}B$ values of these carbonates should be several permil higher than observed if anything like the proportion of trigonal B was indeed sourced from aqueous boric acid. $\delta^{11}B$ in the centre of calcification of *L. pertusa* may form an exception here, however, as described above, SIMS and P-TIMS analytical techniques yield conflicting data evidence (Table 2.1), such that this question cannot be settled at this time.

Synthetic calcites provide a strong constraint here, because their $\delta^{11}B$ falls above $\delta^{11}B_{borate}$ at low pH (Figure 2.4), but not high enough to match the supposed boric acid incorporation from lattice coordination constraints, and no biological effects can be invoked. Consequently, at least some if not all of the trigonal B present in marine carbonates must have undergone a coordination change from original tetrahedral borate adsorbed at the crystal surface to boron incorporated into the calcite lattice (Balan et al. 2016; Klochko et al. 2009; Sen et al. 1994). It is therefore not possible to determine from ^{11}B MAS NMR measurements whether the original adsorbed species was borate or boric acid, and this method is consequently not very useful for improving the mechanistic understanding of boron proxies. However, it would be interesting to perform ^{11}B MAS NMR studies on carbonates that were grown under controlled conditions across a wide pH range, such as those experiments performed for Figure 2.4. Such experiments should produce a greater fraction of tetrahedral to trigonal B in carbonates precipitated at higher pH, and any coherency or deviation of NMR predicted $\delta^{11}B$ from observations may shed some light on the incorporation process and associated coordination changes. Mavromatis et al. (2015) and Noireaux et al. (2015) have performed such analyses but their results do not follow this simple prediction. However, their experiments cover a wide range of elemental compositions and several parameters were varied simultaneously, which leaves this question inconclusive.

In summary, NMR studies establish the significant presence of trigonal boron in marine carbonates, but some, if not all of that trigonal boron must have undergone a coordination change from originally adsorbed borate. A potential surface mechanism involving boric acid has been suggested by Tossell (2006), who considered the potential presence of $B(OH)_2CO_3^-$ at the crystal surface. $B(OH)_2CO_3^-$ is a hypothetical molecule that may occur as the product of the reaction between HCO_3^- and either $B(OH)_3$ or $B(OH)_4^-$ (McElligott and Byrne 1997) at the crystal surface. Once incorporated into the carbonate structure, this molecule would break down into boron species of trigonal or tetrahedral coordination, and thus explain the coordination state within the $CaCO_3$ crystal, as well as partial uptake of $B(OH)_3$ at low pH, where boric acid is more abundant (Tossell 2006). In contrast, Balan et al. (2016) used first principles quantum mechanical calculations to identify the geometry of boron species in marine carbonates and suggest that singly protonated $BO_2(OH)^{2-}$ groups are the predominant trigonal B species in calcites, whereas tetrahedral B occurs mostly as $B(OH)_4^-$ groups in aragonites, possibly associated with concomitant substitution of Na^+ for Ca^{2+} to facilitate charge balance. Balan et al. (2016) further suggest that any $B(OH)_3$ molecules in biogenic aragonites may have been scavenged directly from solution, whereas other trigonal species like $BO_2(OH)^{2-}$ and BO_3^{3-}, which are not abundant in seawater, more likely result from coordination change and deprotonation of boron species adsorbed at the surface. In summary, although much progress has been made by coupling NMR observations to theoretical constraints on electron density and overall energy of the mineral structure (Balan et al. 2016), it remains uncertain how much of the observed trigonal boron in marine carbonates has undergone a coordination change. To resolve this issue, new methods are required to identify the boron species adsorbed at the crystal surface as calcification progresses.

To conclude this discussion, it is interesting to consider how the $\delta^{11}B_{CaCO3}$ vs. pH relationship would look like if one assumes a constant distribution between boric acid and borate in solution and $CaCO_3$ (Figure 2.8). McCulloch et al. (2012b) calculated "Kd" = (boric acid/borate)$_{solution}$/(boric acid/borate)$_{CaCO3}$, using values of 0.2 and 0.02. This approach produces strongly curvilinear relationships that disagree with the consistently positive

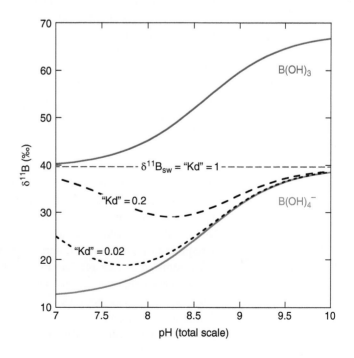

Figure 2.8 Boron isotope partitioning between dissolved boric acid and borate at T = 25 °C, S = 35, p = 1 bar, $\delta^{11}B_{sw}$ = 39.61 ‰ (Foster et al. 2010), $[B_T]$ =432.6 µmol kg^{-1} (Lee et al. 2010) and α_{B3-B4} = 1.0272 (Klochko et al. 2006), as well as $\delta^{11}B_{CaCO3}$ predicted assuming a constant distribution coefficient ("Kd") between boric acid and borate in solution and $CaCO_3$. Figure modified after McCulloch et al. (2012b). Note that the curvilinear relationships are inconsistent with the empirical carbonate calibrations shown, e.g. in Figure 2.4.

empirical carbonate relationships shown in Figure 2.4. Following this exercise, "Kd" cannot be constant and boric acid incorporation would need to obey to more complicated patterns of boric acid/borate incorporation, with less boric acid uptake at low pH (i.e. smaller "Kd") and more boric acid uptake at high pH (i.e. larger "Kd"). It is difficult to conceive why this would happen while boric acid is more abundant at low pH.

2.3.6   The Effect of Precipitation Rate in Synthetic $CaCO_3$

Precipitation rate has also been discussed as a potential reason for boron isotopic disequilibrium between $CaCO_3$ and the parent fluid (Gabitov et al. 2014; Kaczmarek et al. 2016). The theory behind such discussions is generally based on the observation that mineralization may be less ion species and less isotope selective when precipitation rates are fast (e.g. Hobbs and Reardon 1999; McConnaughey 1989). Zeebe et al. (2001) studied the kinetics of boron dissociation and isotope exchange in seawater and calculated ~95 µs as the time required to reach boric acid-borate chemical equilibrium, and ~125 µs to reach boron isotopic equilibrium under standard surface ocean conditions of T = 25 °C and S = 35. Based on these practically instantaneous equilibration times, Zeebe et al. (2001) concluded that kinetic isotope fractionation is unlikely to affect $\delta^{11}B$ in marine carbonates. However, rapid isotopic equilibration between aqueous B species does not necessarily preclude additional kinetic fractionation at the crystal surface, where molecules containing light isotopes may advance faster than molecules containing heavier isotopes (DePaolo 2011; Gabitov et al. 2014; Gussone et al. 2003; Kaczmarek et al. 2016). Applying this theory to explain the more negative $\delta^{11}B_{CaCO3}$ observed by Kaczmarek et al. (2016) upon rapid precipitation (Figure 2.9) we would therefore have to assume that molecules containing ^{10}B are generally faster at diffusing towards the crystal surface. In contrast, Figure 2.4 reveals that $\delta^{11}B_{CaCO3}$ approaches $\delta^{11}B_{borate}$ at higher pH and everything else being equal, higher pH is accompanied by higher Ω and thus precipitation rates (e.g. Allen et al. 2016; Bijma et al. 2002; Langdon et al. 2000; Uchikawa et al. 2015). The precipitation rates reported by Sanyal et al. (2000) follow this pattern, with 100 mg of calcite precipitated within 24, 12, and 3 hours at pH_{TS} 7.77, 8.17 and 8.47, respectively. In contrast to the observation of Kaczmarek et al. (2016), this suggests carbonates approach B isotopic equilibrium at higher precipitation rates, although some $\delta^{11}B_{CaCO3}$ data (Noireaux et al. 2015) fall below $\delta^{11}B_{borate}$ at high pH (Figure 2.4). This may indicate that competing effects are at play across a range of pH, with greater boric acid incorporation at low pH elevating $\delta^{11}B_{CaCO3}$ compared to $\delta^{11}B_{borate}$, and higher precipitation rates at high pH allowing more rapid advance of $^{10}B(OH)_4^-$ and therefore lower $\delta^{11}B_{CaCO3}$ compared to $\delta^{11}B_{borate}$. Given that Kaczmarek et al. (2016) base their interpretation on only two experiments, and that their observed temperature effect on $\delta^{11}B_{CaCO3}$ displays a some-

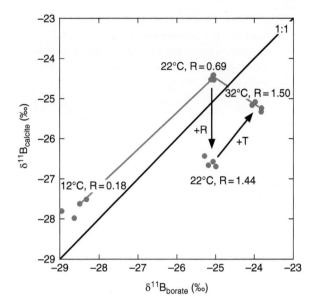

Figure 2.9 Temperature and precipitation rate effects on $\delta^{11}B_{calcite}$ in inorganic precipitation experiments. Precipitation temperatures and rates R (in mg h^{-1}) are indicated next to the respective $\delta^{11}B_{calcite}$ data. Although temperature experiments were performed from 12 to 32 °C and display a non-linear relationship with $\delta^{11}B_{calcite}$, Kaczmarek et al. (2016) focused their interpretation on precipitates grown at the same temperature (i.e. 22 °C, where $\delta^{11}B_{calcite}$ decreases at higher R), and at similar R (R = 1.44–1.5 mg h^{-1}, where $\delta^{11}B_{calcite}$ increases at higher T). The 1 : 1 line is shown for comparison and indicates that none of the $\delta^{11}B_{calcite}$ data are in isotopic equilibrium with $\delta^{11}B_{borate}$.

what complicated pattern across their full temperature range (Figure 2.9), we caution that firm conclusions on this matter are premature at this time. It would be desirable to confirm the inferred precipitation rate effect with several experiments across a range of [Ca^{2+}]. Unfortunately, the data of Gabitov et al. (2014) cannot be used to confirm or negate the findings of Kaczmarek et al. (2016) because they did not monitor the $\delta^{11}B$ of the parent fluid, and the analytical uncertainty of their $\delta^{11}B_{CaCO_3}$ data (i.e. ±2.2 ‰ on average) precludes conclusive interpretation. The general findings of this section and consequent recommendations for future studies on boron proxy systematics in synthetic CaCO$_3$ is given in Box. 2.2.

Box 2.2 Outlook for Studying Boron Proxy Systematics in Synthetic CaCO$_3$

This section highlights that identification of the boron and carbon species adsorbed at the crystal surface is one of the most important puzzles that needs to be resolved before we can fully understand the mechanisms of boron incorporation into marine carbonates. Measuring these interactions is challenging, and probably not possible with existing methods. However, improvements in the spatial resolution and sensitivity of low-energy X-ray and electron spectroscopy techniques, coupled with the use of micro-fluidic reaction cells (De Yoreo and Sommerdijk 2016) could provide useful new insights into this dynamic environment. Ideally, such analyses would be performed over a range of pH, so that surface adsorption, abundance and isotopic composition of incorporated boron can be compared over a range of conditions. Given the observed variability in experimental studies available to date, it seems premature to accept any of these observations as the inorganic baseline against which to compare biogenic carbonates. As we will see later, boron proxy systematics in biogenic carbonates are generally consistent with the observations made in synthetic carbonates, however, some patterns differ greatly, or are even opposite to trends summarized in this section. Ideally, inorganic precipitation experiments should aim to mimic the seawater elemental composition and concentration that living marine organisms experience during secretion of their skeletons and shells. The rate effects observed by many of these experiments further suggest that calcification rates should be monitored in biological systems as much as in inorganic precipitation experiments, and that inorganic experiments should ideally include slower precipitation rates similar to biological systems.

2.4   Boron Isotopes in Biogenic Marine Carbonates

2.4.1   *Boron Isotope Partitioning into Biogenic Marine Carbonates – Evidence from Laboratory Culture Experiments*

The first study calibrating $\delta^{11}B$ in a biogenic marine carbonate was performed by Sanyal et al. (1996), who grew the symbiont-bearing planktic foraminifer *Orbulina universa* in the laboratory from pH_{TS} 7.57 to 8.87 (Figure 2.10a). Although pH in these experiments was originally measured on the NBS scale, we converted these pH values to the total scale (TS), so they can be directly compared to the relative abundance and isotopic composition of borate in seawater (where pK^*_B has been determined on the total scale, Dickson 1990b). At the time when these experiments were performed, α_{B3-B4} had not yet been determined experimentally and Sanyal et al. (1996) compared their results with the $\delta^{11}B_{borate}$ predicted from Kakihana and Kotaka's (1977) theoretical fractionation factor. The empirical $\delta^{11}B_{o.\ universa}$ data match the shape of the theoretically predicted $\delta^{11}B_{borate}$ and show a clear increase with culture water pH, thus confirming the basic $\delta^{11}B$-proxy hypothesis. However, the empirical data fall ~4.4 ‰ below the predicted $\delta^{11}B_{borate}$ of Kakihana and Kotaka (1977). To further investigate this issue, the authors compared their culture data to coretop $\delta^{11}B$ data measured in shells of the symbiont-bearing foraminifer species *Globigerinoides sacculifer* and found that $\delta^{11}B_{o.\ universa}$ is isotopically lighter than $\delta^{11}B_{G.\ sacculifer}$. Based on these comparisons, Sanyal et al. (1996) concluded that a physiological vital effect must be responsible for the low $\delta^{11}B$ values recorded by *O. universa*.

Similar calibration studies have subsequently been performed with the planktic foraminifers *G. sacculifer* (Sanyal et al. 2001), *Globigerinoides ruber* (Henehan et al. 2013), the benthic foraminifer *Amphistegina lobifera* (Rollion-Bard and

Erez 2010) and the tropical coral species *Acropora* sp. (Reynaud et al. 2004), *Porites cylindrica* and *Acropora nobilis* (Hönisch et al. 2004), *Stylophora pistillata* and *Porites* sp. (Krief et al. 2010), and *Cladocora caespitosa* (Trotter et al. 2011), all of which share a similar shape and inflection point with $\delta^{11}B_{borate}$ predicted from Kakihana and Kotaka's (1977) α_{B3-B4}, but they are offset from each other in their absolute $\delta^{11}B$ values (Figure 2.10), with $\delta^{11}B_{coral}$ falling close to predicted $\delta^{11}B_{borate}$ but foraminifers recording lighter boron isotopic compositions than the theoretically predicted $\delta^{11}B_{borate}$.

It is perhaps noteworthy that the three *G. sacculifer* calibration data points generated by Sanyal et al. (2001) describe a slightly steeper $\delta^{11}B$ vs. pH relationship compared to other laboratory calibrations (Figure 2.10a, online Table A2.2), and this is the only biogenic pH-calibration that has been established under 10x boron enriched conditions (cf. Box 2.1). Although increasing the boron concentration in experimental seawater creates a substantial increase in total alkalinity of the experimental seawater, this approach was applied

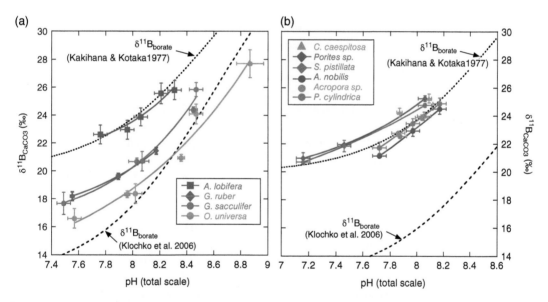

Figure 2.10 Empirical calibrations of $\delta^{11}B$ versus pH in (a) foraminifers and (b) corals. All data are from laboratory culture experiments (see online Table A2.2 for details). Squares indicate SIMS analyses (Rollion-Bard and Erez 2010), circles N-TIMS analyses (Dyez et al. submitted; Hönisch et al. 2004, 2009; Reynaud et al. 2004; Sanyal et al. 1996, 2001), diamonds MC-ICP-MS analyses (Henehan et al. 2013; Krief et al. 2010) and triangles P-TIMS analyses (Trotter et al. 2011). Note that the *Globigerinoides sacculifer* calibration of Sanyal et al. (2001) has been corrected by −1.1 ‰ to match $\delta^{11}B_{G.\ sacculifer}$ data measured on the Thermo TRITON by Hönisch et al. (2009) and Dyez et al. (submitted). Also shown is $\delta^{11}B_{borate}$ calculated for the sea surface at T = 25 °C and S = 35, α_{B3-B4} = 1.0272 (dashed line, Klochko et al. 2006), α_{B3-B4} = 1.0194 (dotted line, Kakihana and Kotaka 1977) and pK^*_B = 8.60 (Dickson 1990b). Least squares exponential regressions have been fit to all calibrations including three or more pH conditions. Note that the calibration for the benthic foraminifer *Amphistegina lobifera* uses the data average for each pH condition, not the individual SIMS data. $\delta^{11}B_{carbonate}$ of all regressions is less sensitive to seawater pH than predicted from $\delta^{11}B_{borate}$ after Klochko et al. (2006).

because planktic foraminifers do not reproduce in the laboratory. Consequently, such planktic foraminifer calibration studies are based on juvenile specimens collected from the open ocean and the portion of the shell grown in the laboratory is then either physically separated (i.e. amputated) from the juvenile shell grown in the ocean (e.g. Sanyal et al. 2001) or a mass balance calculation is performed to determine the shell weight added during the experiment (e.g. Henehan et al. 2013). Amputating foraminifer chambers grown in culture is an accurate but time intensive process and the increase of the experimental seawater [B_T] allowed Sanyal et al. (2001) to limit the number of specimens required for boron isotope analyses. However, this leaves us with the question whether $\delta^{11}B_{G.\ sacculifer}$ is indeed more sensitive to pH and whether experiments performed under boron-enriched conditions accurately reflect the $\delta^{11}B$ of shells grown under natural seawater [B_T]. Fortunately, additional pH experiments performed by B. Hönisch (Figure 2.10a, online Table A2.2, Dyez et al. submitted) in natural seawater support a similar $\delta^{11}B$ vs. pH sensitivity in *G. sacculifer* compared to other biogenic marine carbonates grown in the laboratory and suggest that the pH_{TS} 8.47 calibration point of Sanyal et al. (2001) may either be an outlier or an artifact of the [B_T] enriched seawater approach. Because the amount of boron required for isotopic analyses has decreased substantially since Sanyal's original investigations, experiments in boron enriched or synthetic seawater are now more rarely applied.

While the first empirical carbonate calibration appeared in decisive agreement with $\delta^{11}B_{borate}$ as predicted from Kakihana and Kotaka's (1977) theoretical fractionation factor, the negative $\delta^{11}B$ offsets from $\delta^{11}B_{borate}$, as well as recent theoretical (Liu and Tossell 2005; Oi 2000; Rustad et al. 2010; Sanchez-Valle et al. 2005; Zeebe 2005) and experimental (Klochko et al. 2006; Nir et al. 2015) determinations of $\alpha_{B3-B4} > 1.0194$ seeded doubt about the ability to capture realistic seawater-pH with the boron isotope proxy (Klochko et al. 2006; Liu and Tossell 2005; Pagani et al. 2005; Tripati et al. 2011). The experimental α_{B3-B4} of 1.026–1.0272 (Klochko et al. 2006; Nir et al. 2015) describes a steeper and more sensitive $\delta^{11}B_{borate}$ versus pH relationship than that observed in marine carbonates (Figure 2.10). Comparing empirical carbonate calibrations to $\delta^{11}B_{borate}$ calculated from $\alpha_{B3-B4} = 1.0272$ (Klochko et al. 2006) appeared to resolve the issue of negative $\delta^{11}B_{carbonate}$ offsets from $\delta^{11}B_{borate}$ at least in some symbiont-bearing foraminifers (i.e. *G. ruber* and *G. sacculifer*, Figure 2.11, Foster 2008), but the offsets were now larger at low pH and smaller at high pH. These discrepancies gave rise to an intense debate about issues including the instrumental techniques used to analyze $\delta^{11}B$ (Foster 2008), vital effects (Rollion-Bard and Erez 2010) and the possibility of increased boric acid incorporation at lower pH (Klochko et al. 2009).

In particular the development of the multi-collector inductively coupled mass spectrometry (MC-ICP-MS) technique with its removal of $CaCO_3$ matrix before analysis was originally believed (Foster 2008; Rae et al. 2011) to resolve the lower $\delta^{11}B_{carbonate}$ vs. pH sensitivity established by calibrations using the conventional negative thermal ionization mass spectrometry

(N-TIMS) technique (Hönisch et al. 2004; Reynaud et al. 2004; Sanyal et al. 1996, 2000, 2001). This is because samples of planktic and benthic foraminifers collected from sediments, and analyzed by MC-ICP-MS fall close to $\delta^{11}B_{borate}$ predicted from the experimental $\alpha_{B3-B4} = 1.0272$ (Klochko et al. 2006) (Figure 2.11).

However, naturally occurring pH variations in the ocean are relatively small and limit the calibration range from natural samples. As visualized in Figure 2.3, modern seawater-pH is relatively uniform within the surface ocean habitat of planktic foraminifers, and within the deep ocean environment of benthic foraminifers. The ~0.3 pH unit difference between surface and deep ocean pH encompasses the total pH range present in the ocean, but species specific $\delta^{11}B$ offsets prevent generalizations from combining planktic and benthic calibrations (Foster 2008; Rae et al. 2011). Calibrations using natural samples from a limited pH range therefore do not permit unequivocal identification of the $\delta^{11}B_{carbonate}$ vs. pH sensitivity, as eventually demonstrated by MC-ICP-MS calibrations of $\delta^{11}B$ in the shallow water corals *Porites* sp. and *S. pistillata* (Krief et al. 2010), and the planktic

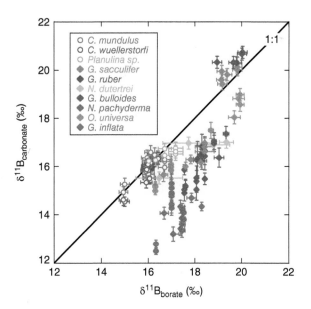

Figure 2.11 Coretop calibrations of $\delta^{11}B$ versus $\delta^{11}B_{borate}$ in planktic (diamonds, Foster 2008; Henehan et al. 2013, 2016; Martínez-Botí et al. 2015a; Yu et al. 2013a) and benthic foraminifera (open circles, Rae et al. 2011). $\delta^{11}B_{borate}$ has been calculated for the respective environmental conditions of the growth habitat of each foraminifer species, where $\alpha_{B3-B4} = 1.0272$ (Klochko et al. 2006) and pK^*_B is temperature, pressure and salinity corrected after Dickson (1990b) and Millero (1995). All calibrations are based on MC-ICP-MS analyses, so relative offsets between calibrations reflect on vital effects, not analytical differences. Because data fall close to the 1 : 1 line between $\delta^{11}B$ and $\delta^{11}B_{borate}$, the respective authors have suggested that $\delta^{11}B$ in planktic foraminifer shells reflects $\delta^{11}B_{borate}$. Later MC-ICP-MS studies with *G. ruber* and corals have confirmed that when grown over a wider pH range than present in the modern ocean, the $\delta^{11}B$ versus $\delta^{11}B_{borate}$ sensitivity is actually smaller than suggested here (see e.g. Figure. 2.10).

foraminifer *Globigerinoides ruber* (Henehan et al. 2013) (Figure 2.10). These biogenic carbonates were grown in the laboratory over a wide range of seawater-pH (i.e. 1 and 0.7 pH units, respectively) and they display $\delta^{11}B_{carbonate}$ vs. pH sensitivities smaller than predicted from $\alpha_{B3-B4} = 1.0272$, as seen in N-TIMS calibrations (Figure 2.10).

The similarity in the $\delta^{11}B_{carbonate}$ vs. pH sensitivity between different techniques can also be seen in Figure 2.12, which shows *O. universa* calibrations based on analyses by N-TIMS, SIMS and MC-ICP-MS. This figure displays the calibration as $\delta^{11}B_{O.\ universa}$ versus $\delta^{11}B_{borate}$ predicted from $\alpha_{B3-B4} = 1.0272$ (Klochko et al. 2006), a technique introduced by Foster (2008) to compare calibrations performed under different environmental conditions (i.e. p, S, T), and thus different pK^*_B. The reason for choosing this relationship over a simple pH-plot is that the calibrations shown in Figure 2.12 were performed in two locations – in Puerto Rico in the Caribbean, where T = 27 °C and S = 34.5, and on Santa Catalina Island in the Pacific, where T = 22 °C and S = 33. The linear slope of the $\delta^{11}B_{carbonate}$ versus $\delta^{11}B_{borate}$ relationship can be

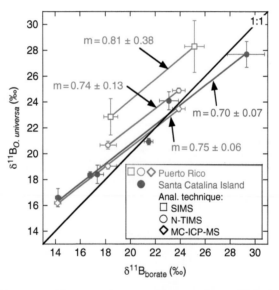

Figure 2.12 Comparison of laboratory calibrations of $\delta^{11}B_{O.\ universa}$ versus $\delta^{11}B_{borate}$ established by different analytical techniques. Squares indicate SIMS analyses (Kasemann et al. 2009), circles N-TIMS analyses (Hönisch et al. 2009; Sanyal et al. 1996) and diamonds MC-ICP-MS analyses (Hönisch and Rae, unpublished data). Different analytical techniques yield different absolute $\delta^{11}B$ values but the calibration slopes (m) are similar for all techniques, such that relative $\delta^{11}B$ differences for a given pH difference are the same. This result demonstrates that different techniques can take advantage of any published calibration for paleoreconstructions, as long as technique specific offsets are accounted for. Note, however, that *Orbulina universa* grown in Puerto Rico (red symbols) records generally higher $\delta^{11}B$ than *O. universa* grown on Santa Catalina Island in California (blue symbols). This offset may be explained by the presence of different *O. universa* genotypes found in California and the Caribbean (Darling and Wade 2008). Such genotype-specific $\delta^{11}B$-difference has not yet been observed in any other foraminifer species and may be unique to *O. universa*.

directly compared to the 1 : 1 line, thus highlighting similarities and differences from aqueous $\delta^{11}B_{borate}$.

There are two important observations to be made from Figure 2.12. First, it is obvious that there are absolute $\delta^{11}B$-offsets between techniques – at the same $\delta^{11}B_{borate}$ (or pH) and in the same measured foraminifer species, SIMS analyses produce higher $\delta^{11}B$ values than N-TIMS, and N-TIMS analyses produce higher $\delta^{11}B$ values than MC-ICP-MS. However, the $\delta^{11}B_{carbonate}$ vs. pH sensitivity (i.e. the slope m of the linear $\delta^{11}B_{foram}$ vs. $\delta^{11}B_{borate}$ regressions) in all these analyses is the same within error (Figure 2.12). We have verified this by calculating York fits of the form $\delta^{11}B_{borate} = (\delta^{11}B_{CaCO3} - c)/m$ for each calibration, where c is the intercept and m the slope of the fit. These calculations were done in Matlab (York et al. 2004), using the 2σ uncertainties in $\delta^{11}B_{CaCO3}$ and $\delta^{11}B_{borate}$. The similarity of the slopes confirms that all three techniques will yield similar paleo-pH reconstructions when analyzing the same material, as long as the technique-specific offsets (i.e. the y-intercepts of the calibration fits, see online Table A2.2) are taken into account. Second, the Caribbean $\delta^{11}B_{O.\ universa}$ culture calibrations yield higher $\delta^{11}B$ values than the Pacific culture calibration. This observation is based on the comparison between N-TIMS analyses performed on Pacific and Caribbean *O. universa* shells, which display an offset, and the fact that Caribbean MC-ICP-MS analyses match Pacific N-TIMS analyses. Because the culture techniques applied for these calibrations were identical (i.e. all experiments were performed in natural seawater), the calibrations are directly comparable, and the observed offsets are significant. Furthermore, because MC-ICP-MS analyses generally yield lower $\delta^{11}B$ values than N-TIMS analyses of the same material (see also the *Cibicidoides wuellerstorfi* offset between analyses done by N-TIMS and MC-ICP-MS in Figure 2.13c and the $\delta^{11}B$ interlaboratory comparison study by Foster et al. 2013), the lack of difference between Caribbean MC-ICP-MS and Pacific N-TIMS data, and the observed offset between Pacific and Caribbean N-TIMS data is consistent with a potential physiological difference (i.e. vital effect) between the two *O. universa* genotypes found in the Caribbean (Type I) and California (Type III) (Darling and Wade 2008; Marshall et al. 2015). A similar deviation in $\delta^{11}B_{O.\ universa}$ has also been observed by Henehan et al. (2016) in specimens from the Gulf of Eilat, which yield ~2 ‰ higher values than other specimens in their calibration data set similar to the offset between N-TIMS calibrations from Pacific and Caribbean specimens observed here. Henehan et al. (2016) argued against a genotype effect, as most specimens in their calibration data set are Type III.

With the currently available data we cannot completely resolve this issue; direct comparison of genotypes and $\delta^{11}B$ data warrants further investigation at least for this species. In contrast, despite the presence of multiple genotypes in most planktic foraminifers species (Darling and Wade 2008), *O. universa* is the only planktic foraminifer for which genotype-based $\delta^{11}B$ offsets may exist. This is borne out by coretop calibrations and paleoreconstructions using shells of *G. sacculifer* and

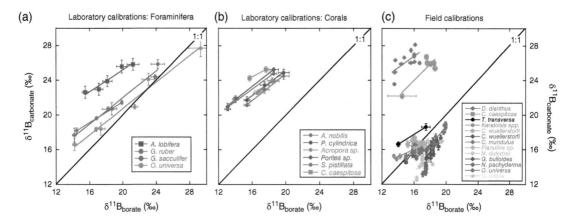

Figure 2.13 Laboratory calibrations of $\delta^{11}B_{carbonate}$ versus $\delta^{11}B_{borate}$ in (a) foraminifers: *Orbulina universa* (Hönisch et al. 2009; Sanyal et al. 1996), *Globigerinoides sacculifer* (Dyez et al. submitted; Sanyal et al. 2001), *G. ruber* (Henehan et al. 2013) and *Amphistegina lobifera* (Rollion-Bard and Erez 2010), (b) corals: *Ac. nobilis* and *Porites cylindrica* (Hönisch et al. 2004), *Acropora* sp. (Reynaud et al. 2004), *Porites* sp. and *Stylophora pistillata* (Krief et al. 2010) and *Cladocora caespitosa* (Trotter et al. 2011). (c) shows field calibrations of biogenic carbonates from coretop sediments: *Cibicidoides wuellerstorfi* (Hönisch et al. 2008; Rae et al. 2011), *C. mundulus* and *Planulina* spp. (Rae et al. 2011), *Neogloboquadrina dutertrei* (Foster 2008), *C. caespitosa* (Trotter et al. 2011), *D. dianthus* (Anagnostou et al. 2012), *Neogloboquadrina*

pachyderma (Yu et al. 2013a), *Terebratalia transversa* (Penman et al. 2013), *Globigerina bulloides* (Martínez-Botí et al. 2015a), *Keratoisis* spp. (Farmer et al. 2015) and *O. universa* and *Globorotalia inflata* (Henehan et al. 2016). All figures are shown across the same 18 ‰ range in $\delta^{11}B$, to highlight the much smaller pH range of field calibrations compared to laboratory calibrations. The smaller pH ranges and greater environmental uncertainty associated with coretop material makes accurate determination of the respective $\delta^{11}B_{carbonate}$ versus $\delta^{11}B_{borate}$ sensitivity difficult. On the other hand, laboratory calibration over a much wider pH range may introduce a growth rate effect on $\delta^{11}B$ (Rae 2018). Squares indicate SIMS analyses, circles N-TIMS analyses and diamonds MC-ICP-MS analyses.

G. ruber, which display consistent patterns between locations and analytical techniques (e.g. Bartoli et al. 2011; Foster and Sexton 2014; Henehan et al. 2013; Hönisch et al. 2009; Seki et al. 2010). Vital effects will be discussed in greater detail in Sections 2.3.1–2.4.7, and here we will return to evaluation of the similarities between the empirical calibrations established to date.

Additional calibrations using specimens grown in their natural environment have meanwhile been established for the planktic foraminifer species *Globigerina bulloides* (Martínez-Botí et al. 2015a), *O. universa* (Henehan et al. 2016), benthic foraminifers (Hönisch et al. 2008; Rae et al. 2011), the shallow water coral *C. caespitosa* (Trotter et al. 2011), deep-sea scleractinian coral *Desmophyllum dianthus* (Anagnostou et al. 2012), and deep-sea octocoral *Keratoisis* sp. (Farmer et al. 2015), as well as brachiopods (Lécuyer et al. 2002; Penman et al. 2013) (Figure 2.13c). Because most of these calibrations used skeletal remains derived from coretop sediments, the covered pH-range is limited, and differences in pressure, temperature, and salinity, and thus

pK^*_B apply. This makes the display shown in Figure 2.10 unsuitable and Figure 2.13c therefore displays $\delta^{11}B_{carbonate}$ versus $\delta^{11}B_{borate}$ predicted from $\alpha_{B3-B4} = 1.0272$ (Klochko et al. 2006). However, even application of sample specific pK^*_B values does not allow unequivocal determination of the individual species' sensitivity to $\delta^{11}B_{borate}$ because the data variability in these natural samples is often large compared to the covered pH range. A notable exception forms the *O. universa* field calibration study of Henehan et al. (2016), which covers a much wider $\delta^{11}B_{borate}$ range than other field calibrations (Figure 2.11). However, the calibration also covers a 19 K-range in temperature, and we will therefore revisit this calibration and potential temperature effects in Section 2.4.8. Although all calibrations conform to the prediction of higher $\delta^{11}B_{CaCO3}$ with higher seawater pH, we recommend caution when applying such natural calibrations to paleoreconstructions. The uncertainty of such calibrations often exceeds the analytical uncertainty (online Table A2.2), and this should be taken into account when inferring paleo-pH.

Because we will discuss species-specific differences in the proxy's sensitivity to $\delta^{11}B_{borate}$ (and pH) throughout this book, and to facilitate comparisons between different biogenic species, we have calculated York fits for each empirical calibration and listed the species-specific coefficients in online Table A2.2. It should be noted that the slopes and intercepts of these fits differ slightly from original publications because we updated the calculations of $\delta^{11}B_{borate}$ and their uncertainty (see empirical calibration data in online Table A2.2) using the recently refined mass balance equations for boron in seawater by Rae (2018). The updated $\delta^{11}B_{borate}$ calculations exert a relatively minor effect on the York fits and the patterns described in the original publications generally remain the same. However, because few calibration studies have reported temperature and salinity uncertainties, our $\delta^{11}B_{borate}$ uncertainty estimates are based only on the reported pH uncertainties. This is not ideal, as the restriction to pH-uncertainties on $\delta^{11}B_{borate}$ underestimates the true uncertainty of each calibration's slope and intercept, but it does not skew the fits significantly. This can be verified by comparing the original and revised *O. universa* field calibration of Henehan et al. (2016), who reported pH, temperature and salinity uncertainties, but calculated $\delta^{11}B_{borate}$ in the same simplified approximation as all other studies before Rae (2018) refined the mass balance equations for aqueous boron in seawater. The original calibration reported by Henehan et al. (2016) used a Monte Carlo approach in R (R-Core-Team 2015), where the calibration uncertainty was determined using a wild bootstrapping approach. This evaluation is statistically more rigorous than the York fits used here, and their reported equation is $\delta^{11}B_{borate} = (\delta^{11}B_{O.\ universa} + 0.42 \pm 2.85)/0.95 \pm 0.17$. After updating the $\delta^{11}B_{borate}$ calculations and including the pH, T and S uncertainties, the simple York fit becomes $\delta^{11}B_{borate} = (\delta^{11}B_{O.\ universa} + 1.03 \pm 1.27)/0.99 \pm 0.07$. Excluding the T and S uncertainties, the York fit becomes $\delta^{11}B_{borate} = (\delta^{11}B_{O.\ universa} + 0.31 \pm 0.86)/0.94 \pm 0.05$. Although the intercept and uncertainties of slope and intercept in this example appear quite variable, slope, and intercept values are within error. The effect of different calibration slopes on the pH-estimate will be discussed in Chapter 3.

In summary, marine carbonates grown in the laboratory over a wide pH range (i.e. >0.5 pH units) display $\delta^{11}B_{carbonate}$ vs. $\delta^{11}B_{borate}$ (or pH) relationships that are less sensitive than predicted from aqueous $\delta^{11}B_{borate}$, assuming borate is the only species incorporated. Using the York fits listed in online Table A2.2, the slopes of all laboratory-calibrated biogenic marine carbonates (including some planktic and benthic foraminifera species, and several species of scleractinian corals) averages m = 0.71 ±0.17. In contrast, field calibrations with symbiont-barren benthic and planktic foraminifera (e.g. *G. bulloides, C. wuellerstorfi, Cibicidoides mundulus*), but also the symbiont-bearing *O. universa* appear as sensitive as predicted from the aqueous boron isotope fractionation (m = 1.11 ±0.43 on average, Figure 2.13c). These differences cannot be explained by differences between analytical techniques, because relative $\delta^{11}B_{carbonate}$ differences between the same type of carbonate (i.e. the same species of coral, foraminifer, etc.) grown at different pH values are comparable between different analytical techniques (e.g. Figure 2.12 and Foster et al. 2013). Consequently, there must be other factors that cause the $\delta^{11}B$ deviation from $\delta^{11}B_{borate}$. Importantly, many paleo-proxies do not conform exactly to theoretical predictions (e.g. Mg/Ca-paleothermometry, Lea 2014), but paleo-reconstructions applying these proxies generate consistent data across ocean basins and time scales. The boron isotope proxy falls remarkably close to theory and observed deviations thus do not preclude accurate reconstruction of seawater pH – as long as carbonate-specific and technique-specific empirical calibrations are available and applied (see also Zeebe 2005). However, because pH-information is also sought to identify and study pH variations at the site of calcification and thus to infer physiological adaptations of marine organisms to ocean acidification (e.g. Blamart et al. 2007; Levin et al. 2015; McCulloch et al. 2012b; Trotter et al. 2011), it is important to identify the cause of these $\delta^{11}B_{carbonate}$ deviations from theory. The following sections will therefore review the evidence for vital effects in biogenic carbonates.

2.4.2 Physiological Modification of the pH Recorded by Marine Calcifying Organisms

Biological controls on boron incorporation and isotope fractionation have been considered since the early studies by Vengosh et al. (1991) and Hemming and Hanson (1992), when it became clear that scleractinian corals record higher $\delta^{11}B$ and B concentrations than e.g. foraminifers. Because modern scleractinian coral skeletons are made of aragonite, and because inorganic aragonite generally incorporates more boron than calcite (Hemming et al. 1995; Kitano et al. 1979; Mavromatis et al. 2015), one might predict that aragonitic skeletons generally record higher boron concentrations than calcitic skeletons. While this prediction appears to be correct in general, there is at least one exception from this rule. The aragonitic benthic foraminifer *Hoeglundina elegans* incorporates much less boron than calcitic species from the same deep ocean environment (Rae et al. 2011; Yu and

Elderfield 2007), and their $\delta^{11}B$ is the lowest of any modern biogenic marine carbonate measured to date (Hönisch et al. 2008; Rae et al. 2011). Clearly, some physiological process must control boron incorporation into this species, but absolute differences in $\delta^{11}B$ and B/Ca occur between most marine calcifying organisms and there are some systematic similarities within certain groups of organisms. We will discuss variations in the total amount of incorporated boron in Section 2.6, but will start our analysis here with the evidence and systematics observed from boron isotopes.

Figure 2.14 summarizes the boron isotope calibrations of the dominant groups of marine carbonates calibrated to date. In general, aragonitic corals record the highest $\delta^{11}B$ values, followed by symbiont-bearing planktic and benthic foraminifers. Benthic foraminifers, deep-sea calcitic corals, brachiopods and symbiont-barren planktic foraminifers record the lowest $\delta^{11}B$ values, and some of them even fall below $\delta^{11}B_{borate}$ (Figure 2.13). Although some exceptions apply (e.g. the aragonitic deep-sea corals *D. dianthus* and *L. pertusa*), there is a general tendency for higher $\delta^{11}B$ to be recorded by those organisms that harbor photosynthetic symbionts. How does photosynthesis affect $\delta^{11}B$, and are there other processes that may play a role as well?

As described in Figure 1.2, pH is a function of the relative proportions of carbon species in seawater. For a given total DIC, the higher the CO_2 concentration, the lower the pH. The dominant physiological processes affecting carbonate chemistry in seawater are the calcification process itself, respiration, and photosynthesis in those organisms that live in symbiosis with photosynthetic algae. Simplified equations for these processes can be described as follows:

$$\text{Calcification}: Ca^{2+} + 2HCO_3^- \rightarrow CaCO_3 + CO_2 + H_2O \quad (2.17)$$

$$\text{Respiration}: C_6H_{12}O_6 + 6O_2 \rightarrow 6CO_2 + 6H_2O \quad (2.18)$$

$$\text{Photosynthesis}: 6CO_2 + 6H_2O \rightarrow C_6H_{12}O_6 + 6H_2O \quad (2.19)$$

The production of CO_2 from calcification and respiration lowers pH in the calcification environment of the organism, whereas photosynthesis consumes CO_2 and thereby elevates pH in the vicinity of the organism. Detailed observations on the calcification process in corals and foraminifers can be found elsewhere (Allemand et al. 2004; Erez 2003; de Nooijer et al. 2014), but a simplified schematic of the physiological processes affecting pH in planktic foraminifers and scleractinian corals is shown in Figure 2.15. According to this schematic, modification of pH in the calcifying fluid from ambient seawater-pH depends on the relative strength of the calcification, respiration, and photosynthesis processes. Fortuitously, the pH modification resulting from these processes has been studied in living organisms using microelectrodes and fluorescent dyes in symbiont-bearing planktic foraminifera (Jørgensen et al. 1985; Köhler-Rink and Kühl 2005; Rink et al. 1998), symbiont-bearing benthic foraminifera (Bentov et al. 2009; Köhler-Rink and Kühl

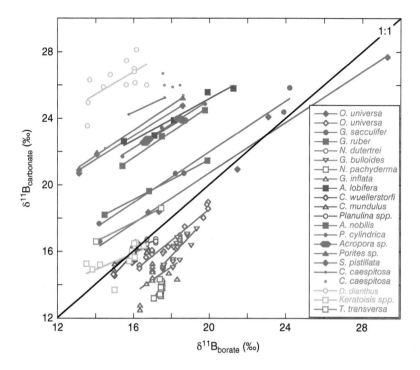

Figure 2.14 Compilation of laboratory (closed symbols) and coretop (open symbols) $\delta^{11}B_{carbonate}$ versus $\delta^{11}B_{borate}$ calibrations as compiled in online Table A2.2 (see table for any corrections and details on analytical techniques, as well as York fits for each calibration and its uncertainties). Colors reflect planktic foraminifers in light blue (Dyez et al. submitted; Foster 2008; Hönisch et al. 2009; Henehan et al. 2013, 2016; Martínez-Botí et al. 2015a; Sanyal et al. 1996, 2001; Yu et al. 2013a), benthic foraminifers in dark blue (Rae et al. 2011), scleractinian corals in red (Hönisch and Hemming 2004; Krief et al. 2010; Reynaud et al. 2004; Trotter et al. 2011), deep-sea corals in green (Anagnostou et al. 2012; Farmer et al. 2015) (although note the difference in phylogenetic order and biomineralization process: *Desmophyllum dianthus* is a scleractinian coral and *Keratoisis* spp. belong to the octocorals), and brachiopods in orange (Penman et al. 2013). Error bars have been omitted for clarity, but can be found in prior figures. Note that laboratory calibrations universally reflect a smaller $\delta^{11}B_{carbonate}$ versus $\delta^{11}B_{borate}$ sensitivity than predicted from aqueous boron isotope fractionation. Some coretop calibrations may share the aqueous fractionation (e.g. planktic foraminifers *Orbulina universa* and *Globigerina bulloides*, deep-sea benthic foraminifers *Cibicidoides wuellerstorfi*, *Cibicidoides mundulus* and *Planulina* sp.), although the scatter in each data set and the small pH range covered by each calibration makes unequivocal comparison difficult. Symbiont-bearing corals and planktic foraminifers tend to record higher $\delta^{11}B$ than $\delta^{11}B_{borate}$, whereas symbiont-barren benthic foraminifers, brachiopods, and the deep-sea coral *Keratoisis* spp. tend to record $\delta^{11}B$ close to or lower than $\delta^{11}B_{borate}$. Note that the *O. universa* field calibration has been established by MC-ICP-MS, which generally reports lower $\delta^{11}B_{carbonate}$ than N-TIMS (used for the *O. universa* laboratory calibration); see Figure 2.12 for details on technique specific analytical offsets. The general tendency of elevated $\delta^{11}B_{carbonate}$ in symbiont-bearing organisms suggests that symbiont-photosynthesis plays a significant role in controlling pH in the calcifying environment of different organisms, although the high $\delta^{11}B$ recorded by the symbiont-barren *D. dianthus* forms a striking exemption from this rule.

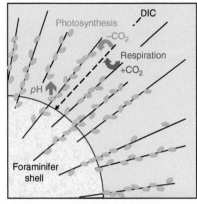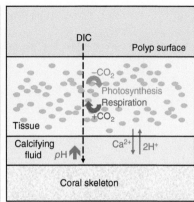

Figure 2.15 Simplified schematic of the calcifying environment in symbiont-bearing planktic foraminifers and scleractinian corals. While all organisms calcify and respire, and thereby lower the pH in their calcifying environment, symbiont photosynthesis sequesters CO_2 and thereby elevates pH in the calcifying environment. DIC has to pass the symbiont halo in planktic foraminifers, and the symbiont-bearing tissue in corals, thus contributing to the elevated pH and $\delta^{11}B_{carbonate}$ recorded by these organisms. Active modification of the calcifying fluid via ion pumping may further contribute to the pH elevation, the reader is referred to studies by de Nooijer et al. (2014), Erez (2003) and Allemand et al. (2004) for further details.

2000) and scleractinian corals (Al-Horani et al. 2003; Cai et al. 2016; Holcomb et al. 2014; Kühl et al. 1995; Venn et al. 2011, 2013). The results of these studies consistently suggest pH elevation to 8.5 and higher, and because of this similarity, only a subset of coral studies is displayed in Figure 2.16 alongside foraminifera studies. Information on the applied pH scales is missing for some studies and instead had to be inferred from reported ambient seawater-pH values, so therefore the absolute pH values (Figure 2.16a) may not be entirely accurate. To account for this uncertainty, and to directly compare the extent of pH modification between different organisms studied under different ambient pH conditions, all data are normalized to their respective ambient pH condition and the pH deviation (ΔpH) is displayed in Figure 2.16b. It should further be noted that microelectrodes cannot puncture the shells of benthic foraminifers and Köhler-Rink and Kühl (2000) therefore measured the pH at the shell surface, which may underestimate the pH of the internal calcification vacuoles. In contrast, Bentov et al. (2009) reported internal pH, which was determined using the fluorescent, pH-sensitive dye SNARF-1. The organisms studied for Figure 2.16 are all symbiont-bearing, so pH is displayed relative to experimental light intensity, and it can clearly be seen that pH at or near the site of calcification is elevated in the light in all species. Whereas all studies report relatively similar microenvironmental pH_{TS} values of ~8.5 under saturating light levels, and lower values in the dark, the extent of pH modification differs somewhat between species. For instance, the corals *Favia* sp. (Kühl et al. 1995) and *Galaxea fascicularis* (Al-Horani et al. 2003) elevate pH by >1.1 pH units in the light compared to the dark, but *Acropora* sp. elevates pH by only

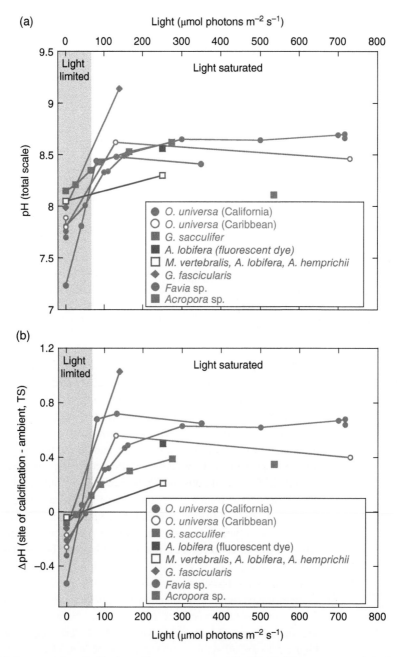

Figure 2.16 (a) Comparison of light-dependent pH-modification in symbiont-bearing planktic (light blue symbols) and benthic foraminifers (dark blue symbols), as well as scleractinian corals (red symbols). All estimates are based on microelectrode studies (Al-Horani et al. 2003; Jørgensen et al. 1985; Köhler-Rink and Kühl 2000, 2005; Kühl et al. 1995; Rink et al. 1998), respectively pH-sensitive dyes (Bentov et al. 2009). All pH data are presented on the total scale, although in some studies the original pH scale has not been defined, in which case the original pH scale has been guessed from reported ambient seawater-pH values. Note that microelectrodes can puncture coral tissue but not benthic foraminifer shells, where calcification is believed to take place inside the shell. Microelectrode studies therefore observe expression of

0.35 units above ambient (Kühl et al. 1995). In comparison, the benthic foraminifer *A. lobifera* elevates pH by ~0.5 pH units (Bentov et al. 2009; Köhler-Rink and Kühl 2000), whereas the planktic foraminifers *O. universa* and *G. sacculifer* elevate pH by ~0.6–1 and ~0.5 units, respectively (Jørgensen et al. 1985; Köhler-Rink and Kühl 2005; Rink et al. 1998). With the data of Cai et al. (2016) and Venn et al. (2011) included, these organisms rank *G. fascicularis* > *Favia* sp./*S. pistillata*/*Orbicella favoleata* > *O. universa* > *G. sacculifer*/*A. lobifera*/*Marginopora vertebralis*/*Amphisorus hemprichii* > *Turbinaria reniformis*/*Acropora* sp./*Acropora millepora* in terms of greatest pH elevation. Microelectrode pH measurements have not yet been performed for *G. ruber*, but Lombard et al. (2009) measured photosynthesis and respiration in *G. ruber*, *O. universa* and *Globigerinella siphonifera* and determined that the photosynthesis-to-respiration ratio decreases from *O. universa* to *G. ruber* to *G. siphonifera*. Assuming that respiration and calcification rates may be similar for all three species (an assumption that needs to be tested), one could therefore predict that pH-elevation in these species should rank *O. universa* > *G. ruber* > *G. siphonifera*. However, it should be noted that the *G. ruber* specimens were much smaller than the *G. siphonifera* and *O. universa* specimen studied by Lombard et al. (2009), which may have introduced a symbiont density effect. These uncertainties leave the ranking of physiological effects on the microenvironmental pH somewhat inconclusive.

While the greatest pH elevation in corals is consistent with highest $\delta^{11}B$ values for this group of calcifiers, *Acropora* species show similar $\delta^{11}B$ to *Porites* and *Stylophora*, but lesser pH elevation (Figure 2.16 and 2.14). Similarly, $\delta^{11}B$ elevation in *A. lobifera* is greater than predicted from actual pH measurements, and $\delta^{11}B$ data would suggest greater pH elevation for *G. sacculifer* and *G. ruber* compared to *O. universa* (e.g. Figure 2.14), which is contrary to the pH observations (Figure 2.16). Before we can discuss the differences with regard to boron isotope fractionation, we must consider whether some of these differences may be due to inaccuracies of the respective boron isotope analytical technique applied. For instance, SIMS analyses (as applied for *A. lobifera*, Rollion-Bard and Erez 2010) appear to produce

Figure 2.16 (*Continued*) CO_2 gas exchange at the foraminifer shell surface only, whereas fluorescent dyes allow observation of internal pH. pH-modification in the dark has not been recorded by all studies. Because of uncertainties in the original pH scale, and differences in ambient pH, (b) presents all data as the deviation from ambient pH, i.e. ΔpH. It is clear from this comparison that scleractinian corals elevate pH in their calcifying fluid to a greater degree than foraminifers, all organisms experience negative ΔpH at light levels that are unsaturating for photosynthesis, suggesting that CO_2 production from respiration and calcification predominates at low light levels and in the dark. Predicting $\delta^{11}B_{carbonate}$ from these observations thus suggests $\delta^{11}B_{coral} > \delta^{11}B_{planktic, symbiotic foraminifer} \geq \delta^{11}B_{A. lobifera}$. This prediction is not entirely met by $\delta^{11}B$ records, although the SIMS technique applied by Rollion-Bard and Erez (2010) to *Amphistegina lobifera* tends to report higher $\delta^{11}B$ values than other techniques (see Figures 2.12 and 2.13 for comparison). Similarly, pH elevation in Caribbean *Orbulina universa* is not greater than in Californian *O. universa*, as one might expect from greater $\delta^{11}B_{O. universa}$ in the Caribbean (see Figure 2.12 for comparison).

generally higher $\delta^{11}B$ values than other techniques (e.g. Figure 2.12 but also an 8 ‰ $\delta^{11}B$ offset between SIMS and P-TIMS analyses on *L. pertusa*, Blamart et al. 2007; McCulloch et al. 2012b). Similarly, the N-TIMS technique applied for the *G. sacculifer* (Dyez et al. submitted; Sanyal et al. 2001) and *O. universa* (Hönisch et al. 2009; Sanyal et al. 1996) calibrations shown in Figure 2.14 produce higher $\delta^{11}B$ values than MC-ICP-MS analyses of the same species (Hönisch and Rae, unpublished data; Martínez-Botí et al. 2015a). However, both MC-ICP-MS and N-TIMS measure $\delta^{11}B_{G.\ sacculifer}$ > $\delta^{11}B_{O.\ universa}$ (compare Figures 2.11, 2.12 and 2.14), and any prediction of pH at the site of calcification for these two species is therefore inconsistent with microelectrode analyses that predict $pH_{O.\ universa}$ > $pH_{G.\ sacculifer}$. Similarly, MC-ICP-MS analyses yield $\delta^{11}B_{G.\ ruber}$ > $\delta^{11}B_{O.\ universa}$ (compare Figures 2.11 and 2.12). Furthermore, *O. universa* cultured in Puerto Rico record higher $\delta^{11}B$ than the same species cultured in California (Section 2.4.1 and Figure 2.12), but microelectrode studies predict a smaller pH elevation for the Caribbean genotype (Köhler-Rink and Kühl 2005) compared to the California genotype (Rink et al. 1998).

Predicting pH at the site of calcification of an organism is apparently not as straightforward as some authors would like to suggest, but studying deviations from the above predictions may help to improve our understanding of the calcification process of different organisms. Potential explanations include uncertainties of local light intensity, different species and genotypes, size, the relative proportion of day-time calcification, when photosynthesis predominates, to night-time calcification, when respiration and calcification are the only processes affecting organismal physiology. In addition, some organisms may actively modify the elemental composition of the calcification fluid (e.g. Allemand et al. 2004; de Nooijer et al. 2014; Erez 2003). The relative importance of physiological processes on carbonate and stable carbon, oxygen and boron isotope systematics in individual planktic foraminifer species has been explored using a diffusion-reaction model (Wolf-Gladrow et al. 1999; Zeebe et al. 1999, 2003) and the reader is encouraged to digress into that literature. Here we will focus on direct observations of living organisms and their fossil remains. Systematic patterns of $\delta^{11}B$ variability in symbiont-bearing organisms have been associated with ambient light conditions, and these patterns have been studied in the laboratory and in the natural environment.

2.4.3   *Light Effects on Symbiont-Bearing Planktic Foraminifers*

The symbiont-bearing planktic foraminifer *O. universa* has been grown in the laboratory under saturating ($\sim$320 µmol photons m^{-2} s^{-1}) and photosynthesis-limiting light conditions ($\sim$20 µmol photons m^{-2} s^{-1}) on Santa Catalina Island in California (Hönisch et al. 2003). This is the same laboratory where the microelectrode studies of Rink et al. (1998) were performed, and isotopic and instrumental data are therefore directly comparable. Hönisch et al. (2003) measured 1.5 ‰ higher $\delta^{11}B$ in shells grown under light-saturating conditions compared to low-light conditions and translated this to an ~0.2 pH unit difference (ΔpH) between the two experiments. This ΔpH~0.2 is much smaller than the ΔpH~0.75 determined by microelectrode analyses for the same light conditions, but taking into account that the symbiont community of this foraminifer species photosynthesizes for only four to six hours per day at the full rate (Spero and Parker 1985), then one third of the foraminifer shell is precipitated at night (Lea et al. 1995). As a result, photosynthesis must be somewhat greater than zero under the experimental low light conditions (Rink et al. 1998). Hönisch et al. (2003) calculated an integrated pH modification due to photosynthesis of ΔpH ~0.24, which is within the uncertainty range of the $\delta^{11}B$-based ΔpH estimate. Using their diffusion-reaction model, Zeebe et al. (2003) predicted a similar light modulated $\delta^{11}B_{O.\ universa}$ difference (i.e. 1.6 ‰) as determined experimentally, but a larger difference for *G. sacculifer* (i.e. 3.2 ‰).

The exact difference depends on the assumed symbiont density associated with the foraminifer, the ratio of day-time to night-time calcification (e.g. *G. sacculifer* secretes ~40% of its shell at night, Anderson and Faber Jr 1984), and the buffering capacity of the experimental seawater, which was elevated in the experiments of Hönisch et al. (2003) due to their use of 10x natural seawater-B concentrations. Nevertheless, the agreement between observations and predictions in *O. universa* is convincing for the boron isotope proxy to reflect pH variations even under physiologically altered conditions.

Additional constraints on light effects on $\delta^{11}B_{foram}$ come from shell sizes of symbiont-bearing foraminifers. Foraminifer shell size effects have been observed for a number of proxies, including carbon and oxygen isotopes (e.g. Oppo and Fairbanks 1989; Spero et al. 2003) and various trace element ratios (e.g. Anand et al. 2003; Ni et al. 2007). Observed differences have often been associated with different habitat depths, where environmental controls differ between the shallowest and deepest growth habitats. Size effects on $\delta^{11}B$ in symbiont-bearing foraminifera have so far been studied for *G. sacculifer* and *G. ruber* (Figure 2.17, Henehan et al. 2013; Hönisch and Hemming 2004; Ni et al. 2007), *O. universa* and, with very few data, in *Globigerinella siphonifera* (Henehan et al. 2016). In coretop sediments from the Ontong-Java Plateau in the Pacific Ocean and the 90°E Ridge in the Indian Ocean, Hönisch and Hemming (2004) observed ~2 ‰ higher $\delta^{11}B$ in the largest *G. sacculifer* shells compared to the smallest shell size class. Translated to $pH_{TS} < 7.7$, the $\delta^{11}B$ data of the smallest shells would suggest a growth habitat depth > 500 m at these locations, which is inconsistent with observations from plankton nets and temperature proxy evidence that place *G. sacculifer*'s growth habitat within the upper 20–100 m of the water column (Fairbanks et al. 1980, 1982; Farmer et al. 2007; Spero et al. 2003). Following observations that *G. sacculifer* grows larger shells in the laboratory when experimental light levels are higher (Spero and Lea 1993), and experimental (Hönisch et al. 2003) and

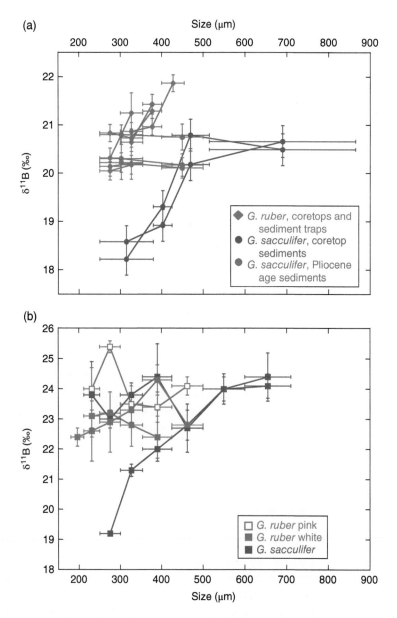

Figure 2.17 Shell size effects on foraminiferal δ¹¹B have been studied in *Globigerinoides sacculifer* (blue symbols) and *G. ruber* (red symbols). (a) compares data from MC-ICP-MS (Henehan et al. 2013) and N-TIMS analyses (Bartoli et al. 2011; Hönisch and Hemming 2004), (b) compares data collected by TE-N-TIMS (Ni et al. 2007). The relative results obtained by these methods are comparable, only the uncertainty and absolute values of the TE-N-TIMS technique are much higher compared to MC-ICP-MS and N-TIMS, which hampers visual comparison in a single graph. Both foraminifers harbor dinoflagellate symbionts of the species *Gymnodinium béei*, but not all samples show a shell size effect and no spatial or temporal pattern can be identified that would allow for prediction of the presence or absence of the effect in a given location or time interval. Where the shell size effect occurs, lower δ¹¹B is recorded in smaller shell size fractions. However, the δ¹¹B of the largest shell size fractions is similar in all samples (observed differences in absolute δ¹¹B-values can be explained by differences in local seawater-pH), suggesting that focusing paleoreconstructions on the larger shell size fractions will yield consistent results between locations and time scales. When using smaller shell sizes, the presence of the shell size effect can be evaluated in discrete samples, and corrected for a constant offset from calibrations if necessary (see also Henehan et al. 2013).

theoretical (Zeebe et al. 2003) predictions of higher $\delta^{11}B$ in symbiont-bearing foraminifers grown under high light conditions, Hönisch and Hemming (2004) suggested the higher $\delta^{11}B$ could therefore reflect on the foraminifers' growth environment: Larger specimens appear to spend relatively more time near the sea surface, where light levels are higher, stronger photosynthesis elevates pH in the foraminiferal microenvironment, specimens grow larger shells and record higher $\delta^{11}B$ compared to specimens that spend relatively more time deeper in the water column. In addition, or alternatively, the symbiont density of larger shells could also be greater (Spero and Parker 1985), and thus contribute to the larger shell growth and elevation of the recorded pH (Henehan et al. 2013; Hönisch and Hemming 2004).

Similar observations have also been made by Ni et al. (2007) and Henehan et al. (2013) for *G. sacculifer* and *G. ruber*, and by Henehan et al. (2016) for *O. universa* and *G. siphonifera*. However, although $\delta^{11}B$ is generally somewhat higher in larger specimens of *O. universa* and *G. siphonifera* analyzed by Henehan et al. (2016), those data are somewhat more scattered than for other species, and in particular for *O. universa* suggest a more limited size effect (see original publication for visual presentation of the data). Although observed in at least some specimens of the symbiont-bearing species studied to date, these shell size effects are not observed everywhere. For instance, although their analytical uncertainty is relatively large, Ni et al. (2007) did not observe systematic shell size effects in *G. ruber*, and one core location in the South Atlantic did not produce the shell size effect in *G. sacculifer* (Figure 2.17b). Ni et al. (2007) therefore concluded that the shell size effect in *G. sacculifer* may instead be related to this species' tendency to secrete a gametogenic calcite layer upon reproduction at greater water depths, and removal of the ontogenetic calcite upon partial shell dissolution would shift the bulk shell $\delta^{11}B$ towards the lower $\delta^{11}B$ of the gametogenic crust. Because *G. ruber* is not known to secrete a gametogenic crust, dissolution effects should not affect this species and $\delta^{11}B$ would therefore be the same in all size classes. However, Henehan et al. (2013) subsequently observed shell size effects in *G. ruber* similar to *G. sacculifer*, and because the authors could exclude any dissolution effect on their studied material, they agreed with Hönisch and Hemming (2004) that the size effect is likely due to photosynthetic pH elevation.

Similarly, Bartoli et al. (2011) did not observe a shell size effect in Pliocene age *G. sacculifer* from the Caribbean (Figure 2.17a), and Penman et al. (2014) found no $\delta^{11}B$ difference between two size classes of their Paleocene and Eocene age, symbiont-bearing *Morozovella velascoensis* (not shown). Dissolution effects will be discussed in Section 2.6.4 and while dissolution can contribute to the shell size effect (Hönisch and Hemming 2004), it is unclear why the shell size effect is absent in some samples. Location specific depth habitat differences may explain this observation, but this hypothesis cannot even be verified with plankton nets, because foraminifers migrate throughout their life cycle (e.g. Hemleben and Bijma 1994) and it is difficult to predict whether a live-collected specimen had already reached its final size at collection or whether it would have grown larger if its life had not been terminated prematurely.

It is also not known if different size fractions of the same foraminifer species share the same $\delta^{11}B$ vs. pH sensitivity, and the limited pH range in the surface ocean will make it difficult to verify this possibility (e.g. Figures 2.11 and 2.13c). For the present time, by relying on the similarity of $\delta^{11}B$ vs. pH sensitivities observed for planktic and benthic foraminifers (Figure 2.13a), it seems reasonable to apply a size-specific correction to the intercept of a species-specific pH calibration. Furthermore, it is important to restrict paleoreconstructions to narrow size fractions and single species, so that any size or species effect on the reconstruction can be excluded (Henehan et al. 2013; Hönisch and Hemming 2004). Additional observations of physiological pH-modification in planktic foraminifers are desirable to corroborate these hypotheses, including instrumental (i.e. microelectrode) analyses of different ontogenetic stages and shell sizes, analysis of additional species such as *G. ruber*, but also symbiont-barren species such as *G. bulloides* and *Neogloboquadrina pachyderma*, whose integrated effect of respiration and calcification on pH in their microenvironment remains unquantified. Some limited data for the symbiont-barren *G. bulloides* and the facultative chrysophyte-bearing *Globorotalia inflata* suggest that $\delta^{11}B$ in these species is relatively invariant with shell size, or possibly slightly decreases in larger specimens (Henehan et al. 2016). Even though any inference of higher respiration rates in larger specimens would be premature in the light of the scattered $\delta^{11}B$ data, the lack of a positive trend in $\delta^{11}B$ with size in *G. bulloides* is consistent with the absence of photosynthetic symbionts. The photosynthetic activity of *G. inflata*, when present, has not yet been studied, but chrysophytes in *G. siphonifera* have been found to be specifically adapted to low light conditions (Takagi et al. 2016), suggesting that the symbiont-density in *G. inflata* must be generally rather low to prevent pH elevation in the microenvironment of this species. Importantly, coretop and downcore $\delta^{11}B$ analyses of the paleoceanographically important species *G. ruber* and *G. sacculifer* are consistent across a wide range of environments and time scales, and we have no evidence for any location-specific deviations from pH-control on $\delta^{11}B$. So, although we do not understand all details of the mechanism by which planktic foraminifers record $\delta^{11}B$, we have no evidence that paleo-pH reconstructions are compromised by vital effects, as long as samples are analyzed within a restricted size fraction. To avoid any unknown size effects, down core studies will ideally use the same size fraction as used in the calibration study.

2.4.4 *Light Effects on Symbiont-Bearing Benthic Foraminifers*

Laboratory calibrations of deep-sea benthic foraminifers remain a challenge and we consequently do not have any benthic foraminifer calibrations that allow us to unequivocally evaluate their $\delta^{11}B$ sensitivity to pH. In contrast, shallow benthic foraminifers can be cultured quite easily, but in contrast to

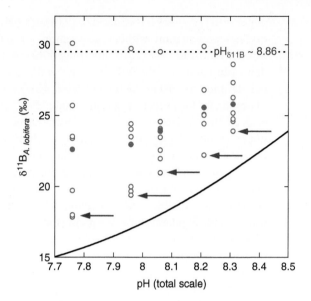

Figure 2.18 Laboratory δ^{11}B versus pH calibration of the symbiont-bearing shallow water benthic foraminifer *Amphistegina lobifera* (Rollion-Bard and Erez 2010). Open symbols indicate individual spot analyses by SIMS, closed symbols reflect averages of individual SIMS measurements under each pH condition. Because δ^{11}B$_{borate}$ has been determined for the experimental conditions on the total scale following Dickson (1990b) and α_{B3-B4} after Klochko et al. (2006), culture water pH (originally on the NBS scale) has been converted to the total pH scale (the gray arrows indicate the extent of the pH correction). Lower bound δ^{11}B$_{A.\ lobifera}$ data trace δ^{11}B$_{borate}$ with a +2.6 ‰ offset, but upper bound δ^{11}B$_{A.\ lobifera}$ data are more or less constant and translate to calcification site pH$_{TS}$ ~ 8.86. Calcification site pH elevation above experimental seawater pH has been interpreted to reflect the activity of a Na$^+$/H$^+$ exchanger that contributes to modifying the chemistry of the biomineralization vacuole relative to ambient seawater (Rollion-Bard and Erez 2010).

their deep-sea relatives they often host symbionts. To date the only laboratory δ^{11}B vs. pH calibration on benthic foraminifers has been performed with the symbiont-bearing *A. lobifera*, and δ^{11}B of these samples has been analyzed by SIMS, revealing large intra-shell variability and the internal range of calcification-pH, in addition to bulk shell responses to external pH (Rollion-Bard and Erez 2010). Figure 2.18 displays both the intra-shell and bulk shell δ^{11}B$_{A.\ lobifera}$ data versus culture water pH. In contrast to the original study, which compared δ^{11}B$_{A.\ lobifera}$ with pH on the NBS scale to δ^{11}B$_{borate}$ on the total scale (Rollion-Bard and Erez 2010), Figure 2.18 shows all data versus pH on the total scale. This correction is necessary because δ^{11}B$_{borate}$ has been determined for the experimental conditions on the total scale following Dickson (1990b) and α_{B3-B4} after Klochko et al. (2006). In contrast to the original presentation by Rollion-Bard and Erez (2010), this revised presentation results in an ~2.6 ‰ offset between the lowest intra-shell δ^{11}B$_{A.\ lobifera}$ data at each pH and the corresponding δ^{11}B$_{borate}$. Importantly, the intra-shell δ^{11}B$_{A.\ lobifera}$ variability is much greater at low pH than at high pH, but the highest δ^{11}B$_{A.\ lobifera}$ ~29.5 ‰ is the same for all pH treatments. Whereas the lower bound on δ^{11}B$_{A.\ lobifera}$ at each pH traces the shape of the δ^{11}B$_{borate}$ curve, translating the upper $\delta^{11}B_{A.\ lobifera}$ ~29.5 ‰ bound to pH yields pH_{TS} ~8.86. Following their original presentation without accounting for differences in pH scales, Rollion-Bard and Erez (2010) interpreted their pH estimates to reflect two dominant controls – the $\delta^{11}B$ of borate as the lower bound, and symbiotic pH elevation to ~8.9 as the upper bound. However, the only peer reviewed pH estimates for *A. lobifera* give internal pH~8.7 on the NBS scale (Bentov et al. 2009), which translates to pH 8.56 on the total scale. Consequently, the upper bound $\delta^{11}B_{A.\ lobifera}$ pH_{TS} estimate exceeds pH observations at the site of calcification by ~0.3 pH units, and the lower bound $\delta^{11}B_{A.\ lobifera}$ exceeds $\delta^{11}B_{borate}$ by ~2.6 ‰. While some of the mismatch between $\delta^{11}B_{A.\ lobifera}$ data and predictions of pH and $\delta^{11}B_{borate}$ may be explained by analytical offsets (see e.g. elevated $\delta^{11}B_{SIMS}$ data compared to other techniques in Figure 2.12), this example highlights how comparing data on different pH scales can bias data interpretation (see also Chapter 1 and Table 1.1 for discussion of the different pH scales).

Irrespective of any arguments regarding the accuracy of these data (see also discussion of $\delta^{11}B$ accuracy obtained by different analytical techniques in Chapter 4), the $\delta^{11}B_{A.\ lobifera}$ range observed by Rollion-Bard and Erez (2010) at different experimental pH values appears to reflect pH variation at *A. lobifera*'s site of calcification. Although Rollion-Bard and Erez (2010) could not associate their data to any specific shell structures (all data were collected on the massive shell "knobs"), Bentov et al. (2009) observed pH elevation in the vacuoles that are believed to supply the calcification process in benthic foraminifers, and their observed pH elevation generally agrees with $\delta^{11}B_{A.\ lobifera} > \delta^{11}B_{borate}$. Whether the elevation in $\delta^{11}B_{A.\ lobifera}$ is a by-product of symbiont photosynthesis or the result of active pH-upregulation remains to be determined, and the study of the actual range of pH in foraminiferal calcification vacuoles (Bentov et al. 2009) might help our understanding. It would be particularly helpful if the SNARF-1 method could be applied to foraminifers maintained at different light levels, so that the observations displayed in Figure 2.16 can be expanded. Furthermore, SIMS analyses of shallow-water symbiont-barren and deep-sea benthic foraminifers would be very useful to evaluate whether the intra-shell variability observed in *A. lobifera* extends to other foraminifer species, and that intra-shell variability is indeed a function of symbiont photosynthesis in *A. lobifera*. Relevant data for the deep-sea benthic foraminifer species *C. wuellerstorfi* have been presented by Sadekov et al. at the International Conference on Paleoceanography in 2016 but remain to be published. Those data suggest a 10 ‰ range in $\delta^{11}B$, suggesting that active pH regulation is present in this species.

Importantly, the averages of $\delta^{11}B_{A.\ lobifera}$ data collected at each pH condition describe a $\delta^{11}B_{A.\ lobifera}$ vs. pH sensitivity similar to other marine carbonates calibrated over a wide pH range (e.g. Figure 2.14). This observation stands in contrast to coretop calibrations of deep-sea benthic foraminifers, and begs the question whether the pH-sensitivity of $\delta^{11}B$ in coretop benthic foraminifera is indeed similar to aqueous boron isotope fractionation (Rae et al. 2011). We will discuss this caveat in Section 2.4.8, but while unequivocal data remain to be generated, we recommend that the benthic foraminiferal $\delta^{11}B$ SIMS calibration should be replicated using bulk shell material of off-spring produced under experimental conditions (i.e. clear separation of shell material grown in culture compared to shell material secreted in natural seawater). While bulk shell techniques like MC-ICP-MS or TIMS do not allow for analysis of intra-shell variability, they average $\delta^{11}B$ of a larger number of shells and thus reduce potential sampling bias. Bulk shell calibration is particularly important for paleoreconstructions, where investigators typically seek to average out individual shell variability. Benthic foraminifer calibrations will likely have to focus on shallow-water species until mass culturing of deep-sea benthic foraminifers can be achieved, but in the meantime experiments with shallow benthic foraminifers would still advance our goal of understanding boron fractionation into marine carbonates.

2.4.5   Vital Effects in Asymbiotic Foraminifers

As indicated above, the dominant physiological processes affecting asymbiotic calcifiers are the calcification process itself and respiration, both of which produce CO_2 and thereby lower pH in the calcifier's microenvironment. Rae et al. (2011) observed that the $\delta^{11}B$ recorded by infaunal benthic foraminifera is depleted by up to ~2 ‰ relative to bottom water-$\delta^{11}B_{borate}$ and also compared to epifaunal benthic foraminifera. A much larger deviation up to -3.5 ‰ was observed in the epibenthic species *H. elegans*. However, as the only aragonitic deep-sea benthic foraminifer species *H. elegans* is rather unusual and appears to bear vital effects that are not easily translated to other species. Because the -2 ‰ deviation from bottom water-$\delta^{11}B_{borate}$ is in reasonable agreement with lower pH and aqueous $\delta^{11}B$ of porewater caused by organic matter degradation and respiratory calcite dissolution in the sediment, Rae et al. (2011) did not consider vital effects a major factor in (infaunal) benthic foraminifera.

In contrast, $\delta^{11}B$ of the symbiont-barren planktic foraminifers *Neogloboquadrina dutertrei*, *G. bulloides* and *N. pachyderma* and the facultative chrysophyte-bearing *G. inflata* falls below $\delta^{11}B_{borate}$ of the seawater these organisms grew in, suggesting their microenvironmental pH is acidified relative to ambient seawater (Figure 2.11, Foster 2008; Hönisch et al. 2003; Martínez-Botí et al. 2015a; Yu et al. 2013a). To the best of our knowledge, pH-microelectrode studies have not yet been performed in symbiont-barren organisms, but microelectrode studies on symbiont-bearing organisms in the dark hint at microenvironmental acidification by order 0.1–0.5 pH units (Figure 2.16), consistent with our prediction for symbiont-barren calcifiers.

Using the example of *N. pachyderma* (Figure 2.11), and assuming that $\delta^{11}B_{N.\ pachyderma}$ reflects $\delta^{11}B_{borate}$ (i.e. applying $\alpha_{B3-B4} = 1.0272$, Klochko et al. 2006) at the site of calcification, we estimate the calcification site pH in *N. pachyderma* is on average ~0.57 ± 0.1 (1sd) units lower than ambient seawater pH. Because Yu et al. (personal communication, 2013a) have assembled ambient carbonate chemistry data for their calibration samples, we can also calculate the corresponding calcite saturation state (Eq. 2.15). Assuming the concentration of dissolved inorganic carbon at the site of calcification is equal to ambient DIC~2060 μmol kg^{-1}, we calculate $\Omega_{calcite} = 1.05 \pm 0.04$ (1sd). However, because calcification consumes carbonate ions, we have also repeated the calculation for ambient DIC – 200 μmol kg^{-1}, which yields $\Omega_{calcite} = 0.95 \pm 0.04$ (1sd). In this case all data fall below $\Omega_{calcite} < 1$. While the actual extent of DIC consumption due to calcification depends on the rate of calcification and potential deliberate carbon concentration mechanisms, the $\delta^{11}B_{N.\ pachyderma}$ suggests this foraminifer species calcifies on the brink of undersaturation with respect to calcite. This is a striking result, as most marine calcifiers thrive only in waters that are several-fold oversaturated with respect to CaCO$_3$.

Neogloboquadrina pachyderma records some of the lowest $\delta^{11}B$ of all marine carbonates studied to date (Figure 2.11), and this is thus an extreme example. Estimating calcification site pH from $\delta^{11}B_{G.\ bulloides}$ (Martínez-Botí et al. 2015a) yields acidification on the order of -0.23 ± 0.07 (1sd) pH units on average, with some pH$_{TS}$ estimates as low as 7.82. If these estimates are accurate for the site of calcification, these foraminifer species could be highly vulnerable to impending ocean acidification. However, in contrast to the low saturation state estimates, we need to consider whether foraminifers may instead exert or experience additional boron isotope fractionation beyond the pH-dependent fractionation in seawater. In particular, a temperature correction on α_{B3-B4} could reduce the negative pH offset dramatically (Section 2.4.8), but the existence of such a temperature effect remains to be verified. While we do not yet have direct observations of boron isotope fractionation at the water-mineral interface, calcification in undersaturated conditions seems questionable, and we therefore caution interpretation of such geochemical data in terms of pH at the organism's site of calcification; additional confirmation from microelectrode or fluorescent probe studies is clearly warranted.

2.4.6   Light Effects on Symbiont-Bearing Corals

Hemming et al. (1998b) analyzed boron and carbon isotopes in the symbiont-bearing coral *Porites lobata*, collected at Fanning Island in the Central Pacific Ocean. The coincidence of heavy $\delta^{13}C$ and $\delta^{11}B$ in high-density bands led Hemming et al. (1998b) to suggest that seasonally varying photosynthesis of the coral-symbiont association and the entire reef community may cause elevated $\delta^{13}C$ and $\delta^{11}B$ when photosynthesis is high, $^{12}CO_2$ is sequestered by the algae, and skeletal growth is consequently sourced from high $\delta^{13}C_{DIC}$ and high-pH seawater. Similarly, Rollion-Bard et al. (2003) observed 12 ‰ $\delta^{11}B$-variation in *Porites lutea*, measured by SIMS. They translated these values to a pH-range ~7.1–9.0 (note that updating their parameterization to

$pK_B^* = 8.59–8.64$, $\delta^{11}B_{sw} = 39.61$ ‰ and $\alpha_{B3–B4} = 1.0272$ would reduce this pH_{TS} -range to ~8.0–9.0) and interpreted this range to be indicative of the influence of variable photosynthesis on the site of calcification.

pH elevation in the calcifying fluid of zooxanthellate corals has been observed similar to pH elevation in symbiont-bearing foraminifers (Figure 2.16) and measured $\delta^{11}B_{coral}$ data are broadly consistent with direct pH measurements (Al-Horani et al. 2003; Holcomb et al. 2014; Venn et al. 2013). However, Hönisch et al. (2004) analyzed $\delta^{11}B$ in corals grown over a range of light intensities (540–1210 µmol photons m^{-2} s^{-1}) and feeding frequency in the laboratory, including a natural depth profile with light intensities as low as 30–70 µmol photons m^{-2} s^{-1}. Although these treatments caused significant variations in $\delta^{13}C$ (Grottoli 1999, 2002), little $\delta^{11}B$ variability was observed (Hönisch et al. 2004). Given the experimental light levels exceeded the zooxanthellae's light saturation, and that long-term growth under low light conditions may be accommodated by a symbiont population that is adapted to limited light availability, the authors concluded that symbiont photosynthetic activity is unlikely to compromise the $\delta^{11}B$ record in zooxanthellate corals. While this conclusion may be correct for healthy corals that are not affected by environmental stress, it remains to be tested whether coral bleaching and subsequent recovery will be recorded in $\delta^{11}B$.

To reconcile the laboratory observations of limited light effects on $\delta^{11}B_{coral}$ and the observation of seasonal $\delta^{11}B$ variability in the *P. lobata* coral studied by Hemming et al. (1998b), Hönisch et al. (2004) suggested that the coral may instead record the chemical difference between the two water masses bathing Fanning Island. The South Equatorial Current and the North Equatorial Counter Current shift seasonally over this site, and these currents differ in their CO_2 concentration and acidity. While the reef community likely contributes to the seasonal changes in reef water chemistry, Hemming et al. (1998b) are likely correct in suggesting that $\delta^{11}B_{coral}$ reflects a combination of vital effects and ambient seawater chemistry variations. To date the vital effects have mostly been interpreted as pH up-regulation at the site of calcification.

2.4.7 pH Up-Regulation in Corals

Until $\alpha_{B3–B4}$ was determined experimentally (Byrne et al. 2006; Klochko et al. 2006; Nir et al. 2015), the similarity of the $\delta^{11}B$ versus pH sensitivity in corals to other empirical calibrations and to $\delta^{11}B_{borate}$ predicted by the theoretical $\alpha_{B3–B4}$ of Kakihana and Kotaka (1977) led some studies to suggest the $\delta^{11}B$ offsets caused by physiological vital effects must be constant across a wide range of seawater pH (e.g. Hönisch et al. 2004; Reynaud et al. 2004). This view has changed now that it has been determined that the $\delta^{11}B_{coral}$ versus pH sensitivity is lower than predicted from $\delta^{11}B_{borate}$ (Figures 2.10b, 2.13b, and 2.14), and this lower sensitivity has also been observed in at least some foraminifera species (Figures 2.10a, 2.14, and Sections 2.4.3–2.4.5). Several studies have explored the difference between $\delta^{11}B_{borate}$ and $\delta^{11}B_{coral}$ and determined the inferred pH-difference ($\Delta pH = pH_{calcifying\ fluid} - pH_{ambient}$) is greater at low than high pH (Allison and Finch 2010; Allison et al. 2014; McCulloch et al. 2012a, 2012b; Trotter et al. 2011). The geochemical observation has now also been confirmed by direct observation of calcifying fluid pH in the zooxanthellate coral *S. pistillata* (Holcomb et al. 2014; Venn et al. 2013), although $\delta^{11}B_{S.\ pistillata}$ continues to reflect higher pH than observed by fluorescent probe (Figure. 2.19, Holcomb et al. 2014; Krief et al. 2010). Holcomb et al. (2014) explained this pH offset by the different portions of the coral analyzed by fluorescent probe and $\delta^{11}B$, because the microscopic observations can only be performed at the lateral growth edge of these corallites, where the coral skeleton is thin and symbionts do not obscure the field of view. Although the observed offset between pH estimated from boron isotopes and fluorescent probe is large (~0.4 units), the similarity of the

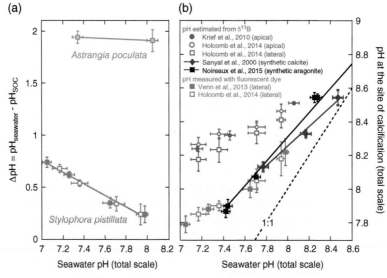

Figure 2.19 (a) Instrumental measurements of pH at the site of calcification (SOC). The tropical coral *Stylophora pistillata* (Holcomb et al. 2014; Venn et al. 2011) up-regulates pH at the SOC specifically at low seawater-pH, whereas the temperate *Astrangia poculata* (Ries 2011) up-regulates pH_{SOC} equally at low and high pH. (b) Comparison of pH_{SOC} by fluorescent probe (red symbols) and inferred from $\delta^{11}B_{S.\ pistillata}$ (blue symbols). All pH_{SOC} are higher than ambient seawater-pH and are interpreted to reflect active pH up-regulation by the coral. pH estimates based on boron isotopes are generally higher than fluorescent probe analyses but because they have been analyzed on different portions of the skeleton, Holcomb et al. (2014) interpret the difference to reflect the greater vulnerability of lateral

(compared to apical) coral extension to ocean acidification. Also shown is the precipitation pH inferred from $\delta^{11}B$ in synthetic aragonite (black symbols, Noireaux et al. 2015, see also Figure. 2.4 for comparison of the data presented herein) and calcite (gray symbols, Sanyal et al. 2000). Whereas synthetic calcite displays "pH-upregulation" similar to corals, the sensitivity of the $\delta^{11}B_{aragonite}$ versus pH relationship is more similar to $\delta^{11}B_{borate}$ (indicated by the similarity in slope to the 1 : 1 relationship between pH in seawater and at the SOC). Because inorganic $CaCO_3$ precipitation is not affected by physiological processes, this comparison highlights the need for a solid understanding of inorganic systematics before physiological processes can be inferred with confidence from proxies.

inferred pH ranges is remarkable and gives confidence to the reliability of the boron isotope technique for estimating differences in pH between different treatments.

Based on collective evidence from coral calibrations, Trotter et al. (2011) and McCulloch et al. (2012a, b) speculated that corals may in fact be resilient to ocean acidification, as they appear to actively up-regulate their pH at the site of calcification (SOC) and thus counteract low-saturation conditions in seawater. Given synthetic aragonite tends to record $\delta^{11}B$ lower than calcite (Figure 2.4) but most shallow and deep-sea corals record $\delta^{11}B$ higher than other biogenic marine carbonates (Figure 2.14), this general observation holds true for most but not all corals. For instance, elevated $\delta^{11}B$ is not only observed in zooxanthellate corals, as for example the asymbiotic corals *D. dianthus* (Figure 2.14, Anagnostou et al. 2012) and *L. pertusa* (Blamart et al. 2007; McCulloch et al. 2012b; Wall et al. 2015) record some of the highest $\delta^{11}B$ values in biogenic marine carbonates measured to date. This indicates that while some $\delta^{11}B$ elevation may be caused by symbiont photosynthetic activity, some $\delta^{11}B$ elevation must also be caused by active pH up-regulation or passage of isotopically heavy boric acid to the site of calcification (i.e. the uncharged $B(OH)_3$ may preferentially pass tissue layers but would likely dissociate into $B(OH)_4^-$ and H^+ at the site of calcification). In contrast, the calcitic bamboo coral *Keratoisis* sp. records $\delta^{11}B$ as low as deep ocean benthic foraminifers (Figure 2.14, Farmer et al. 2015; Hönisch et al. 2008; Rae et al. 2011) indicating that this species does not modify pH and/or the isotopic composition of the calcifying fluid at all, or at least to a much smaller extent than *D. dianthus* and *L. pertusa*. Some of the differences may be related to the respective biomineralization processes of different calcifiers (e.g. scleractinian corals and octocorallia), and the interested researcher should look for specific details in the respective literature.

Trotter et al. (2011) and McCulloch et al. (2012a, b) specifically considered the greater $\Delta\delta^{11}B_{coral-borate}$ at low compared to high ambient pH, and argued pH up-regulation is greater at low ambient pH, and effectively counteracts the low $\Omega_{aragonite}$ at low pH. While this argument has been confirmed for *S. pistillata* (Holcomb et al. 2014; Venn et al. 2013), we urge caution in interpreting coral resilience to ocean acidification based on $\delta^{11}B_{coral}$ observations. There is no doubt that $\delta^{11}B_{coral}$ is less sensitive to seawater pH than predicted from $\delta^{11}B_{borate}$, however, microelectrode pH_{SOC} measurements in the temperate coral *Astrangia poculata* (Figure 2.19, Ries 2011) show this species up-regulates pH_{SOC} equally at low and high ambient pH, and the saturation state at the site of calcification must be lower at low ambient pH in *A. poculata*. pH_{SOC} clearly behaves differently in different coral species, yet all $\delta^{11}B_{coral}$ calibrations established to date display similar pH-sensitivities (Figures 2.13b and 2.14, and York fits in online Table A2.2).

Ideally, we should be able to compare $\delta^{11}B_{coral}$ to $\delta^{11}B_{synthetic\ CaCO3}$, however, evidence from synthetic carbonate precipitation experiments yields conflicting results. While the $\delta^{11}B_{aragonite}$ calibration of Noireaux et al. (2015) appears to conform to the pH sensitivity of $\delta^{11}B_{borate}$ (Figure 2.4), the same does not hold true for the $\delta^{11}B_{calcite}$ calibrations of Noireaux et al. (2015) and Sanyal et al. (2000), which show a lesser pH sensitivity compared to that predicted from $\delta^{11}B_{borate}$ (Figure 2.4). Using the constraints applied by Trotter et al. (2011) and McCulloch et al. (2012a, b), we can calculate the pH inferred from synthetic calcite precipitated across a range of pH in the laboratory (Figure 2.19). According to the low pH sensitivity, synthetic calcite seems to "up-regulate" pH by ~0.4 units at ambient pH_{TS} ~7.8, and this ΔpH decreases towards zero near $pK^*_B = 8.6$. There are no vital effects to explain active pH up-regulation in synthetic calcite, which begs the question whether observations from synthetic $CaCO_3$ precipitation experiments are actually comparable to biological calcification in natural seawater. At this point we conclude boron isotope fractionation into inorganic carbonates requires more detailed study, and that evidence for greater pH up-regulation by corals at low pH should be considered with caution.

2.4.8   Temperature Effects on Boron Isotope Fractionation

Following Urey (1947), the thermodynamic isotope exchange between two molecules relates to the energy state of the molecules involved in the reaction, and can be explained almost exclusively by differences in the vibrational frequencies of isotopically different atoms. As a general rule, the heavy isotope concentrates in the compound in which the element is bound most strongly, and this is typically the compound with the highest vibrational frequency. In the case of boron, boric acid is characterized by higher vibrational frequencies than borate (see Zeebe 2005), and ^{11}B is more concentrated in $B(OH)_3$ than in $B(OH)_4^-$ (Figure 2.2). Thermodynamic principles predict that isotope fractionation effects are temperature dependent and fractionation factors approach unity at very high temperatures. For boron isotopes, however, accurate experimental determination of the temperature effect on α_{B3-B4} in solution has not yet been accomplished. Palmer et al. (1987) studied boron adsorption onto marine clays at variable pH and over a range of temperatures (5–40 °C), but their data exhibit too much variability to discern a systematic temperature effect (Figure 2.20). Similarly, Klochko et al. (2006) studied only two temperatures (25 and 40 °C) and their experimental uncertainty is very large compared to theoretical predictions (Figure 2.21a). We will therefore combine experimental and modeling studies to explore the current status of boron isotope fractionation systematics in response to temperature variations.

Figure 2.21a and b display several attempts to constrain the temperature effect on α_{B3-B4}. It is important to note that both the absolute value of α_{B3-B4} and its sensitivity to temperature depend on the accuracy of the vibrational frequency and the numerical method chosen to determine α_{B3-B4} (Zeebe 2005); Figure 2.21 should therefore be regarded as a sensitivity study. Rustad et al. (2010) explored a similar range of methods with overall similar results, but for Figure 2.21a we only selected the one that best matches the experimental

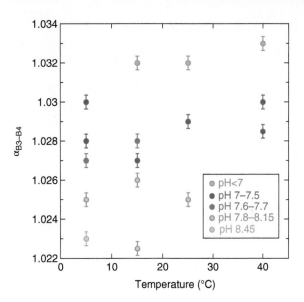

Figure 2.20 Experimental study of boron isotope fractionation during adsorption onto marine clays. Experiments were done over a range of pH and temperature. Note that the boron isotope fractionation onto clays decreases with pH but within a narrow pH range, the data scatter is too large to discern a systematic trend with temperature. Data are from Palmer et al. (1987).

data of Klochko et al. (2006) for pure water. The experimental α_{B3-B4} estimates for pure water and artificial seawater at 25 and 40 °C (Klochko et al. 2006) are also shown, along with the early theoretical estimate of Kotaka and Kakihana (1977). The general pattern resulting from the theoretical approaches necessarily agrees with theory – the temperature sensitivity (i.e. the slope of the regressions of α_{B3-B4} versus T) is greater when the absolute value of α_{B3-B4} is larger (Figure 2.21b), and α_{B3-B4} is larger at lower temperatures (Figure 2.21a). In contrast, the experimental data of Klochko et al. (2006) yield somewhat conflicting results. While the pure water results agree with the theoretical prediction of a smaller α_{B3-B4} at higher temperature, the artificial seawater results suggest the opposite. However, the uncertainties of the 40 °C experiments are too large to confirm or negate a temperature effect on the experimental α_{B3-B4} with confidence. Judging from biogenic carbonates, the ~1 : 1 $\delta^{11}B_{borate}$ sensitivity of $\delta^{11}B_{CaCO3}$ in symbiont-barren foraminifera species such as *C. wuellerstorfi, C. mundulus, G. bulloides*, but also the field calibration of the symbiont-bearing *O. universa* (Figure 2.13c, online Table A2.2) has been used to argue against the existence of a significant temperature effect on α_{B3-B4} (e.g. Henehan et al. 2016; Martínez-Botí et al. 2015a; Rae et al. 2011). This is because respiration rates of deep-sea benthic foraminifera are low (Nomaki et al. 2007), and Henehan et al. (2016) suggested that their *O. universa* may live deeper in the water column, where photosynthetic pH-elevation may be more limited. While these are reasonable arguments, they do not unequivocally prove that there is no temperature effect on α_{B3-B4}. As the arguments against a temperature effect on α_{B3-B4} have already been presented in the published literature, we apply a different approach in this book, and explore the approximate theoretical magnitude of a temperature effect on α_{B3-B4} in seawater and its possible consequences for paleo-pH reconstructions.

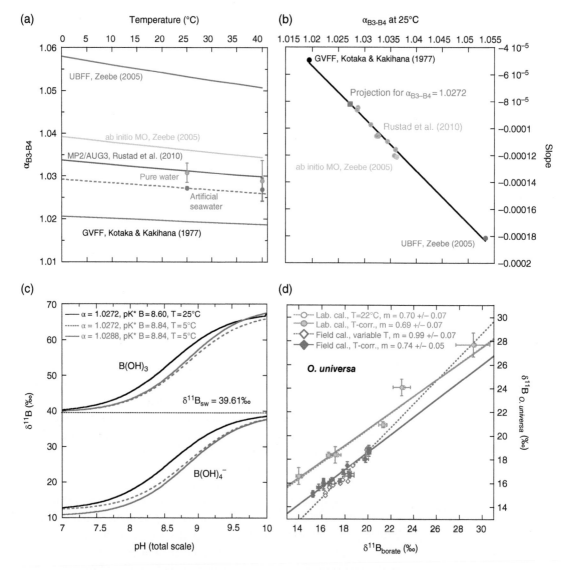

Figure 2.21 (a) Temperature effect on the boron isotope fractionation factor α_{B3-B4} as estimated from modeling (Kotaka and Kakihana 1977; Rustad et al. 2010; Zeebe 2005) and isotopically induced differences in aqueous boron speciation (orange symbols reflect experiments performed in pure water, red symbols artificial seawater experiments, Klochko et al. 2006). UBFF stands for "Urey-Bradley force fields," MO for "molecular orbital theory," MP2 for "second order Møller-Plesset correlated molecular orbital calculations," and GVFF for "generalized valence force fields" (see original studies for details). (b) Slope values of linear regressions of data shown in (a), but also including approximate values for Rustad et al. (2010). The general tendency of a stronger temperature effect with larger α_{B3-B4} is a consistent prediction in all modeling studies. Because the analytical uncertainties associated with the experimental determination of α_{B3-B4} (Klochko et al. 2006) are large, we projected the temperature sensitivity for $\alpha_{B3-B4} = 1.0272$ (dashed red line in (a)) from the regression shown in (b). See text for details. (c) Aqueous boron isotope partitioning predicted for a range of pH at T = 25 °C and $\alpha_{B3-B4} = 1.0272$ (black lines) and 5 °C and $\alpha_{B3-B4} = 1.0288$ (red lines). The dashed red line uses $\alpha_{B3-B4} = 1.0272$ and reflects the temperature effect on pK^*_B only (Dickson 1990b). (d) Comparison of *Orbulina universa* culture (green circles and lines, Hönisch et al. 2009; Sanyal et al. 1996) and field (blue diamonds and lines, Henehan et al. 2016) calibrations. Open symbols and dashed lines reflect original data with $\delta^{11}B_{borate}$ calculated using $\alpha_{B3-B4} = 1.0272$. Closed symbols and solid lines show $\delta^{11}B_{borate}$ calculated using temperature-dependent α_{B3-B4}. The slopes of the calibrations are significantly different without temperature correction, but align after temperature correction of α_{B3-B4}. The offset along the y-axis is due to the analytical techniques used for each calibration – N-TIMS for the laboratory calibration and MC-ICP-MS for the field calibration.

Because $\alpha_{B3-B4} = 1.0272$ (Klochko et al. 2006) is the fractionation factor assumed to correctly reflect aqueous boron isotope fractionation in seawater, but the experimental uncertainty is large (Figure 2.21a), we choose not to regress the temperature sensitivity from the two available temperature conditions, but instead project an approximate temperature sensitivity for $\alpha_{B3-B4} = 1.0272$ at 25 °C (dashed line in Figure 2.21a) using the regression shown in Figure 2.21b. The projected temperature sensitivity for $\alpha_{B3-B4} = 1.0272$ thus becomes $\alpha_{B3-B4} = 1.0292 - 0.00008188 \times T$ (°C, see also online Table A2.1 and Figure 2.21a and b). Based on these estimates, between 25 and 0 °C the temperature effect on α_{B3-B4} amounts to ~2 ‰ when $\alpha_{B3-B4} = 1.0272$ (Klochko et al. 2006).

It should be noted that the regression shown in Figure 2.21b excludes the data of Rustad et al. (2010) because we could not obtain the actual data and only estimated the slopes from reading the data off their Figure 2.2. Including those estimated slopes into the regression would increase the approximate temperature sensitivity for $\alpha_{B3-B4} = 1.0272$ from -0.00008188 (as shown in Figure 2.21a and b) to -0.00008258. This is well within the variability of the different model estimates (Figure 2.21b) and given that it is based on merely approximated data, we prefer to use the smaller sensitivity instead. However, we reiterate again that this projection of the temperature sensitivity on α_{B3-B4} is only a rough approximation, not only because of the uncertainties described above, but also because the temperature sensitivity for each α_{B3-B4} is only approximately linear with temperature (Figure 2.21a, cf. Zeebe 2009). The actual temperature sensitivity will eventually have to be verified and determined experimentally, but for the time being we can explore its potential effect on boron isotope studies with this approximation.

Figure 2.21c displays $\delta^{11}B_{borate}$ and $\delta^{11}B_{boric\ acid}$ estimated for 25 and 5 °C, using the projected temperature sensitivity for $\alpha_{B3-B4} = 1.0272$ (Klochko et al. 2006). In response to the larger pK^*_B and larger α_{B3-B4} at lower temperature, both $\delta^{11}B_{borate}$ and $\delta^{11}B_{boric\ acid}$ are lower at 5 °C compared to 25 °C. However, much of the difference is given by adjusting pK^*_B, illustrated by the black and red dashed curves in Figure 2.21c. The temperature effect on α_{B3-B4} generally exerts a small decrease in $\delta^{11}B_{borate}$, illustrated by the difference between the dashed and solid red lines in Figure 2.21c, but this decrease is largest within the oceanic pH_{TS} range of 7.5–8.3.

As described above, the temperature effect on $\delta^{11}B_{borate}$ displayed in Figure 2.21 is based on theory, and experimental studies are urgently needed to study and quantify this effect in seawater, either by extending the studies of Byrne et al. (2006) and Klochko et al. (2006), or by performing synthetic $CaCO_3$ precipitation experiments across a wide range of temperatures at constant pH and precipitation rate. As mentioned before, inorganic precipitation experiments have been done by Sanyal et al. (2000), Noireaux et al. (2015), Kaczmarek et al. (2016) and Farmer et al. (2016), but those experiments were all performed under different experimental seawater compositions and at different growth rates, which precludes unambiguous identification of a temperature effect on boron isotope partitioning in seawater. Wara et al.

(2003) used foraminiferal $\delta^{11}B$ and Mg/Ca (a temperature proxy) data from a downcore sediment record to suggest that $\delta^{11}B_{foram}$ decreases with warmer growth temperatures. However, this observation contradicts the direction of a temperature effect on $\delta^{11}B$ (Figure 2.21), and instead correlates with higher seawater-pH predictions for glacial times, when temperatures were lower (Hönisch et al. 2007). Due to simultaneous variation of multiple parameters through time, downcore sediment records such as done by Wara et al. (2003) are therefore unsuitable for investigating temperature effects.

The only possibly meaningful comparison of a marine carbonate grown in seawater but across a range of temperature conditions is given by laboratory and field calibrations of the symbiont-bearing foraminifer *O. universa*, where this species was either grown across a range of $pH_{TS} = 7.57–8.87$ ($\delta^{11}B_{borate} = 14.2–19.3$ ‰, online Table A2.2) at constant T = 22 °C (Hönisch et al. 2009; Sanyal et al. 1996), or across a range of $pH_{TS} = 8.056–8.20$ ($\delta^{11}B_{borate} = 16.3–19.9$ ‰) and temperature (9–27.9 °C) (Henehan et al. 2016). As indicated in online Table A2.2, the sensitivity of the culture calibration is m = 0.70±0.07 for the culture calibration and m = 0.99±0.07 for the field calibration. This difference in sensitivity was recently addressed by Rae (2018), who suggested that the lesser sensitivity of laboratory compared to field calibrations (see online Table A2.2 for a summary) maybe be caused by a growth rate effect in the laboratory, where $\Omega_{calcite}$ can be as high as 20 at $pH_{TS} = 8.87$, whereas ocean values typically range between 1 and 7 (Takahashi et al. 2014). Rae (2018) suggests that rapid calcification under such conditions may decrease microenvironment pH at high ambient pH and decrease $\delta^{11}B_{CaCO3}$ of organisms grown under such conditions. Although estimates of planktic foraminiferal calcification rates do not necessarily support this interpretation (Allen et al. 2016; Holland et al. 2017), the change in saturation state between cultures and the natural ocean is significant and Rae (2018) suggests that field calibrations may in fact more accurately reflect the true sensitivity of the proxy in biogenic carbonates. Here we explore the potential of a temperature effect on the field calibration of Henehan et al. (2016) and find that field and laboratory calibrations can in fact be aligned when the theoretical temperature effect on α_{B3-B4} (Figure 2.21a–c) is applied to the *O. universa* calibrations (Figure 2.21d). Although this comparison is not proof that the applied temperature effect is valid, the similarity of the two calibrations after application of this simple effect indicates that further experimental investigation of the effect is indeed warranted.

It is clear that more data are needed to conclude whether and to which extent temperature affects aqueous boron isotope fractionation and thus $\delta^{11}B_{carbonate}$. The question now is, should we correct $\delta^{11}B_{carbonate}$ data for precipitation temperatures? To discuss this question, we can evaluate published $\delta^{11}B$ data from benthic and planktic foraminifers that were grown across a range of temperatures.

As discussed above, the $\delta^{11}B$ of epifaunal deep-sea benthic foraminifers falls on or close to (depending on the applied analytical technique) $\delta^{11}B_{borate}$ predicted by $\alpha_{B3-B4} \sim 1.0272$ (Hönisch et al. 2008; Klochko et al. 2006; Rae et al. 2011) (Figure 2.14, online Table A2.2). The agreement is indeed remarkable and one could argue that deep-sea benthic foraminifers should fall close to inorganic theory (Rae et al. 2011) because they are not affected by symbiont-photosynthetic pH modification as described in Section 2.4.4. However, the deep-sea benthic foraminifer calibrations cover only a small pH range, and no calibrations exist from laboratory cultures over a wide pH range or under controlled temperature conditions.

Hönisch et al. (2008) and Rae et al. (2011) explored the effect of different α_{B3-B4} and temperature on their benthic foraminifer $\delta^{11}B$ data, using the temperature sensitivities shown in Figure 2.21. We have extended this evaluation by considering all available field calibrations to date, including benthic foraminifera (Rae et al. 2011), *N. pachyderma* (Yu et al. 2013a), and *G. bulloides* (Martínez-Botí et al. 2015a). These data sets cover a wide range of temperatures (i.e. 15 K for *G. bulloides*, 5 K for *N. pachyderma*, 6 K for *C. wuellerstorfi*, 9 K for *C. mundulus* and 4 K for *Planulina* sp.) and the studied foraminifer species are symbiont-barren, thus limiting physiological vital effects to calcification and respiration only. Using the estimated temperature sensitivity (Figure 2.21) for $\alpha_{B3-B4} = 1.0272$ (Klochko et al. 2006), we recalculate temperature-corrected $\delta^{11}B_{borate}$ for the calibrations shown in Figure 2.13c and display the comparison with the original data in Figure 2.22 alongside the *O. universa* field calibration of (Henehan et al. 2016). Because most of the foraminifera samples in this comparison grew at T < 25 °C, the temperature corrected calibrations shift towards lower $\delta^{11}B_{borate}$. This is because the temperature correction applies $\alpha_{B3-B4} > 1.0272$ to carbonates grown at T < 25 °C. Importantly, the temperature correction is largest at the lowest temperatures. Because the lowest temperatures typically prevail in the deeper ocean, where respired CO_2 accumulates, lowest temperatures tend to coincide with low bottom water pH and consequently low $\delta^{11}B_{foram}$. The temperature correction therefore tends to affect $\delta^{11}B_{benthic\ foram}$ data the most. This correction results in a relaxation of the $\delta^{11}B_{foram}$ versus $\delta^{11}B_{borate}$ calibration slopes and thus brings them into closer agreement with the low pH (i.e. $\delta^{11}B_{borate}$) sensitivities of marine carbonates grown at higher temperatures and across a wider pH range in the laboratory (cf. Figure 2.14 and online Table A2.2).

The above comparison leaves us with the question: which interpretation we should adopt? The disappointing answer is that we cannot yet make a strong recommendation for or against the existence of a temperature effect. It has been suggested that the low metabolic rates on deep-sea benthic foraminifera (Nomaki et al. 2007) should minimize any vital effects in these species and $\delta^{11}B_{benthic\ foram}$ should equal $\delta^{11}B_{borate,}$ arguing against a temperature effect (Henehan et al. 2016; Rae 2018; Rae et al. 2011). On the other hand, deep-sea benthic foraminifera typically live in ocean bottom waters that are often minimally saturated or even undersaturated with respect to calcite. If there are any marine calcifiers that should upregulate their micro-environmental pH to facilitate calcification, then benthic foraminifera should surely be prime candidates. The temperature-corrected calibrations

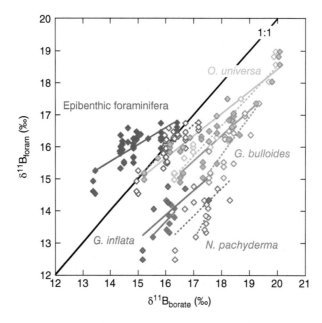

Figure 2.22 Comparison of field $\delta^{11}B_{foram}$ versus $\delta^{11}B_{borate}$ relationships, where $\delta^{11}B_{borate}$ is calculated either with constant $\alpha_{B3-B4} = 1.0272$ (open symbols and dashed lines, Klochko et al. 2006) or with a temperature corrected α_{B3-B4} according to Figure 2.21 (closed symbols and solid lines). Epibenthic foraminifera data are from Rae et al. (2011), planktic *Neogloboquadrina pachyderma* data from Yu et al. (2013a), *Globigerina bulloides* data from Martínez-Botí et al. (2015a) and *Orbulina universa* and *Globorotalia inflata* from Henehan et al. (2016). All data were generated by MC-ICP-MS. Note that the temperature correction is largest at the lowest temperatures, which are typically found deeper in the water column, where accumulated CO_2 is typically higher and pH correspondingly lower. The temperature correction therefore tends to shift samples from colder environments to lower $\delta^{11}B_{borate}$ and thus decreases the pH-sensitivity of these calibrations.

shown in Figure 2.22 would be consistent with the existence of a vital effect, where e.g. *C. wuellerstorfi* would elevate its microenvironmental pH above $\delta^{11}B_{borate}$. This is furthermore consistent with observations of up to 10 ‰ variability in $\delta^{11}B$ of individual *C. wuellerstorfi* studied by laser ablation (Sadekov et al. 2016: poster presentation at the International Conference of Paleoceanography, Utrecht). However, we do not have any direct pH measurements for the microenvironment of deep-sea benthic foraminifers that would allow us to confirm physiological pH up-regulation in these organisms. The only studies available to date are based on shallow ocean benthic foraminifera, which show a range of results. Whereas the symbiotic *Marginopora vertebralis*, *Amphistegina radiata*, *Heterostegina depressa* and *Peneroplis* sp. display significant pH elevation consistent with symbiont photosynthesis (Glas et al. 2012), the asymbiotic *Quinqueloculina* sp. and *Miliola* sp. show little or no pH upregulation (Glas et al. 2012), but the asymbiotic *Ammonia* sp. upregulates pH in its microenvironment actively and significantly (Toyofuku et al. 2017). This dichotomy of observations makes it difficult to predict the behavior of specimens living in the deep-sea, but until these knowledge gaps are filled, we can consider how temperature may affect boron isotope fractionation in cold environments.

Continuing the thought experiment, when the hypothetical temperature effect correction is applied to planktic foraminifera, sensitivities in the $\delta^{11}B_{CaCO3}$ versus $\delta^{11}B_{borate}$ relationships converge, and the large negative off-sets from $\delta^{11}B_{borate}$ observed in *G. bulloides* and *N. pachyderma* are reduced after applying the temperature correction, which makes it easier to interpret the residual offsets in terms of foraminiferal respiration and calcification. In comparison to Section 2.4.5, where we showed the $\delta^{11}B_{N.\ pachyderma}$ calibration established by Yu et al. (2013a) implies calcification at $\Omega = 0.85–1.10$, we can now recalculate the saturation state in the calcifying fluid of *N. pachyderma* with a temperature corrected pH estimate. Assuming that DIC equals ambient $DIC_{seawater}$, $\Omega = 3.87 \pm 0.12$ (1sd); when assuming DIC at the calcifi-cation site is reduced by $200\,\mu mol\,kg^{-1}$ due to the calcification process, $\Omega = 3.49 \pm 0.11$ (1sd). This comparison reveals these foraminifers may actu-ally calcify at supersaturation, if our assumed temperature effect is indeed applicable.

In summary, a temperature effect on α_{B3-B4} is predicted by thermodynamic laws, but the effect's existence and magnitude remain to be tested in targeted experiments. Ideally, such experiments should focus directly on the aqueous fractionation in solution, as differences in precipitation rates of inorganic carbonates may bias the result. The uncertainties of temperature-corrected field calibrations are essentially the same as those applying constant $\alpha_{B3-B4} = 1.0272$ (online Table A2.2), but applying the temperature correction shifts the $\delta^{11}B_{foram}$ calibrations towards higher $\delta^{11}B_{borate}$ and calcification pH. If correct, this scenario would suggest benthic foraminifers up-regulate their calcification pH to increase the saturation state of their calcifying fluid relative to ocean bottom water. Similarly, symbiont-barren planktic foramin-ifers would also calcify at supersaturation with respect to calcite, instead of the low calcifying fluid saturation suggested by non-temperature corrected calibrations (Section 2.4.5). Applying the temperature correction also relaxes the pH sensitivity observed in field calibrations of samples collected from different temperature conditions, and creates greater consistency with warm water calibrations, many of which have been established over a much wider pH range (Figures 2.10 and 2.14). If correct, the similarity in the sensitivity of various biogenic calcifiers to $\delta^{11}B_{borate}$ will facilitate calibration of now extinct species that can no longer be calibrated over a known range of $\delta^{11}B_{borate}$. We will explore this feature in greater detail in Chapter 3. However, given current knowledge of the temperature dependence of α_{B3-B4}, these considerations remain hypothetical. Experiments are clearly needed to investigate the pH and temperature sensitivity of benthic and symbiont-bar-ren planktic foraminifers in laboratory culture, where environmental param-eters can be studied in isolation and across much wider ranges than present in the ocean. While $\delta^{11}B_{borate}$ estimates at $T \ll 25\,°C$ (e.g. relevant for deep ocean benthic calcifiers) will be most affected by a temperature correction, it should be noted that paleotemperature changes within a given surface or deep-ocean environment are typically no larger than 5 K, which entails only a small temperature correction. We will investigate the effect of a temperature correction on paleo-pH and pCO_2 reconstructions in Chapter 3.

2.4.9 Diagenesis

At the time when boron concentrations in marine sediments were still considered as a potential means to infer paleosalinity, Cook (1977) investigated the boron content of fossil oyster shells and observed decreasing boron concentrations with age. Although the distribution of boron within the shells was not known, the most likely reason for the decreasing concentration was sub-aerial weathering. Similarly, Stewart et al. (2015) studied B/Ca and $\delta^{11}B$ in Marinoan-age (i.e. ca. 640 Ma) bulk sediments from the Great Bahamas Bank. Glaciation led to a sealevel lowstand during the Marinoan and exposed these sediments to meteoric waters and subaerial diagenesis. Stewart et al. (2015) observed B/Ca ratios decreasing by 90% and $\delta^{11}B$ by 6 ‰ across the interval of subaerial exposure, suggesting that recrystallization and carbonate precipitation in the presence of meteoric waters can bias original boron proxy signals significantly. Furthermore, Gaillardet and Allègre (1995) studied $\delta^{11}B$ and B concentrations in tropical coral reefs and observed lower values for both parameters in the oldest, glacial samples, in particular those that were collected from now submerged reef terraces. Because $\delta^{11}B$ of the ancient, recrystallized corals also correlated with the reciprocal of the boron concentration, Gaillardet and Allègre (1995) interpreted these data to reflect post-depositional alteration of the original geochemical signals. Although these observations are intriguing, the study is based on various unidentified coral species. It is now clear that different coral species record different geochemical signatures (data presented in Figures 2.10b and 2.14, but also Wei et al. 2014), and it therefore appears that there is some uncertainty in the study of Gaillardet and Allègre (1995). However, evidence for diagenetic alteration of boron proxy records in live collected corals has recently been suggested by Lazareth et al. (2016), who observed variable skeletal recrystallization in a live collected *Porites* specimen, and $\delta^{11}B_{coral}$ decreased by −0.001 ‰ per additional % calcite present.

Spivack and You (1997) studied bulk sediments from a deep-sea sediment core in the eastern equatorial Pacific and observed $\delta^{11}B_{bulk\,carbonate}$ data varying between +23 ‰ in coretop material and − 5 ‰ in late Miocene aged material. Because such large $\delta^{11}B$ variations had previously not been observed in similar aged deep-sea sediments (Spivack et al. 1993), Spivack and You (1997) interpreted these data to reflect varying proportions of recrystallized carbonate with $\delta^{11}B_{rec.\,CaCO3}$ ~ −5 ‰. This observation remains unique and it should be noted that the samples with the lowest $\delta^{11}B_{bulk}$ values coincide with elevated concentrations of diatomaceous ooze and low magnetic susceptibility in the studied sediment core (Mayer et al. 1991). $\delta^{11}B$ values of −15 to +5 ‰ have been reported for siliceous oozes and cherts (Ishikawa and Nakamura 1993), and although Spivack and You (1997) cleaned their bulk samples by ultrasonication and separated acid soluble and insoluble fractions by centrifugation, it is possible boron leached from the siliceous fraction may have contaminated the carbonate geochemical signature. Recent Miocene studies of $\delta^{11}B$ in planktic foraminifers (Badger et al. 2013; Foster et al. 2012; Greenop et al. 2014) yield consistent results across several sediment cores, suggesting Spivack and You's (1997) observation may be unique to bulk sediment samples. Similarly, Edgar et al. (2015) studied ~40 Myr old glassy planktic foraminifers from clay-rich sediments off Tanzania and recrystallized specimens from typical open ocean carbonaceous sediments in the western equatorial Pacific. They find recrystallization at the seafloor has little effect on $\delta^{11}B$ in planktic foraminifers, suggesting B loss to pore fluids does not impose any problems. Diagenetic effects on $\delta^{11}B$ and B/Ca in Jurassic limestones have been evaluated (Paris et al. 2010b), but the study of such ancient material bears many uncertainties and does not warrant detailed discussion herein.

Systematic studies of partial dissolution have only been performed on planktic foraminifers (Hönisch and Hemming 2004; Seki et al. 2010), where shells were selected from defined size fractions and sediment cores recovered from different water depths and various levels of bottom water saturation (Eq. 2.15). Hönisch and Hemming (2004) studied $\delta^{11}B_{G.\ sacculifer}$ in sediments from the Ontong Java Plateau (OJP) in the western equatorial Pacific and the 90°East Ridge in the Indian Ocean. The idea of such studies is that planktic foraminifer shells grown in the same region should record the same geochemical signature from their sea surface habitat, and any geochemical deviation from well-preserved material must be due to partial dissolution at the seafloor. In sediment cores spanning Ω = 1.25–0.95 and 1780–3260 m water depth they observed systematically decreasing $\delta^{11}B_{G.\ sacculifer}$ with increasingly corroded samples (Figure 2.23). The preservation state was independently determined by shell weight analyses, where any weight deficit compared to well-preserved specimens indicates partial dissolution. Whereas the dissolution trend on the OJP correlates with water depth and bottom water $\Omega_{calcite}$, the 90° East Ridge receives large amounts of organic matter rain at relatively shallow depths, leading Hönisch and Hemming (2004) to suggest excessive respiration of organic matter creates greater calcite dissolution in shallow compared to intermediate depth sediments. Correlation of $\delta^{11}B_{G.\ sacculifer}$ with corresponding shell weights bears this out (Figure 2.23b) and highlights that water depth and bottom water saturation are not always good indicators of a sample's preservation state. Importantly, the dissolution effect on the largest samples is smallest, which is consistent with a relatively smaller surface area-to-volume ratio in larger compared to smaller shells, and consequently relatively less surface area exposed to corrosive bottom waters. Hönisch and Hemming (2004) therefore

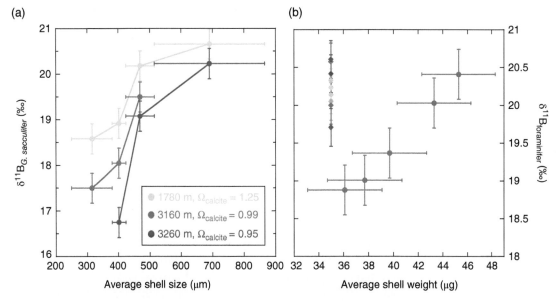

Figure 2.23 (a) Comparison of $\delta^{11}B_{G.\ sacculifer}$ in four different shell size classes from a sediment depth transect on the Ontong-Java Plateau in the western equatorial Pacific. Data replotted from Hönisch and Hemming (2004) but for consistency with other data shown in this book, 1.1 ‰ has been removed from $\delta^{11}B_{G.\ sacculifer}$ to account for the interlaboratory difference between the 15 cm radius of curvature NBS-design TIMS at SUNY Stony Brook and the modern Thermo Triton at LDEO. $\Omega_{calcite}$ reflects the saturation state of ocean bottom water at each core site and indicates corrosive conditions when $\Omega_{calcite} < 1$. Similar to Figure 2.17, there is an original size effect between the largest and smallest size fraction in undissolved samples (turquoise symbols) but the size difference increases with dissolution in the smaller size fractions in deeper sediment cores. Note that $\delta^{11}B_{G.\ sacculifer}$ in the largest shell size fraction is within analytical uncertainty in undissolved and partially dissolved shells. (b) Comparison of $\delta^{11}B_{G.\ sacculifer}$ from the 90° East Ridge in the Indian Ocean (380–425 μm shell size fraction, blue circles, Hönisch and Hemming 2004), and $\delta^{11}B_{G.\ sacculifer}$ (blue diamonds) and $\delta^{11}B_{G.\ ruber}$ (pink diamonds) from Ceara Rise in the western equatorial Atlantic (300–355 μm shell size fraction, Seki et al. 2010). Seki et al. (2010) did not determine shell weights; their data are plotted versus an estimated undissolved weight but the inferred dissolution state of their samples is indicated by symbol shading, where darker colors reflect deeper cores. Although the magnitude of the dissolution effect on Ceara Rise is overall smaller than in the Indian Ocean, Seki et al. (2010) assume that $\delta^{11}B_{G.\ ruber}$ is unaffected by partial dissolution.

recommend paleoreconstructions focus on the larger size fractions to minimize any potential dissolution bias.

Following similar lines of argument, Ni et al. (2007) also considered partial shell dissolution to explain their observed size effect in $\delta^{11}B_{G.\ sacculifer}$. Because they only observed a size effect for $\delta^{11}B_{G.\ sacculifer}$ but not $\delta^{11}B_{G.\ ruber}$ (Figure 2.17b), and because *G. sacculifer* secretes a gametogenic calcite layer towards the end of its life cycle (Bé 1980), Ni et al. (2007) speculated that the size effect in *G. sacculifer* may be a dissolution effect, where the ontogenetic portion of the shell is preferentially dissolved relative to the gametogenic portion. Because gametogenic calcite is secreted at greater depth and after symbionts have been expelled, it may have a lower $\delta^{11}B$ than the ontogenetic calcite and its preferential preservation would lower bulk shell $\delta^{11}B_{G.\ sacculifer}$. However, more recent observations indicate a pronounced size effect is not necessarily related to partial shell dissolution but is an original signal in some samples (see also Section 2.4.2). Specifically, it is now clear that *G. ruber* also records higher $\delta^{11}B$ in larger shells of some well-preserved samples (Henehan et al. 2013) and this size effect occurs despite the lack of a gametogenic crust in this species. Observation of a shell size effect alone therefore does not provide evidence for partial dissolution. Furthermore, the sediment cores studied by Ni et al. (2007) were all bathed in bottom waters with $\Omega_{calcite} > 1.13$, and because the authors did not provide any shell weight data to prove potential respiratory dissolution within the sediment, we can only assume these shells were preserved in these oversaturated bottom waters. Finally, Hönisch and Hemming (2004) provided evidence from scanning electron microscopy that the gametogenic crust was present in well preserved shells but not in partially dissolved shells, suggesting the model of greater ontogenetic calcite solubility (Brown and Elderfield 1996) needs to be revisited.

A subsequent study by Seki et al. (2010) analyzed $\delta^{11}B_{G.\ sacculifer}$ and $\delta^{11}B_{G.\ ruber}$ from a sediment profile at Ceara Rise, western equatorial Atlantic (Figure 2.23b). Similar to Ni et al. (2007), this study did not provide shell weights, and the preservation state is inferred from the range of bottom water $\Omega_{calcite} = 1.47{-}0.98$, with only the deepest core bathed in undersaturated water. Due to the lack of shell weights, Figure 2.23b assigns all their data at an arbitrary average weight of $35\,\mu g$, which may approximate undissolved shell weights. Similar to other studies, $\delta^{11}B_{G.\ ruber}$ is generally higher than $\delta^{11}B_{G.\ sacculifer}$, but whereas $\delta^{11}B_{G.\ sacculifer}$ decreases systematically with $\Omega_{calcite}$ and water depth, $\delta^{11}B_{G.\ ruber}$ displays some variability between cores but no systematic trend with depth. Seki et al. (2010) inferred from this comparison that $\delta^{11}B_{G.\ ruber}$ is not affected by dissolution and recommended this species be preferred for paleoreconstructions. Similar to Ni et al. (2007), the authors argued that the lack of a gametogenic crust in this species must distribute $\delta^{11}B$ homogenously throughout the shell and $\delta^{11}B_{G.\ ruber}$ should therefore not be susceptible to dissolution. While this appears reasonable based on the data of Seki et al. (2010), whether it is sufficient to confidently exclude a dissolution effect on $\delta^{11}B_{G.\ ruber}$ for all locations is less clear. First of all, the dissolution effect on $\delta^{11}B_{G.\ sacculifer}$ observed at Ceara Rise is smaller than that observed on the OJP and the 90° East Ridge by Hönisch and Hemming (2004). The small effect observed on Ceara Rise could simply reflect minimal dissolution at this location, but this cannot be confirmed without independent quantitative constraints such as shell weights. Furthermore, Mg/Ca ratios in *G. ruber* have been shown to be susceptible to dissolution (Dekens et al. 2002; Regenberg et al. 2006; Rosenthal and Lohmann 2002), indicating that the lack of a gametogenic crust in this species does not prevent post-depositional modification of the original bulk shell composition. To solidify confidence in $\delta^{11}B$ reconstructions using this species, dissolution studies should therefore be expanded to other regions, and actual dissolution verified by shell weight and/or foraminiferal fragmentation analyses.

In summary, several studies hint at post-depositional modification of original $\delta^{11}B$ records. Existing analyses are sparse and should be expanded to all carbonates used for $\delta^{11}B$ reconstructions, and ideally supported by microanalysis of $\delta^{11}B$ heterogeneity within individual carbonates, e.g. by SIMS or Laser Ablation techniques. The magnitude of the dissolution effect observed in $\delta^{11}B_{G. sacculifer}$ is significant for paleoreconstructions, but any potential bias of the original geochemical signature can be minimized by restricting paleoreconstructions to large shell sizes (Hönisch and Hemming 2004). Because the surface area-to-volume ratio argument applies to all samples, the large shell-size approach is recommended for all reconstructions, including those species that may appear resistant to dissolution. Diagenetic bias is a problem observed in most paleo-archives and screening for diagenetic overprints should always be considered. The general findings of this section and consequent recommendations for future studies on boron isotope systematics in marine carbonates are given in Box. 2.3.

Box 2.3 Secondary Effects of Organismal Physiology, Temperature and Diagenesis on Boron Isotope Signatures in Marine Carbonates

Boron isotope signatures in marine carbonates are characterized by a high degree of coherency. Empirical $\delta^{11}B_{carbonate}$ calibrations display fundamental similarities across a wide range of seawater-pH and a wide range of calcifiers, including foraminifera, tropical and deep-sea corals (aragonitic and calcitic), and brachiopods. Although species-specific calibrations need to be applied, the similarities are remarkable because these organisms differ greatly in their calcification processes and rates of calcification. This is a strong argument against the critics of the boron isotope proxy who have argued that the complexities of vital effects and the incorporation process are too enigmatic to apply the proxy with confidence. $\delta^{11}B$ differences between organisms are broadly consistent with physiological processes such as photosynthesis, respiration and calcification, although some parts of the observed signals remain unexplained. Direct observations of organismal pH modification at the site of calcification will help to improve our understanding of physiological processes and inorganic boron isotope systematics, but until the inorganic theory behind the boron isotope proxy is understood in its entirety, use of the $\delta^{11}B$ proxy for interpretations of physiological processes should be made with caution. To further improve our understanding of the proxy, synthetic $CaCO_3$ precipitation experiments in seawater should be expanded over a range of chemical and physical conditions, aqueous boron isotope fractionation should be studied over a range of temperatures, and general calibration and validation of the boron isotope proxy in modern and fossil marine carbonates should be continued. For techniques that sample carbonate skeletons at the microscale (e.g. SIMS or laser ablation techniques), the minimal sampling area or data density required to overcome geochemical heterogeneity should be assessed before comparisons with bulk shell techniques (e.g. TIMS, MC-ICP-MS) are made. As with most paleo-proxies, measurement of numerous shells/specimens is advised so as to average out any intra-shell heterogeneity caused by variability in physiological control on pH. These refinements of the boron proxy's theoretical basis will build on existing evidence, which already provides a strong basis for applying the boron isotope proxy to the geological record, so that paleoreconstructions can be approached with confidence.

2.5   Secular Evolution of [B$_T$] and δ^{11}B in Seawater

Faithful application of the δ^{11}B and B/Ca proxies to the paleorecord requires not only accurate calibrations but also accurate knowledge of the seawater isotopic composition and boron concentration at the given time of interest in Earth history. [B$_T$] and δ^{11}B$_{sw}$ in the modern ocean have been quantified as [B$_T$] = 432.6 μmol kg^{-1} in seawater with S = 35 (Lee et al. 2010) and δ^{11}B$_{sw}$ = 39.61 ± 0.04‰ (Foster et al. 2010). These estimates are based on many replicates from all ocean basins and these data are accepted as the best estimates available. However, the chemistry of seawater evolves over time and past fluxes of boron into and out of the ocean must therefore be quantified. Quantifying natural boron fluxes is now complicated by significant anthropogenic fluxes, which have increased the natural river flux by more than 50% (Schlesinger and Vengosh 2016). The reader is referred to Schlesinger and Vengosh (2016) for further details on anthropogenic fluxes, here we will focus on the natural fluxes that determine the boron inventory and isotopic composition in seawater on geological time scales.

The dominant natural sources of boron to the ocean have been identified as terrestrial weathering, fluids expelled at subduction zones and hydrothermal vents. On the other hand, the dominant natural sinks for boron removal from the ocean are adsorption onto clays, co-precipitation in marine carbonates, as well as chemical alteration of newly formed oceanic crust (e.g. Harriss 1969; Lemarchand et al. 2002; Spivack and Edmond 1987; Vengosh et al. 1991). Several estimates have been published for these fluxes and a selection of the first and most recent ones is listed in Table 2.2. Assuming steady state in the sources and sinks of boron, the residence time of boron in the ocean can be calculated as

$$\tau_B = \left[B_T \right] \times M_O / B_{\text{source (or sink)}} \qquad (2.20)$$

where $M_O \sim 1.4 \times 10^{24}$ g is the mass of the ocean, and [B$_T$] = 4.68 × 10^{-6} g g^{-1} of seawater in the modern ocean (Lee et al. 2010). By analyzing the boron chemical and isotopic composition of the world's major rivers Lemarchand et al. (2000, 2002) greatly improved estimates of the riverine boron inputs and in combination with other flux estimates (Table 2.2), they estimated τ_B at 14–15 Myr in the modern ocean. In comparison to the residence times of many other elements in the ocean (e.g. 1.1 Myr for calcium, Broecker and Peng 1982), 14 million years is a very long and seemingly comforting residence time for a proxy. However, this does not mean that [B$_T$] and δ^{11}B$_{sw}$ were constant over the past 14 million years. Lemarchand et al. (2000, 2002) modeled past changes in δ^{11}B$_{sw}$ and [B$_T$] (Figure 2.24) and estimated that δ^{11}B$_{sw}$ changed by ~0.1‰/Myr. With δ^{11}B analytical uncertainty being ≥0.2‰ at best, this means that δ^{11}B$_{sw}$ can really only be considered constant for the past two million years, and paleoreconstructions going further back in time must either constrain δ^{11}B$_{sw}$ to accurately estimate past pH variations, or resign to relative pH estimates across sufficiently short time intervals.

Table 2.2 Oceanic boron budgets.

	Harriss (1969)	Spivack and Edmond (1987) and Vengosh et al. (1991)	Lemarchand et al. (2000, 2002)	$\delta^{11}B$ (‰) after Lemarchand et al. (2002) and references therein
Sources ($\times 10^{14}$ g yr^{-1})				
River runoff	40	30	38	10
Hydrothermal fluids	–	0.9	4	6.5 ± 8
Fluids expelled at subduction zones	–	–	2	25 ± 5
Sum of all inputs	40	31	44	
Residence Time (Myr)	16.4	21	14.9	
Sinks ($\times 10^{14}$ g yr^{-1})				
Oceanic crust alteration	0.6	14	27	4
Adsorption onto clays	0.3	10	13	15 ± 1
Co-precipitation in marine carbonates	3.3	6.4	6	20 ± 5
Sum of all outputs	40	30.4	46	
Residence Time (Myr)	16.4	21.6	14.2	

Applying published estimates of oceanic crust production, clastic sediment flux, marine $CaCO_3$ production and riverine runoff, Lemarchand et al. (2000, 2002) modeled $\delta^{11}B_{sw}$ as low as ~36‰ 60 million years ago (Figure 2.24a), and Cretaceous [B_T] ~ 25% lower compared to the modern ocean. Unfortunately, these estimates are poorly constrained because estimates of the controlling processes are fraught with uncertainty. Simon et al. (2006) performed a sensitivity study where they tested the effect of uncertainties on these controlling processes and estimated that $\delta^{11}B_{sw}$ could really vary by as much as ±10‰ on 10-million year time scales. Such uncertainties are clearly not good enough for accurate reconstructions of seawater acidity. The model of Lemarchand et al. (2000, 2002) predicts that the B concentration of seawater was higher ~10–30 Ma compared to today, but lower >40 Ma. The B concentration is heavily influenced by estimates of global river runoff (Figure 2.24b), and in contrast to $\delta^{11}B_{sw}$, the evolution of [B_T] has not yet been estimated by alternative approaches.

Experimental constraints on $\delta^{11}B_{sw}$ have been attempted using analyses of fluid inclusions in ancient halite evaporites (Paris et al. 2010a), and mixed fossil foraminifers (Spivack et al. 1993), planktic (Palmer et al. 1998; Pearson and Palmer 1999), benthic (Raitzsch and Hönisch 2013) and paired planktic/benthic foraminifers (Greenop et al. 2017). While the pioneering study of Spivack et al. (1993) is now outdated due to their use of mixed foraminifer samples (cf. Figure 2.14 for significant species effects within and between planktic and benthic foraminifera), the other studies deserve individual attention.

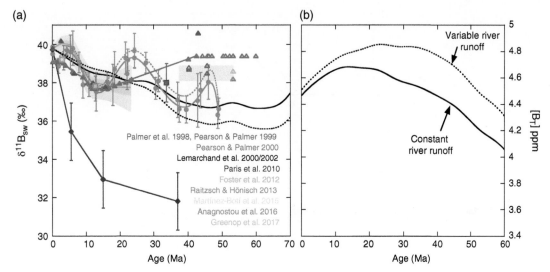

Figure 2.24 (a) Cenozoic evolution of $\delta^{11}B_{sw}$ as estimated from modeled boron fluxes in and out of the ocean (constant river flux = solid black line, variable river flux = dashed black line, Lemarchand et al. 2000, 2002); surface ocean pH profiles and corresponding $\delta^{11}B$ in planktic foraminifers (dark blue triangles, Palmer et al. 1998; Pearson and Palmer 1999); $\delta^{11}B$ in benthic foraminifers (orange symbols and solid line, assume variable pH scenario and $\delta^{11}B_{benthic\ foram.}$ vs. pH sensitivity similar to laboratory calibrated marine carbonates; dashed line assumes variable pH scenario and $\delta^{11}B_{benthic\ foram}$ vs. pH sensitivity similar to $\delta^{11}B_{borate}$ (Raitzsch and Hönisch 2013)); fluid inclusions in halites (gray diamonds, Paris et al. 2010a); and surface-to-deep ocean pH and $\delta^{13}C$ profiles: green square – Foster et al. (2012), turquoise line and shading – Greenop et al. (2017), green triangles and shading – Anagnostou et al. (2016) (where closed and open triangles display estimates based on the assumption that $\delta^{11}B_{Eocene\ foram} = \delta^{11}B_{borate}$, or $\delta^{11}B_{Eocene\ foram} = \delta^{11}B_{T.\ sacculifer}$, respectively; estimates at 40.3 and 45.6 Ma reflect upper bounds and estimates at 53.2 Ma lower bounds for the entire interval 37–53 Ma, see text for details). Note that the model of Pearson and Palmer (2000) is based on analyses of Pleistocene to middle Eocene aged foraminifers (Palmer et al. 1998; Pearson and Palmer 1999), and all data beyond 43 Ma are mere assumptions based on the oldest Eocene reconstruction (open blue triangles). Yellow and turquoise squares display interpolation between modern and Miocene estimates (Martínez-Botí et al. 2015b), and the assumed best estimate for the Eocene/Oligocene Boundary (Pearson et al. 2009), respectively. Halite inclusions may be compromised by stronger boron isotope fractionation in response to higher Ca^{2+} and lower Mg^{2+} concentrations in past seawater. See text for details. (b) Modeled $[B_T]$ evolution from Lemarchand et al. (2000, 2002).

Palmer et al. (1998) and Pearson and Palmer (1999) analyzed contemporaneous sets of four to ten mono-specific planktic foraminifer samples, covering the age range from 53 Ma (Eocene) to 85 ka (Pleistocene). Based on the argument that these species inhabited different water depths in the surface ocean and that the upper ocean pH gradient must have been relatively similar at different times in ocean history, the approach explores the non-linearity of the $\delta^{11}B$ versus pH relationship and its variation in response to different $\delta^{11}B_{sw}$ (Figure 2.25). Because the $\delta^{11}B$ versus pH relationship is steeper when $\delta^{11}B_{sw}$ is lower, a given $\delta^{11}B_{CaCO3}$ difference between shallow and deeper living foraminifera translates into a smaller pH difference at e.g. $\delta^{11}B_{sw} = 37‰$ compared to $\delta^{11}B_{sw} = 40‰$. Knowledge of the

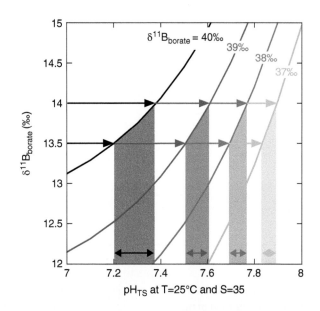

Figure 2.25 Conceptual model of the surface-to-deep ocean pH gradient technique to infer paleo-$\delta^{11}B_{sw}$. Because the $\delta^{11}B_{borate}$ versus pH relationship is curved and the curve is steeper when $\delta^{11}B_{sw}$ is lower, a given range in $\delta^{11}B_{borate}$ (here: 13.5–14‰) reconstructed from surface and deep dwelling foraminifera (ideally benthics, to remove any uncertainty on the depth habitat of now extinct planktic species) translates into a progressively smaller pH gradient as $\delta^{11}B_{sw}$ decreases. Knowledge of the pH and surface-to-deep $\delta^{11}B_{borate}$ gradients thus allows estimation of paleo-$\delta^{11}B_{sw}$. Studies applying this technique have either assumed the modern surface-to-deep pH gradient remained constant in the past, or inferred it from the planktic-benthic $\delta^{13}C$ difference, see text and original studies for details. Figure modified after Anagnostou et al. (2016).

pH gradient in addition to the $\delta^{11}B_{CaCO3}$ difference between shallow and deeper living foraminifera can therefore be used to infer $\delta^{11}B_{sw}$ for a given study interval. Palmer et al. (1998) and Pearson and Palmer (1999) used this approach to reconstruct upper ocean pH gradients, but they applied the same $\delta^{11}B_{foraminifer}$ vs. pH calibration to all species. Although we now know the use of a universal calibration to all foraminifer species is flawed (e.g. Figure 2.14), Palmer et al. (1998) noted $\delta^{11}B$ of their deepest-living Miocene species falls below the sensitivity of the proxy and those data could not be translated to pH. Similar to Spivack et al. (1993), Palmer et al. (1998) inferred variations in $\delta^{11}B_{sw}$, and assuming that both pH and temperature were the same as in the modern ocean, their Miocene foraminifers would reflect $\delta^{11}B_{sw} \sim 38‰$. The subsequent study of Pearson and Palmer (1999) followed the same approach for their Eocene-aged foraminifers but their surface pH gradient estimates were refined by estimates of biological productivity. Following various lines of argument, they estimated $\delta^{11}B_{sw} = 40.6‰$ 43 million years ago. Pearson and Palmer (2000) later used this approach to model $\delta^{11}B_{sw}$ throughout the past 60 million years (Figure 2.24), so they could reconstruct pH and atmospheric pCO_2 across the Cenozoic. Their data broadly agree with the earlier estimates, but were generated following the strategy of Pearson and Palmer (1999) to improve the $\delta^{11}B_{sw}$ estimates of Palmer et al. (1998). The assumed best estimate for the Eocene ($\delta^{11}B_{sw} = 40.6‰$, Pearson and Palmer 1999), however, was not applied to their reconstruction (Pearson and Palmer 2000).

Paris et al. (2010a) analyzed $\delta^{11}B$, Mg/Ca and $^{87}Sr/^{86}Sr$ in fluid inclusions in cloudy halites and interpreted their data to be indicative of the seawater composition at the time of evaporation. Because modern evaporites reflect $\delta^{11}B_{fluid\ inclusion} = 39.7 \pm 0.3‰$, the authors assumed that boron is incorporated into fluid inclusions without further fractionation. This assumption was corroborated by laboratory evaporation experiments by Liu et al. (2000), who evaporated NaCl solutions with varying concentrations of Ca^{2+} and SO_4^{2-} across a range of pH and temperature, and found boron is included in fluid inclusions with relatively little isotopic fractionation. Similarly, Vengosh et al. (1992) evaporated seawater and found that precipitated halite contains no boron in systems that have only evolved to precipitate halite, but not Mg- and/or K-sulfates. Salts containing Mg- and K-sulfates can incorporate small quantities of B, and their $\delta^{11}B$ is depleted by 20–30‰ compared to the parent seawater solution, but such precipitates are rare in the geological record (Vengosh et al. 1992). Paris et al. (2010a) measured halites from five time intervals during the Cenozoic, and their results from all samples were replicated multiple times and in some cases from different locations, suggesting they reflect a global ocean signal. However, pairing their Eocene $\delta^{11}B_{sw}$ estimates (i.e. ~31.8‰) with $\delta^{11}B_{foram}$ data from Pearson et al. (2009), results in estimated Eocene/Oligocene surface ocean pH_{TS} values of ~8.3. This result is inconsistent with independent evidence that atmospheric pCO_2 was approximately twice as high as today (Figure 1.4), and model estimates of Eocene/Oligocene surface ocean acidity which suggest pH_{TS} ~7.9 (Ridgwell and Schmidt 2010). Similar comparisons were made by Raitzsch and Hönisch (2013), who highlighted the massive oversaturation of surface and deep ocean water implied by such high pH values. These comparisons cast doubt on the halite $\delta^{11}B_{sw}$ estimates and suggest $\delta^{11}B_{fluid\ inclusion}$ is compromised in a way that affects globally distributed halite samples uniformly. Although we cannot identify the source of this discrepancy unambiguously at this time, the explanation is likely related to the boron isotope fractionation effect observed by Liu et al. (2000) in solutions containing Ca^{2+}, and sulfates (Vengosh et al. 1992). Liu et al. (2000) observed $\delta^{11}B_{fluid\ inclusion}$ is up to 5‰ lower than the parent brines in solutions containing ~1 g [Ca] per liter, and Vengosh et al. (1992) observed B residing in evaporites containing Mg and K-sulfates, with $\delta^{11}B_{salt}$ depleted by 20–30‰ compared to the parent solution. The likely explanation for this fractionation is co-precipitation of B in gypsum and anhydrite during the evaporation sequence. While 1 kg of modern seawater contains ~10 mmol $[Ca^{2+}]$ and thus ~0.4 g Ca^{2+}, the seawater calcium concentration was approximately twice as high 40 million years ago (Brennan et al. 2013), which brings $[Ca^{2+}]$ within the range where it affects boron isotope fractionation (Liu et al. 2000). Importantly, Paris et al. (2010a) observed a positive correlation between $\delta^{11}B_{fluid\ inclusion}$ and Mg/Ca analyzed in their samples. This is consistent with the higher [Ca^{2+}] and lower [Mg^{2+}] values reported for the early Cenozoic (Lowenstein et al. 2014), and suggests greater ^{10}B depletion when [Ca^{2+}] was higher. Because Ca^{2+} and Mg^{2+} are well mixed in the ocean, a Ca-effect on boron isotope fractionation would affect halites around the globe uniformly. We recommend the experiments of Liu et al. (2000) be expanded to test the effect of gypsum and anhydrite mineral precipitation under ancient seawater conditions. The conclusion that $\delta^{11}B_{fluid\ inclusion} = \delta^{11}B_{sw}$ (Paris et al. 2010b) does not appear tenable without accounting for [Ca^{2+}]-modulated boron isotope fractionation. If the 5‰ boron isotope fractionation observed by Liu et al. (2000) applies to the ancient halites studied by Paris et al. (2010a), their values would range ~37–38 ‰, which would be consistent with other observations (Figure 2.24).

Raitzsch and Hönisch (2013) approached estimating $\delta^{11}B_{sw}$ by analyzing $\delta^{11}B$ in epifaunal benthic foraminifera. The idea behind these estimates is that deep-ocean pH variations on geological time scales are muted relative to the surface ocean (Hönisch et al. 2008), and that $\delta^{11}B_{benthic\ foram.}$ variations simultaneously observed in the Atlantic and Pacific Ocean and across a range of depths should reflect variations in past $\delta^{11}B_{sw}$. This approach is not entirely independent, as whole ocean pH variations have been suggested for the Cenozoic (e.g. Ridgwell and Schmidt 2010) and must consequently be removed from the $\delta^{11}B_{benthic\ foram.}$ signal. Applying independent pH estimates from Earth system models (Ridgwell and Schmidt 2010; Tyrrell and Zeebe 2004), Raitzsch and Hönisch (2013) inferred the estimates shown in Figure 2.24, where the pH sensitivity of $\delta^{11}B_{benthic\ foram.}$ has been estimated according to aqueous $\delta^{11}B_{borate}$ (Klochko et al. 2006) and the lower sensitivity observed in laboratory calibrated marine carbonates (Figure 2.13a and b). Habitat and/or vital effects (i.e. species-specific offsets) between different modern and extinct benthic foraminifer species have thereby been cross-calibrated and normalized across the entire record (Raitzsch and Hönisch 2013). The results are in close agreement with the model estimates of Lemarchand et al. (2000, 2002) and for at least the past 16 million years also with the pH-gradient estimates of Pearson and Palmer (2000), albeit with oscillations the previous studies did not detect. These oscillations are similar to the δ^7Li_{sw} record of Misra and Fröhlich (not shown, 2012), which further supports the validity of these estimates because the B and Li cycles are affected by similar processes (Raitzsch and Hönisch 2013).

Foster et al. (2012) estimated Miocene $\delta^{11}B_{sw}$ from the $\delta^{11}B_{foram}$ difference between contemporaneous planktic and benthic foraminifer samples, using (i) the modern surface-to-deep ocean pH gradient as a constant, and (ii) the surface-to-deep ocean pH gradient inferred from the $\delta^{13}C$ gradient between planktic and benthic foraminifera. Similar to Palmer et al. (1998) and Pearson and Palmer (1999, 2000), this approach explores the non-linear relationship between $\delta^{11}B_{CaCO3}$ and pH (Figure 2.25), and the systematic differences between carbonate chemistry patterns in the surface and deep ocean, which are ultimately connected by the biological pump. Based on these two pH-gradients, Foster et al. (2012) estimated $\delta^{11}B_{sw} = 37.6‰$ for the constant pH-gradient and 38.0‰ from the pH-gradient inferred from $\delta^{13}C_{planktic-benthic}$, and averaged them to $37.82 \pm 0.35‰$ (Figure 2.24).

Anagnostou et al. (2016) applied a similar approach to the Eocene, where $\delta^{11}B_{sw}$ was estimated based on the surface-to-deep $\delta^{11}B_{borate}$-pH relationship (Figure 2.25). Because this relationship is non-linear, there is a minimum $\delta^{11}B_{CaCO3}$ for which pH can be calculated and this minimum $\delta^{11}B_{CaCO3}$ therefore provides the upper limit of $\delta^{11}B_{sw}$ in a given time interval. Using the lowest $\delta^{11}B_{CaCO3}$ values analyzed in their study, Anagnostou et al. (2016) therefore determined an upper $\delta^{11}B_{sw}$ limit of 39.5‰ for the Eocene. To refine this estimate, Anagnostou et al. (2016) assumed the Eocene surface-to-300 m pH-gradient was similar to the modern gradient in the tropical oceans. Because Eocene surface-deep dwelling planktic foraminiferal $\delta^{13}C$ gradients were larger, a higher metabolic rate and larger surface-deep pH gradient can be assumed for the Eocene, such that the assumption of the modern gradient would place a lower limit on $\delta^{11}B_{sw}$ (Figure 2.25). With this approach, they estimated the minimum Eocene $\delta^{11}B_{sw} = 38.2$–38.6‰, depending on the $\delta^{11}B_{CaCO3}$ versus $\delta^{11}B_{borate}$ sensitivity assumed for their now extinct foraminifera species (see also online Table A2.2). Finally, Anagnostou et al. (2016) paired their boron-based pH estimates (with $\delta^{11}B_{sw} = 38.2$–39.6‰) with model-derived $\Omega_{calcite}$ to determine the surface-to-deep DIC profile, and via the Redfield Ratio of 106 CO_2/138 O_2 the Apparent Oxygen Utilization of the water column. Combining this estimate with temperature-dependent oxygen solubility in the surface ocean and the minimum oxygen tolerance of the modern planktic foraminifera *Hastigerinella digitata*, they estimate the upper $\delta^{11}B_{sw}$ bound as 38.5–38.8‰ at 45.6 Ma and as 38.7–38.9‰ at 40.3 Ma (Figure 2.24). These estimates were then applied to all their data within the interval 37–53 Ma.

Finally, Greenop et al. (2017) extended the combined planktic-benthic $\delta^{11}B$ and $\delta^{13}C$ approach over the entire Neogene (Figure 2.24) and found data that are broadly consistent with the estimates of Pearson and Palmer (2000), Lemarchand et al. (2000, 2002) and Raitzsch and Hönisch (2013). Further agreement comes from Pearson et al. (2009), who examined their Eocene-Oligocene foraminiferal $\delta^{11}B$ across a range of possible $\delta^{11}B_{sw} = 37$–39‰ (using the estimates of Lemarchand et al. 2000, 2002), and found that $\delta^{11}B_{sw} \sim 38‰$ produced pCO_2 estimates that match modeled pCO_2 thresholds for Antarctic ice sheet growth (DeConto and Pollard 2003; DeConto et al. 2008) (Figure 2.24). Lastly, Martínez-Botí et al. (2015b) interpolated between modern $\delta^{11}B_{sw}$ and the Miocene estimate of Foster et al. (2012) to estimate Pliocene $\delta^{11}B_{sw} \sim 39.2 \pm 0.2‰$ (Figure 2.24), and although this is not an independent constraint, the data are shown for completeness.

In summary, as more $\delta^{11}B_{sw}$ estimates are being generated, the data shown in Figure 2.24 appear to converge and thus increase the confidence in absolute pH values and surface ocean pCO_2 estimates inferred from the boron isotope proxy. However, differences between records still amount to $\sim 1‰$ and increase further back in time. Consequently, paleoreconstructions should continue to explore the sensitivity of their interpretations to uncertainties in $\delta^{11}B_{sw}$ (e.g. Anagnostou et al. 2016; Bartoli et al. 2011; Pearson et al. 2009). For the B/Ca proxy we can unfortunately only rely on the $[B_T]$ model estimates of Lemarchand et al. (Figure 2.24, 2000, 2002) and the sensitivity studies of Simon et al. (not shown, 2006). These estimates range from $[B_T]$ ~1–6 ppm at 20 Ma, highlighting the significant uncertainty in model parameterizations. Culture studies have not yet determined whether variable $[B_T]$ systematically affects the sensitivity of B/Ca to pH. Deep time studies using the B/Ca proxy are therefore currently hampered by uncertainty in this and other parameters, such that reconstructions can only describe qualitative trends in carbonate chemistry. As applications of the B/Ca proxy to the geological record are expanded, additional constraints on $[B_T]$ are needed as much as improved estimates of $\delta^{11}B_{sw}$.

2.6   The B/Ca Proxy in Foraminifera

Following original observations of Hemming and Hanson (1992) that boron concentrations in marine carbonates are sensitive to pH, Wara et al. (2003) were among the first to evaluate the potential of B/Ca in planktic foraminifers, but the proxy only started gaining momentum with targeted calibration studies by Yu et al. (2007) and Yu and Elderfield (2007), who discovered that the environmental controls on the proxy differ between planktic and benthic foraminifers. In contrast to foraminifers, B/Ca in corals is more sensitive to temperature than to marine carbonate chemistry (e.g. Dissard et al. 2012; Fallon et al. 1999; Sinclair et al. 1998; Trotter et al. 2011), and B incorporation in coccoliths is heavily modified by the internal biomineralization process of the coccolithophorid algae (Stoll et al. 2012). Because this book targets boron proxies in the context of marine carbonate chemistry, we will therefore neglect corals and coccolithophorids here, and focus on the B/Ca proxy systematics in planktic and benthic foraminifers.

2.6.1   *B/Ca in Planktic Foraminifers – Environmental Controls and the Partition Coefficient K_D*

As described in Section 2.3.2, boron partitioning into planktic foraminifer shells (Eq. 2.14) was initially believed to be controlled by the relative abundance of $B(OH)_4^-$ and HCO_3^- ions in seawater. Because the modern ratio of borate to bicarbonate ion in seawater is proportional to pH (Figure 2.26), the residence time of boron in seawater is long (i.e. 14–15 Myr, Figure 2.24a, Lemarchand et al. 2000, 2002), and $[B_T]$ is proportional to salinity (i.e. $[B_T] = 432.6 \, \mu mol \, kg^{-1} \times S/35$, Lee et al. 2010), Yu et al. (2007) suggested that pH may be inferred from B/Ca if the partition coefficient K_D can be quantified.

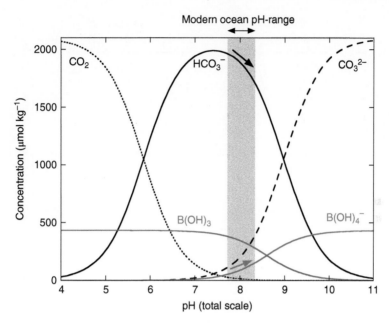

Figure 2.26 This Bjerrum plot displays the concentrations of dissolved carbon (black lines) and boron (red and blue lines) species versus seawater-pH. Individual species concentrations were calculated using the CO2SYS program by Pierrot et al. (2006) with K_1 and K_2 according to Lueker et al. (2000), K_{SO4} according to Dickson (1990a), total [B] after Lee et al. (2010), and using T = 25 °C, S = 35, p = 1 bar and DIC = 2100 µmol kg^{-1}. The modern range of seawater-pH is indicated by the gray bar, and the arrows highlight that $[B(OH)_4^-]$ increases and $[HCO_3^-]$ decreases as seawater pH increases. Boron incorporation into marine carbonates should theoretically increase with the increasing ratio of $[B(OH)_4^-]$ to $[HCO_3^-]$ at higher pH (Yu et al. 2007).

Quantification and calibration of K_D has, however, proven difficult. Sediment coretop (Foster 2008; Hendry et al. 2009; Henehan et al. 2013; Yu et al. 2007, 2013a) and laboratory calibration studies (Allen et al. 2011, 2012; Henehan et al. 2015) determined that $K_D = [B/Ca]_{foram}/[B(OH)_4^-/HCO_3^-]_{seawater}$ as defined by Yu et al. (2007) is species specific and not a constant. Among symbiont-bearing foraminifers, *G. ruber* records higher B/Ca than *G. sacculifer* and *O. universa*; *Globorotalia truncatulinoides* covers a range of B/Ca values (Figure 2.27). Among symbiont-barren species, *G. bulloides* records the lowest B/Ca, whereas *N. pachyderma*, *G. inflata*, *N. dutertrei* and *N. incompta* record B/Ca values similar to *O. universa* (Figure 2.27). Furthermore, pH is unlikely the only factor controlling boron incorporation. While Yu et al. (2007) originally suggested K_D in *G. inflata* and *G. bulloides* is sensitive to temperature (Figure 2.27d), Foster (2008) observed consistently inverse relationships between K_D and temperature and K_D and $[CO_3^{2-}]$ in *G. sacculifer*, *G. ruber* and *N. dutertrei* (Figure 2.27c). A later study by Tripati et al. (2009) found again a positive relationship between K_D and temperature in *G. ruber* and *G. sacculifer*, albeit using down-core sediments for their calibration. Applying their respective K_D calibrations

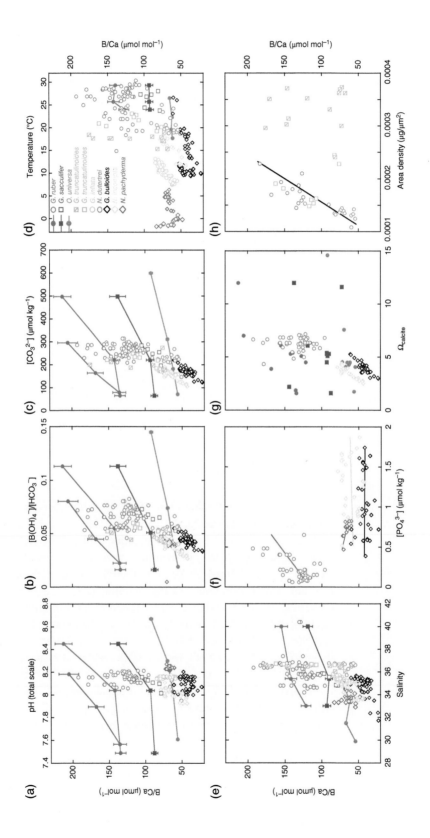

to downcore records of B/Ca in planktic foraminifers, all three studies succeeded in replicating ice core pCO_2, despite the fact that most B/Ca records do not reflect the higher glacial values predicted from higher glacial pH (i.e. lower pCO_2, Figure 2.28). The discrepancy between calibrations, and the apparent disconnect between original B/Ca data and their translation to seawater boron/carbonate chemistry and pCO_2 has raised concern about the functioning of the B/Ca proxy in planktic foraminifers, and solicited further probing of B/Ca proxy systematics.

The use of K_D in these paleoreconstructions was subsequently questioned by Allen and Hönisch (2012), who demonstrated the resulting pH and pCO_2 estimates calculated according to the temperature-dependent K_D approach are essentially independent of the primary B/Ca data. This is because the relative concentrations of carbonate, bicarbonate, and borate ion are interrelated by equilibrium constants (Figure 2.26), and also because temperature, salinity, alkalinity, and atmospheric CO_2 are closely related in the ocean-climate system, both past and present. In cases where only K_D shows a relationship with temperature or carbonate ion, but not the original B/Ca data, the relationship is driven solely by the denominator, $[B(OH)_4^-]/[HCO_3^-]$, which itself is linked to temperature and carbonate ion via equilibrium constants (see also Figure 2.7 for the denominator driven temperature effect on B/Ca in synthetic calcite). Consequently, K_D, temperature and carbonate ion are not independent quantities, and paleoreconstructions that apply such denominator-driven K_D calibrations are the result of relationships artificially built into the calibration rather than a new product that reflects variations in the original B/Ca data. For example, if a Pleistocene K_D-SST calibration is applied to an Eocene record, this forces temperature into a driving role in the reconstruction, automatically projecting the Pleistocene temperature-pCO_2 sensitivity onto the Eocene. Allen and Hönisch (2012) demonstrated the severity of this linkage by recalculating the pCO_2 reconstructions of Yu et al. (2007) and Tripati et al. (2009) with a constant (average) B/Ca value for all samples, and obtained essentially the

Figure 2.27 Summary of B/Ca calibrations in planktic foraminifers. Data are from Allen et al. (2011, 2012), Yu et al. (2007, 2013a), Foster (2008), Hendry et al. (2009), Henehan et al. (2015), Quintana Krupinski et al. (2017), and Salmon et al. (2016). Laboratory culture studies are displayed with closed symbols and coretop studies with open symbols. Example analytical uncertainties are displayed for culture studies. To facilitate direct comparison of studies performed in different laboratories, all data are normalized to analyses performed at Rutgers University, using the conversion equation described by Allen et al. (i.e. B/Ca = 1.085 x B/Ca$_{Cambridge, Bristol, Southampton}$ – 1.09, 2012). Data of Quintana Krupinski et al. (2017), and Salmon et al. (2016) have not been corrected due to unknown interlaboratory offsets. Whereas laboratory culture studies confirm the predicted B/Ca sensitivity to pH, coretop studies are rather scattered and suggest B/Ca in natural samples is subject to multiple environmental controls, none of which seem to predominate the B incorporation behavior. Species differences are relatively robust; the symbiont-bearing *G. ruber* and *G. sacculifer* exhibit the highest values, whereas symbiont-barren species record lower values. *Orbulina universa* forms again a notable exception, by recording low B/Ca values despite its association with symbiotic algae (cf. also Figure 2.14 for a similar pattern in $\delta^{11}B$). (f) displays linear regressions for individual species, (h) the multi-species regression of Salmon et al. (2016), which excludes the encrusted *Globorotalia truncatulinoides* data.

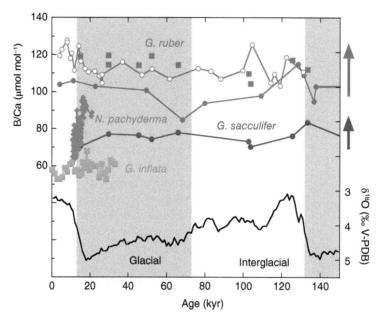

Figure 2.28 Comparison of planktic B/Ca records over the past 150 kyr. Data are from Yu et al. (red and turquoise squares, and green diamonds, 2007, 2013a), Foster (open red circles, 2008), and Tripati et al. (red and blue circles, 2009). The benthic oxygen isotope stack (Lisiecki and Raymo 2005) is given as a reference to indicate glacial and interglacial stages, and the red and blue arrows on the right display the glacial B/Ca increase predicted from laboratory pH calibrations (Allen et al. 2012) with *G. ruber* (red) and *G. sacculifer* (blue). Note that only the *Neogloboquadrina pachyderma* record conforms with the prediction, all other records are either flat or even show lower values during glacial times.

same pCO_2 estimates as with the original B/Ca data (Figure 2.29). Because these denominator-driven temperature and carbonate ion relationships cannot provide any new paleo-information, Allen and Hönisch (2012) recommend paleoreconstructions should focus on primary relationships between original B/Ca data and environmental parameters, and that the use of K_D should be avoided until the aqueous species involved in the calcification process have been identified.

Fortunately, the B/Ca sensitivity to pH (Figure 2.27a), $[B(OH)_4^-]/[HCO_3^-]$ (Figure 2.27b), and/or $[CO_3^{2-}]$ (Figure 2.27c) has been confirmed in laboratory calibrations with the planktic foraminifer species *O. universa*, *G. sacculifer* and *G. ruber* (Allen et al. 2011, 2012; Henehan et al. 2015). With all other environmental parameters being constant in laboratory culture, these three parameters are coupled in seawater, and a positive correlation with B/Ca is expected. In contrast, temperature does not exert significant control on B/Ca in laboratory culture (Figure 2.27d), but B/Ca is somewhat sensitive to salinity (Figure 2.27e). Despite this encouraging result, the sensitivity of the B/Ca proxy to seawater pH in planktic foraminifers is relatively low and ranges from ~3.5 μmol mol⁻¹ (*O. universa*) to ~5.9 μmol mol⁻¹ (*G. sacculifer*) to ~9.3 μmol mol⁻¹ (*G. ruber*) per 0.1 pH unit, and from 4.5 μmol mol⁻¹

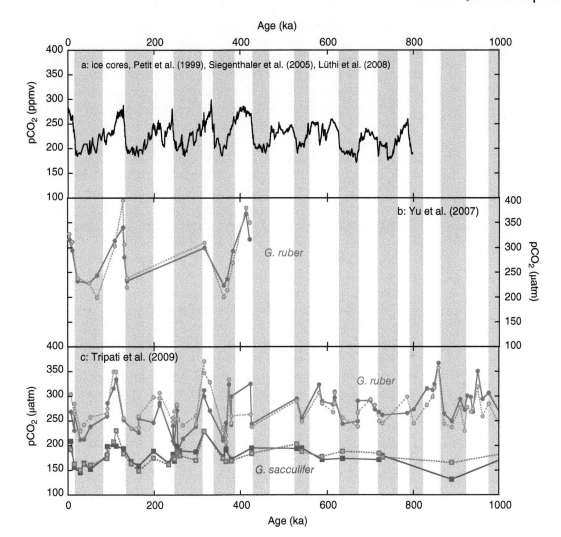

Figure 2.29 Comparison of paleo-pCO$_2$ measured (a) in ice cores (Lüthi et al. 2008; Petit et al. 1999; Siegenthaler et al. 2005), and (b, c) reconstructed from B/Ca in planktic foraminifer shells based on the temperature-dependent K$_D$ calibrations applied in the original studies (Tripati et al. 2009; Yu et al. 2007). B/Ca estimates apparently capture the pCO$_2$ variation observed in ice cores, but detailed analysis (Allen and Hönisch 2012) revealed that the estimates are predominantly driven by other environmental parameters involved in the calculation: (b) and (c) compare pCO$_2$ estimates based on actual measured B/Ca data (dark red and blue symbols, solid lines), and constant B/Ca values that correspond to the average B/Ca values of each data set (pink and light blue symbols, dashed lines). The similarity of the resulting pCO$_2$ estimates demonstrates that B/Ca values play a subordinate role in the reconstructions. Note that the pCO$_2$ difference between *G. ruber* and *G. sacculifer* estimates in (c) is due to species-specific differences in the calibrations. Gray bars indicate glacial stages. Figure modified from Allen and Hönisch (2012).

(*O. universa*) to 5.1 μmol mol^{-1} (*G. ruber*) per salinity unit. For comparison, the analytical uncertainty of B/Ca is typically on the order of ±3%, or ± 2–10 μmol mol^{-1} (depending on the foraminifer species studied). Boron isotope analyses have determined glacial surface ocean pH was ~0.15 units higher than Holocene surface ocean pH (Henehan et al. 2013; Hönisch and Hemming 2005), which translates to glacial B/Ca ~5.3–14 μmol mol^{-1} higher than interglacials, but downcore studies generally do not conform to this prediction (Figure 2.28). Because glacial surface seawater was characterized by both higher pH and higher salinity, the salinity signal also has to be removed from glacial B/Ca data, further diminishing the small change observed in B/Ca data of glacial age.

Field calibration studies of B/Ca in various planktic foraminifera species (Foster 2008; Hendry et al. 2009; Henehan et al. 2015; Quintana Krupinski et al. 2017; Salmon et al. 2016; Yu et al. 2007, 2013a) are also shown in Figure 2.27, where B/Ca data are presented versus the environmental parameters reported by the original studies, and/or could be calculated from reported values (e.g. $\Omega_{calcite}$, Figure 2.27g). Whereas most of these comparisons show much larger variability than observed in laboratory culture, some strong relationships with phosphate concentrations (Henehan et al. 2015), $\Omega_{calcite}$ (Quintana Krupinski et al. 2017) and the degree of calcification (Salmon et al. 2016) were observed in field-grown specimens. However, there is not a single parameter that consistently explains B/Ca in all foraminifera studied in different environmental settings. For instance, while $\Omega_{calcite}$ appears to exert strong control on B/Ca in *G. bulloides* and *N. incompta* (Figure 2.27g, Quintana Krupinski et al. 2017), and area density estimates on B/Ca in *O. universa*, *G. bulloides* and non-encrusted *G. truncatulinoides* (Figure 2.27h, Salmon et al. 2016), laboratory-grown foraminifera do not suggest a saturation state control on B/Ca (Figure 2.27g). This is particularly striking for *G. sacculifer*, where $\Omega_{calcite}$ was varied separately by pH and by addition of DIC, and whereas B/Ca and $\Omega_{calcite}$ both increase at higher pH and thus confirm to the suggestion of $\Omega_{calcite}$ control of Quintana Krupinski et al. (2017), $\Omega_{calcite}$ similarly increased in the DIC experiments, but B/Ca decreased under those conditions, thus contradicting a $\Omega_{calcite}$ control (Allen et al. 2012). These comparisons reinforce the notion that B/Ca is not a very sensitive proxy in planktic foraminifers, and in particular on timescales with relatively small variations in seawater acidity.

2.6.2   *Elemental Species Competition and Potential pH up-Regulation Effects on B/Ca in Planktic Foraminifers*

In addition to the low proxy sensitivity, boron partitioning (i.e., K_D as defined by Yu et al. 2007) also decreases at higher pH (Figure 2.30, Allen et al. 2011, 2012). The same pattern has been observed by He et al. (2013) in synthetic calcite precipitation experiments (Figure 2.5a) and it means that although B incorporation increases at higher pH in absolute terms, B uptake is less than the pH-dependent increase in borate concentration in seawater. Figure 2.30 compares experimental results of B/Ca uptake in response to higher $[B(OH)_4^-]/[HCO_3^-]$ at higher pH (Allen et al. 2012; Allen et al. 2011), to experiments at constant pH but with higher $[B(OH)_4^-]/[HCO_3^-]$ due to

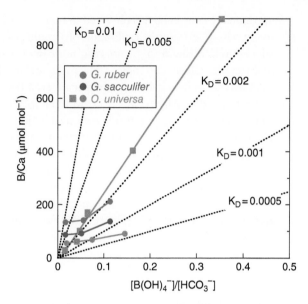

Figure 2.30 Comparison of experimental planktic foraminiferal B/Ca versus aqueous $[B(OH)_4^-]/[HCO_3^-]$ with lines of constant K_D. Data are from Allen et al. (2011, 2012), where circles display data from pH experiments ($[B(OH)_4^-]$ increases but $[HCO_3^-]$ decreases with pH), and squares present data from boron concentration experiments ($[B(OH)_4^-]$ increases but $[HCO_3^-]$ remains constant). K_D values between 0.0005 and 0.01 indicate that foraminifers heavily discriminate against B incorporation, although the discrimination increases at higher $[B(OH)_4^-]/[HCO_3^-]$, as indicated by decreasing K_D at higher pH. In contrast, B/Ca data from the boron concentration experiment fall close to a constant $K_D{\sim}0.0025$. This comparison indicates that although more B is incorporated at higher $[B(OH)_4^-]/[HCO_3^-]$, B uptake is not proportional to the $[B(OH)_4^-]/[HCO_3^-]$ ratio in pH experiments, and other boron or carbon species must be involved in the calcification stoichiometry.

higher $[B_T]$ (Allen et al. 2011; Haynes et al. 2017). Whereas K_D in the pH experiments ranges from ~0.0006 to 0.01 in these three foraminifer species, B/Ca in the B concentration experiment falls close to a constant $K_D{\sim}0.0022$ in *O. universa*; noting other foraminifer species have not yet been calibrated for variable $[B_T]$ in laboratory culture. Allen et al. (2011) discussed potential reasons for the difference in B/Ca uptake between pH and $[B_T]$ experiments, including reduced borate adsorption affinity of the increasingly negatively charged calcite surface at higher pH (van Cappellen et al. 1993; Morse 1986), and potential competition with other pH-sensitive ions. While reduced adsorption efficiency of borate ion at higher pH remains to be tested, Allen et al. (2011) evaluated boric acid incorporation at low pH, which could potentially elevate B incorporation beyond the prediction from borate ion abundance in seawater. However, using evidence from the laboratory calibration of $\delta^{11}B$ in *O. universa* (Sanyal et al. 1996) and its observed reduced pH sensitivity compared to boron isotope fractionation between aqueous boric acid and borate ion (Section 2.4.1, Klochko et al. 2006; Nir et al. 2015), Allen et al. (2011) estimated boric acid incorporation at low pH no greater than 3–8%, which is a small contribution compared to the large quantities of trigonally coordinated B observed in marine carbonates (Table 2.1). The recent MC-ICP-MS field calibration of *O. universa* (Henehan et al. 2016) falls entirely below $\delta^{11}B_{borate}$, making boric acid incorporation even more problematic. Given the comparably large analytical uncertainty of B/Ca analyses (i.e. ~±5% at 2σ, see Chapter 4), even the small addition of boric acid estimated by Allen et al. (2011) would have a negligible effect on B/Ca and K_D. Finally, it is theoretically conceivable that foraminifers up-regulate pH in their calcifying fluid at low ambient pH and thereby elevate B/Ca, but it remains unknown whether planktic foraminifers vacuolize seawater in a fashion similar to benthic foraminifers (Bentov et al. 2009).

Alternatively, Allen et al. (2011) hypothesized that borate incorporation may compete with carbonate ion incorporation, as both ions increase in abundance at higher pH (Figure 2.26), and borate incorporation may become less favorable as more carbonate ions become available for calcification. This hypothesis was subsequently supported by Allen et al. (2012) and Haynes et al. (2017), who compared B/Ca in *G. sacculifer* and *O. universa* grown (i) across a range of pH at constant DIC, and (ii) across a range of DIC at constant pH (Figure 2.31). Whereas B/Ca increases with pH when $[B(OH)_4^-]$ increases, B/Ca decreases when $[HCO_3^-]$ and $[CO_3^{2-}]$ increase due to addition of DIC at constant pH. Similar results have been obtained by Holland et al. (2017), who investigated a handful of *O. universa* specimens

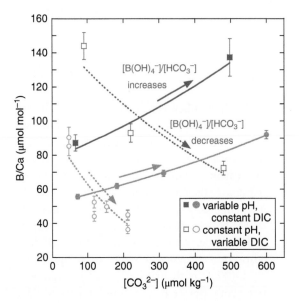

Figure 2.31 In laboratory culture experiments B/Ca increases with pH, when the $[B(OH)_4^-]/[HCO_3^-]$ ratio increases due to increasing $[B(OH)_4^-]$ and decreasing $[HCO_3^-]$ (closed symbols and solid line), but B/Ca decreases when DIC (i.e. $[HCO_3^-]$ and $[CO_3^{2-}]$) is added at constant $[B(OH)_4^-]$ (open symbols and dashed line). These results suggest that boron uptake in planktic foraminifers competes with bicarbonate and/or carbonate ion uptake during calcification. Data are from Allen et al. (2012), growing *G. sacculifer*, and Haynes et al. (2017), growing *Orbulina universa*.

by laser ablation (not shown here due to the low number of shells analyzed in that study), and by Howes et al. (2017), whose data are not shown because they grew their foraminifera under 10x elevated boron concentrations. This does not diminish the results, but makes them more difficult to display alongside foraminifera grown under natural boron concentrations.

The carbonate competition result contrasts with synthetic calcite precipitation experiments across a range of DIC, which display a positive relationship between B/Ca in synthetic calcite and DIC when precipitation rate is allowed to increase at higher DIC (Uchikawa et al. 2015, 2017). However, B/Ca decreases with DIC when precipitation rate is held constant by manipulating $[Ca^{2+}]$, supporting competition of carbonate and borate species in the absence of growth rate effects (Uchikawa et al. 2017) (Figure 2.5). While synthetic $CaCO_3$ precipitation experiments thus appear to be strongly affected by precipitation rate (Figure 2.6), we have few constraints on the precipitation rate of planktic foraminifer shells. Kısakürek et al. (2011) used the weight difference of cultured foraminifera between the time of capture and gametogenesis to estimate planktic foraminiferal calcification rates as logR $(mol/m^2/s) = -6.26$ for *G. siphonifera* and -6.08 for *G. ruber* at 25 °C, respectively. Allen et al. (2016) approximated growth rates in µg/day for *O. universa*, *G. sacculifer*, *G. ruber* and *G. bulloides*, using the final weight of individual specimens grown in culture by the number of days they lived in culture. Although those approximations are not very accurate because a portion of the individual shell weights was grown in the ocean and is not included in the time estimate, it is clear that growth rates decrease from *O. universa* > *G. sacculifer* > *G. ruber* > *G. bulloides*. In comparison, Holland et al. (2017) estimated linear growth rates for *O. universa* by dividing the diurnal banding width by 24 hours and estimated logR $(mol\,m^{-2}\,s^{-1}) = -6.5$ to -6.1 for their experimental conditions. Based on these preliminary constraints, planktic foraminifera are estimated to calcify ~100x more slowly than inorganic calcites grown in similarly saturated solutions, suggesting that a growth rate effect on B/Ca, even if observed in inorganic experiments, may not exert strong influence on foraminifera. Foraminifera instead appear to maintain slow, controlled growth rates in spite of ambient seawater conditions that could allow faster inorganic growth (Allen et al. 2016).

While more and better constrained calcification rate data are needed across a range of experimental conditions, including pH, temperature, DIC and $[Ca^{2+}]$, these estimates suggest planktic foraminifers precipitate their shells at rates that fall towards the low end of precipitation rates studied by Uchikawa et al. (2015) and by Gabitov et al. (2014), where K^*_B changes little. Even without more detailed foraminiferal calcification rate estimates, the culture experiments of Allen et al. (2012), Howes et al. (2017) and Haynes et al. (2017) do not allow us to distinguish between bicarbonate or carbonate ion control on calcification, as both increase by addition of DIC. Nonetheless, it is clear that borate ion is not the only control on B/Ca, and B/Ca uptake is thus more sensitive to increase of borate via addition of $[B_T]$ (at constant pH and DIC) than to pH-driven $[B(OH)_4^-]/([HCO_3^-] + [CO_3^{2-}])$ changes alone (Figure 2.30).

2.6.3   Light Effects on B/Ca in Symbiont-Bearing Planktic Foraminifers

While B/Ca controls are easily identified in laboratory culture, where environmental controls can be modified far beyond conditions observed in the modern ocean and with all other parameters maintained constant, B/Ca observations from plankton nets, sediment traps, and coretop sediments reveal less systematic patterns (Figure 2.27), with even the pH and $[B(OH)_4^-]/[HCO_3^-]$ controls being difficult to distinguish. Open-ocean studies are important to developing our understanding of proxies since they involve natural growth and depositional conditions. However, as described above, part of the difficulty with open-ocean calibrations is the low sensitivity of the proxy, in addition to potential complex interactions with other controls. For instance, Babila et al. (2014) studied *G. ruber* from sediment traps, and observed a strong seasonal cycle in B/Ca that correlated most strongly with seasonal variations in light intensity, and less with temperature. Similar to light effects on $\delta^{11}B$ in symbiont-bearing foraminifers (Section 2.4.3, Hönisch et al. 2003; Jørgensen et al. 1985; Rink et al. 1998; Zeebe et al. 2003), Babila et al. (2014) suggested symbiotic CO_2 sequestration would elevate pH in the microenvironment of the foraminifer, and consequently elevate the recorded B/Ca ratio in response to its pH sensitivity.

This argument may be consistent with a laser ablation study by Allen et al. (2011) and Holland et al. (2017), who studied individual *O. universa* shells and observed an ~50% increase in B/Ca from the first calcite secreted during chamber formation, to the final calcite precipitated right before gametogenesis. Incremental increase in symbiont density over the course of its life cycle (Spero and Parker 1985) can be expected to be associated with an increase in carbon sequestration and pH elevation, and may thus explain the intra-shell variability. In addition, B/Ca is elevated in growth bands secreted during the day compared to growth bands precipitated at night, a feature that is consistent with photosynthesis-dependent pH elevation during the day (Holland et al. 2017). However, bulk shell B/Ca analyses of *O. universa* grown in low light conditions display little to no change in B/Ca with light: Using the ~0.2-unit pH-elevation inferred from boron isotopes measured in *O. universa* grown under saturating and non-saturating light levels (Hönisch et al. 2003), Haynes et al. (2017) estimated B/Ca in *O. universa* should differ by $7–11\ \mu mol\ mol^{-1}$, depending on the sensitivity of the B/Ca versus pH relationship observed on Santa Catalina Island (Allen et al. 2011) and in Puerto Rico (Haynes et al. 2017). While this difference was observed in Puerto Rico, none was observed for the specimens grown on Santa Catalina Island (Haynes et al. 2017), leaving this evidence somewhat inconclusive.

In contrast, Babila et al. (2014) used their B/Ca data to estimate microenvironmental pH-elevation up to 0.2 units during high illumination summer months, and pH-lowering up to −0.3 units during the dimmer winter months, resulting in significantly different bulk shell B/Ca composition. This estimate by far exceeds the integrated boron isotope-based pH estimate from light experiments in *O. universa* (Hönisch et al. 2003), however, light experiments have yet to be undertaken for *G. ruber* and other symbiont-bearing species. Until such experimental data are available to confirm the extent of the light effect, B/Ca studies on symbiont-bearing foraminifers from extra-tropical sediments (where seasonal variations in environmental parameters are largest) should be interpreted with caution.

Henehan et al. (2015) studied B/Ca in *G. ruber* shells from coretop sediments collected in all ocean basins, and from latitudes 38°N to 42°S. The study revealed B/Ca variation much greater than predicted from pH sensitivity alone (open red symbols in Figure 2.27), and found 25% of the B/Ca variation in *G. ruber* covaries with dissolved phosphate (PO_4^{3-}) concentrations (Figure 2.27f). Similarly, Naik and Naidu (2014) studied B/Ca in *G. ruber* from Arabian Sea sediments and found higher B/Ca values during times of increased upwelling and hence higher nutrient concentrations but lower pH. While *G. ruber*, as a symbiont-bearing species adapted to oligotrophic conditions, may not be the best choice for proxy validation in an upwelling area (Naik and Naidu 2014), Henehan et al. (2015) discussed several possible explanations for their specific observation, including greater food supply in high nutrient environments, higher calcite precipitation rate, and higher symbiont-photosynthesis rates. While there is support for and against each of these parameters (e.g. symbiont photosynthesis sequesters CO_2 and thus elevates pH and B/Ca; higher respiration may lead to higher calcification rates and reduced discrimination against impurities, but both respiration and calcification produce CO_2 and thereby lower pH in the foraminiferal microenvironment), the most striking argument is that B/Ca only covaries with $[PO_4^{3-}]$, but not with other nutrients. This conundrum led Henehan et al. (2015) to suggest that the reason for the covariation may instead be crystallographic, where greater $B(OH)_4^-$ uptake might compensate crystal strain or charge imbalances due to incorporation of PO_4^{3-} ions. A similar observation of a $[PO_4^{3-}]$ effect on B/Ca has recently been made in inorganic calcite (Uchikawa et al. 2017). However, evidence from other planktic foraminifera species does not support a universal $[PO_4^{3-}]$ control: B/Ca in the symbiont-barren foraminifer *N. pachyderma* (Yu et al. 2013a) shows a negative covariation with $[PO_4^{3-}]$, and *G. bulloides* and *N. incompta* show no correlation at all (Figure 2.27f, Quintana Krupinski et al. 2017). In addition, Haynes et al. (2017) performed a suite of culture experiments in partially synthetic seawater, in which they inadvertently halved the PO_4^{3-} concentration but found no difference in B/Ca compared to specimens grown under normal $[PO_4^{3-}]$. In summary, evidence from other foraminifera species suggests the positive relationship with $[PO_4^{3-}]$ observed in *G. ruber* (Henehan et al. 2015; Naik and Naidu 2014) may not be systematic nor a controlling factor.

2.6.4 Dissolution Effects on B/Ca in Planktic Foraminifers

Several studies have investigated the effect of partial dissolution on B/Ca in planktic foraminifer shells (Figure 2.32, Coadic et al. 2013; Dai et al. 2016; Seki et al. 2010; Yu et al. 2007), however, not all of these studies are comparable in their rigor of sample selection and verification of the preservation state of their studied shells. For instance, the *G. inflata* and *G. bulloides* samples of Yu et al. (2007) cover 18° of latitude and the carbonate chemistry of the surface ocean growth habitat likely varied independent of bottom water saturation. The *G. sacculifer* samples of Coadic et al. (2013) and Seki et al. (2010) were selected from a small size fraction (300–355 μm), which is more susceptible to dissolution (Berger and Piper 1972) and therefore not preferable for paleoreconstructions using boron proxies (Hönisch and Hemming 2004). Finally, Yu et al. (2007) and Seki et al. (2010) do not provide any evidence from shell weights, which makes the dissolution extent difficult to gauge. In contrast, the studies of Coadic et al. (2013) and Dai et al. (2016) both look at

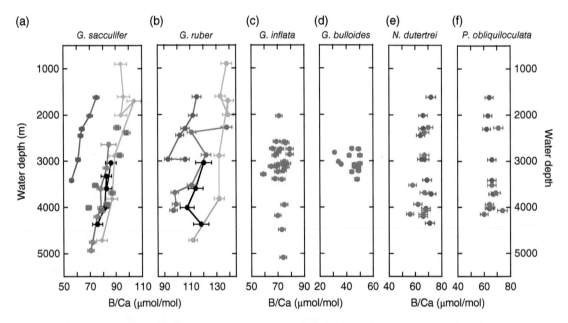

Figure 2.32 Evidence for a dissolution effect on planktic foraminiferal B/Ca from core top sediments. The foraminifer species are indicated at the top of each graph and individual studies are color coded: black symbols – Seki et al. (2010), green symbols – Coadic et al. (2013), red symbols – Yu et al. (2007), blue symbols – Dai et al. (2016), where dark blue symbols reflect data from the Ontong-Java Plateau, medium blue symbols data from the Southwestern Indian Ocean, and light blue symbols data from the Caribbean. Note that the *Globorotalia inflata* (c) and *Globigerina bulloides* (d) data are derived from samples covering a wider geographic range with variable surface ocean chemical conditions. The *G. sacculifer* (a), *G. ruber* (b), *Neogloboquadrina dutertrei* (e) and *Pulleniatina obliquiloculata* (f) data are from depth transects within restricted geographical areas and chemically homogenous surface ocean conditions. Significant B/Ca decreases in *G. sacculifer* (a) and *G. ruber* (b) suggest partial shell dissolution may affect symbiont-bearing foraminifer species more readily than symbiont-barren species (c–f).

defined depth transects and a range of water depths below the calcite saturation horizon. The common feature in these observations is that symbiont-barren species (*G. inflata*, *G. bulloides*, *N. dutertrei* and *Pulleniatina obliquiloculata*, Figure 2.23c–f) tend to retain their B/Ca signature even in undersaturated bottom water conditions, whereas the symbiont-bearing *G. sacculifer* and *G. ruber* (Figure 2.32a and b) show consistent and significant B/Ca decreases even shallower than 3000 m water depth, where bottom waters are typically oversaturated with respect to calcite. B/Ca decreases by ~20 μmol mol^{-1} (*G. ruber*), respectively by ~15 μmol mol^{-1} (*G. sacculifer*) between the shallowest and just undersaturated core depths, which is similar in magnitude to glacial interglacial variations projected from culture calibrations (Figures 2.27 and 2.28). Partial dissolution therefore must be considered a further potential pitfall of the proxy, and shell preservation should be considered when selecting samples and interpreting data for paleoreconstructions. A summary of the incorporation and preservation controls on B/Ca in planktic foraminifera is given in Box 2.4, including recommendations for the use of this proxy in paleoreconstructions.

Box 2.4 Summary of B/Ca systematics in planktic foraminifers

In summary, B/Ca in planktic foraminifers is sensitive to carbonate and borate speciation in seawater, but this sensitivity is relatively weak and predominantly observed in laboratory cultures spanning a wide range of experimental pH. Studies using material from coretop sediments and sediment traps find much larger B/Ca variability than predicted from sensitivity to pH, however, some of that variability may be due to partial shell dissolution in coretop sediments (Figure 2.32). Potential alternative controls include salinity, light intensity, $[PO_4^{3-}]$ and possibly the calcite precipitation process itself (Figure 2.27). Similar to the large variability observed in modern samples, downcore records differ in their absolute values, but within individual sediment cores planktic B/Ca values are relatively consistent (Figure 2.28). This consistency is promising for studies of ancient events with much larger pH variations, e.g. ocean acidification associated with the Paleocene-Eocene Thermal Maximum, where Penman et al. (2014) found parallel shifts in both B/Ca and δ^{11}B. However, quantitative comparisons of B/Ca values between different core sites and time intervals should not be attempted until the reasons for B/Ca offsets between core sites have been identified. Similarly, until the actual carbon and boron species involved in the calcification process are unambiguously identified, studies should only interpret primary B/Ca values, and not K_D.

2.6.5 B/Ca in Benthic Foraminifers

Following observations in planktic foraminifers (Sections 2.6.1–2.6.4), the B/Ca proxy in benthic foraminifers has been studied since Yu and Elderfield 2007, when Yu and Elderfield presented the first calibration from coretop sediments (Figure 2.33). Their initial assessment has subsequently been confirmed and augmented by additional coretop studies by Rae et al. (2011),

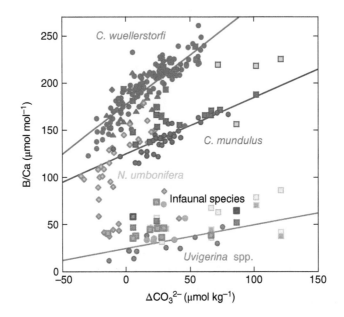

Figure 2.33 Coretop calibration of B/Ca versus bottom water carbonate saturation (i.e. ΔCO_3^{2-}) in various benthic foraminifer species. Different studies are depicted by different symbols, but data of individual species are shown in the same color. Species often used for paleoreconstructions are shown with linear regressions. The epifaunal species *Cibicidoides wuellerstorfi*, *Cibicidoides mundulus* and *Nuttallides umbonifera* are characterized by highest B/Ca values and greatest sensitivity to ΔCO_3^{2-}, whereas infaunal species show lowest B/Ca values and little sensitivity to ΔCO_3^{2-}. Data are from Yu and Elderfield (2007), Yu et al. (circles, 2013b), Rae et al. (squares, 2011), Brown et al. (diamonds, 2011), and Raitzsch et al. (triangles, 2011).

Raitzsch et al. (2011), Brown et al. (2011) and Yu et al. (2013b). Duplicating the [CO_3^{2-}] experiments of Allen et al. (2012) with the symbiont-bearing benthic foraminifer species *A. lessonii*, a recent culture study by Kaczmarek et al. (2015) also found that B/Ca increases with experimental seawater pH, but decreases upon addition of DIC at otherwise constant pH. This shallow benthic species thereby exhibits boron systematics that are comparable to planktic foraminifers (Figures 2.27 and 2.31, Allen et al. 2012; Haynes et al. 2017; Holland et al. 2017), however, deep-sea benthic foraminifers do not necessarily display coherent variation between B/Ca and pH, [B(OH)$_4^-$]/ [HCO$_3^-$] or temperature (Yu and Elderfield 2007). Instead, B/Ca in deep-sea benthic foraminifers shows consistent sensitivity to the carbonate ion saturation of ocean bottom water, ΔCO_3^{2-} (Figure 2.34), which is defined as

$$\Delta CO_3^{2-} = \left[CO^3{}_{2-} \right]_{in\ situ} - \left[CO_3^{2-} \right]_{saturation} \qquad (2.21)$$

and

$$\left[CO_3^{2-} \right]_{saturation} = \left[CO^3{}_{2-} \right]_{in\ situ} / \Omega_{calcite} \qquad (2.22)$$

$\Omega_{calcite}$ has already been defined in Eq. 2.15. Benthic B/Ca invariably follows linear correlations with ΔCO_3^{2-}, although the individual sensitivity varies between foraminifer species and is a function of their respective B affinity

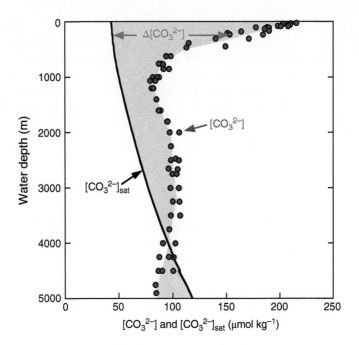

Figure 2.34 The shaded area in this graph highlights ΔCO_3^{2-}, which is defined as the difference between the carbonate ion concentration *in situ* (circles) and the carbonate ion concentration at saturation (solid black line). ΔCO_3^{2-} is positive where $[CO_3^{2-}]_{in\ situ}$ is greater than $[CO_3^{2-}]_{sat}$, and negative where $[CO_3^{2-}]_{in\ situ} < [CO_3^{2-}]_{sat}$. The saturation concentration is a function of temperature, salinity and pressure, and changes little on glacial/interglacial time scales. This figure is based on hydrographic data from the South Atlantic Ocean.

(Figure 2.33), as is also observed in planktic foraminifers. For instance, B/Ca ratios in the shells of the epifaunal species *C. wuellerstorfi* and *C. mundulus* are similar to or higher than B/Ca in the least B-discriminating planktic foraminifer species *G. ruber* (Figure 2.27). Similar to the high pH sensitivity in *G. ruber*, the B/Ca sensitivity to ΔCO_3^{2-} is also highest among these two epifaunal benthic foraminifer species. On the other hand, infaunal benthic foraminifers such as *Uvigerina* spp. and *Oridorsalis umbonatus* incorporate much less boron (~10–80 μmol mol^{-1}). In this regard, they are similar to symbiont-barren planktic foraminifer species, and their sensitivity to ΔCO_3^{2-} is relatively weak (Figures 2.27 and 2.33). The following calibration equations from Yu et al. (2013b) and Yu and Elderfield (2007) describe the sensitivities of the dominant benthic foraminifer species used in deep ocean paleoceanography:

$$\text{B/Ca}_{C.wuellerstorfi}\left(\mu\text{mol mol}^{-1}\right)=1.14\pm0.04\times\Delta CO_3^{2-}\left(\mu\text{mol kg}^{-1}\right)+176.6\pm1.0$$

$$(2.23)$$

$$\text{B/Ca}_{C.mundulus}\left(\mu\text{mol mol}^{-1}\right)=0.69\pm0.04\times\Delta CO_3^{2-}\left(\mu\text{mol kg}^{-1}\right)+119.9\pm1.7$$

$$(2.24)$$

$$\text{B/Ca}_{Uvigerina\ spp.}\left(\mu\text{mol mol}^{-1}\right)=0.27\pm0.076\times\Delta CO_3^{2-}\left(\mu\text{mol kg}^{-1}\right)+19.6\pm3.0$$

$$(2.25)$$

Importantly, the sensitivity of B/Ca in benthic foraminifers to ΔCO_3^{2-} does not diminish at or below saturation (i.e. $\Delta CO_3^{2-} \leq 0$, Figure 2.33), which distinguishes the proxy from the generally reduced pH sensitivity of the boron isotope proxy at low pH (Figures 2.2 and 2.10) and of B/Ca in planktic foraminifers grown at pH < 8 (Figure 2.27). Furthermore, to date there is no indication of a dissolution effect on benthic B/Ca (Yu and Elderfield 2007), which makes the proxy an attractive candidate for deep ocean carbonate chemistry reconstructions.

The question of why B/Ca in benthic foraminifers should respond to ΔCO_3^{2-}, however, remains unresolved, and the correlations shown in Figure 2.33 are entirely empirical. Because $K_D << 1$ and is variable in benthic foraminifers, Yu and Elderfield (2007) argued that Rayleigh fractionation from an internal calcifying pool (Elderfield et al. 1996) unlikely explains the observed sensitivity to ΔCO_3^{2-}. Given inorganic $CaCO_3$ exhibits ~5–10x lower B than benthic foraminiferal calcite at similar seawater boron concentration (Uchikawa et al. 2015), benthic foraminifers may concentrate B in their calcifying fluid, and any modification of the carbon concentration and/or pH in the calcifying fluid would affect its $[B(OH)_4^-]/[HCO_3^-]$ ratio and thus B/Ca in the foraminifer shell. Such a connection might explain the B/Ca proxy systematics if there were a link between ambient seawater ΔCO_3^{2-} and the size of calcifying fluid's carbon pool or pH (Yu and Elderfield 2007). Notwithstanding, we have to learn a lot more about the biomineralization process in deep ocean benthic foraminifers before we can answer this question with certainty.

In summary, the species-specific boron uptake and sensitivity to ΔCO_3^{2-} requires species-specific calibrations; and in this regard the B/Ca proxy in benthic foraminifers is similar to the boron isotope proxy and B/Ca in planktic foraminifers. Epifaunal species are suited best for bottom water reconstructions, because infaunal species are subject to porewater modification of carbonate and boron chemistry (e.g. Rae et al. 2011). Fortuitously, B/Ca ratios in the epifaunal species *C. wuellerstorfi* and *Ch. mundulus* are the most sensitive to ΔCO_3^{2-}, and uncertainties in their respective calibration regressions suggest that paleoreconstructions can be achieved with an uncertainty better than $\pm 10 \, \mu mol \, kg^{-1}$ in ΔCO_3^{2-} (Yu and Elderfield 2007). However, Rae et al. (2011) caution that additional uncertainty could arise from individual morphotype selection, as B/Ca differences up to $40 \, \mu mol \, mol^{-1}$ occur between coexisting morphotypes of the same foraminifer species. Rae et al. (2011) provide helpful images of the morphotypes that are known to differ in their B/Ca composition, and the reader is referred to that publication for further details. Box 2.5 summarizes the incorporation controls on B/Ca in synthetic, planktic and benthic foraminifera, and makes recommendations for further validation studies to improve our understanding of the boron incorporation controls.

Box 2.5 Summary of B/Ca Systematics in Foraminifers and Synthetic $CaCO_3$

B/Ca in $CaCO_3$ secreted by living organisms is subject to a number of vital and environmental effects. Temperature sensitivity of B/Ca in foraminifers and synthetic $CaCO_3$ is low if not nil (Figures 2.7 and 2.27), but correlations with carbonate chemical parameters have been observed in all three. Observations to date suggest B/Ca in planktic foraminifers is sensitive to the pH-dependent relative abundance of borate and bicarbonate ions (Figure 2.27), whereas B/Ca in benthic foraminifers is more sensitive to ocean bottom water saturation state (Figure 2.33). If evidence from synthetic calcification experiments (Figure 2.5) is applicable to foraminiferal calcification, these observations are ultimately due to kinetic effects associated with higher precipitation rates at higher carbonate saturation (Figure 2.6). However, limited evidence from cultured planktic foraminifera suggests much lower precipitation rates and active control over the biomineralization process. This hints at significant vital effects during foraminiferal calcification, and begs the question whether observed effects are constant across a wide range of seawater chemical compositions. B/Ca in planktic foraminifers and synthetic $CaCO_3$ increases with $[B_T]$ of the experimental seawater (Figures 2.5a and 2.29), but the positive correlation observed between B/Ca and $[Ca^{2+}]$ has only been observed in synthetic calcite (Figure 2.5d), and does not occur in foraminiferal calcite. In order to fill the gap between B/Ca systematics in inorganic and biogenic $CaCO_3$, progress is required in constraining calcification rates in planktic foraminifers and exploring controlled cultures with deep-ocean benthic foraminifers. Given the observed sensitivity of B/Ca to $[B_T]$ and DIC, paleoceanographers aiming to apply this proxy have to consider secular variations in $[B_T]$ and DIC in paleo-seawater. Paleoreconstructions are further complicated by species-specific proxy sensitivities and discrimination against boron incorporation (Figures 2.27 and 2.33), which requires careful cross-calibration of coexisting species when going back in time and studying now-extinct species. As daunting as the requirements outlined by this summary may seem, published paleo-reconstructions in particular with benthic foraminifers have revealed consistent and reasonable variations in ocean carbonate chemistry. Similarly, planktic B/Ca sometimes covaries with $\delta^{11}B$ in downcore studies. This makes further probing of the proxy well worth the effort.

References

Al-Horani, F.A., Al-Moghrabi, S.M., and de Beer, D. (2003). The mechanism of calcification and its relation to photosynthesis and respiration in the scleractinian coral *Galaxea fascicularis*. *Marine Biology 142*: 419–426.

Allemand, D., Ferrier-Pagès, C., Furla, P. et al. (2004). Biomineralisation in reef-building corals: from molecular mechanisms to environmental control. *Comptes Rendus Palevol 3* (6–7): 453–467.

Allen, K.A. and Hönisch, B. (2012). The planktic foraminiferal B/Ca proxy for seawater carbonate chemistry: a critical evaluation. *Earth and Planetary Science Letters 345–348* (0): 203–211.

Allen, K.A., Hönisch, B., Eggins, S.M., and Rosenthal, Y. (2012). Environmental controls on B/Ca in calcite tests of the tropical planktic foraminifer species *Globigerinoides ruber* and *Globigerinoides sacculifer*. *Earth and Planetary Science Letters 351–352* (0): 270–280.

Allen, K.A., Hönisch, B., Eggins, S.M. et al. (2011). Controls on boron incorporation in cultured tests of the planktic foraminifer *Orbulina universa*. *Earth and Planetary Science Letters 309* (3–4): 291–301.

Allen, K.A., Hönisch, B., Eggins, S.M. et al. (2016). Trace element proxies for surface ocean conditions: a synthesis of culture calibrations with planktic foraminifera. *Geochimica et Cosmochimica Acta 193*: 197–221.

Allison, N. and Finch, A.A. (2010). $\delta^{11}B$, Sr, Mg and B in a modern *Porites* coral: the relationship between calcification site pH and skeletal chemistry. *Geochimica et Cosmochimica Acta 74* (6): 1790–1800.

Allison, N., Cohen, I., Finch, A.A. et al. (2014). Corals concentrate dissolved inorganic carbon to facilitate calcification. *Nature Communications 5*.

Anagnostou, E., Huang, K.F., You, C.F. et al. (2012). Evaluation of boron isotope ratio as a pH proxy in the deep sea coral *Desmophyllum dianthus*: evidence of physiological pH adjustment. *Earth and Planetary Science Letters 349–350*: 251–260.

Anagnostou, E., John, E.H., Edgar, K.M. et al. (2016). Changing atmospheric CO_2 concentration was the primary driver of early Cenozoic climate. *Nature 533* (7603): 380–384.

Anand, P., Elderfield, H., and Conte, M.H. (2003). Calibration of Mg/Ca thermometry in planktonic foraminifera from a sediment trap time series. *Paleoceanography 18* (2): 15.

Anderson, O.R. and Faber, W.W. Jr. (1984). An estimation of calcium carbonate deposition rate in a planktonic foraminifer *Globigerinoides sacculifer* using ^{45}Ca as a tracer: a recommended procedure for improved accuracy. *Journal of Foraminiferal Research 14*: 303–308.

Babila, T.L., Rosenthal, Y., and Conte, M.H. (2014). Evaluation of the biogeochemical controls on B/Ca of *Globigerinoides ruber* white from the oceanic flux program, Bermuda. *Earth and Planetary Science Letters 404*: 67–76.

Badger, M.P.S., Lear, C.H., Pancost, R.D. et al. (2013). CO_2 drawdown following the middle Miocene expansion of the Antarctic ice sheet. *Paleoceanography 28* (1): 42–53.

Balan, E., Pietrucci, F., Gervais, C. et al. (2016). First-principles study of boron speciation in calcite and aragonite. *Geochimica et Cosmochimica Acta 193*: 119–131.

Bartoli, G., Hönisch, B., and Zeebe, R.E. (2011). Atmospheric CO_2 decline during the Pliocene intensification of Northern Hemisphere glaciations. *Paleoceanography 26* (4): PA4213.

Bé, A.W.H. (1980). Gametogenic calcification in a spinose planktonic foraminifer, *Globigerinoides sacculifer* (BRADY). *Marine Micropaleontology 5*: 283–310.

Bentov, S., Brownlee, C., and Erez, J. (2009). The role of seawater endocytosis in the biomineralization process in calcareous foraminifera. *Proceedings of the National Academy of Sciences 106* (51): 21500–21504.

Berger, W.H. and Piper, D.J.W. (1972). Planktonic foraminifera: differential settling, dissolution, and reproduction. *Limnology and Oceanography 17* (2): 275–287.

Bijma, J., Hönisch, B., and Zeebe, R.E. (2002). Impact of the ocean carbonate chemistry on living foraminiferal shell weight: comment on "carbonate ion concentration in glacial-age deep waters of the Caribbean Sea" by W.S. Broecker and E. Clark. *Geochemistry, Geophysics, Geosystems 3* (11): doi: 10.1029/2002GC000388.

Blamart, D., Rollion-Bard, C., Meibom, A. et al. (2007). Correlation of boron isotopic composition with ultrastructure in the deep- sea coral *Lophelia pertusa*: implications for biomineralization and paleo-pH. *Geochemistry, Geophysics, Geosystems 8*: 11.

Branson, O., Kaczmarek, K., Redfern, S.A.T. et al. (2015). The coordination and distribution of B in foraminiferal calcite. *Earth and Planetary Science Letters 416*: 67–72.

Brennan, S.T., Lowenstein, T.K., and Cendón, D.I. (2013). The major-ion composition of Cenozoic seawater: the past 36 million years from fluid inclusions in marine halite. *American Journal of Science 313* (8): 713–775.

Broecker, W.S. and Peng, T.-H. (1982). *Tracers in the Sea*. Palisades, New York: Lamont Doherty Earth Observatory, Columbia University.

Brown, R.E., Anderson, L.D., Thomas, E., and Zachos, J.C. (2011). A core-top calibration of B/Ca in the benthic foraminifers *Nuttallides umbonifera* and *Oridorsalis umbonatus*: a proxy for Cenozoic bottom water carbonate saturation. *Earth and Planetary Science Letters 310* (3–4): 360–368.

Brown, S. and Elderfield, H. (1996). Variations in Mg/Ca and Sr/Ca ratios of planktonic foraminifera caused by postdepositional dissolution: evidence of shallow Mg-dependent dissolution. *Paleoceanography 11* (5): 543–551.

Byrne, R.H. and Kester, D.R. (1974). Inorganic speciation of boron in seawater. *Journal of Marine Research 32* (2): 119–127.

Byrne, R.H., Yao, W., Klochko, K. et al. (2006). Experimental evaluation of the isotopic exchange equilibrium $^{10}B(OH)_3+^{11}B(OH)_4^-=^{11}B(OH)_3+^{10}B(OH)_4^-$ in aqueous solution. *Deep Sea Research Part I: Oceanographic Research Papers 53* (4): 684–688.

Cai, W.-J., Ma, Y., Hopkinson, B.M. et al. (2016). Microelectrode characterization of coral daytime interior pH and carbonate chemistry. *Nature Communications 7*: 11144.

van Cappellen, P., Charlet, L., Stumm, W., and Wersin, P. (1993). A surface complexation model of the carbonate mineral-aqueous solution interface. *Geochimica et Cosmochimica Acta 57*: 3505–3518.

Coadic, R., Bassinot, F., Dissard, D. et al. (2013). A core-top study of dissolution effect on B/Ca in *Globigerinoides sacculifer* from the tropical Atlantic: potential bias for paleo-reconstruction of seawater carbonate chemistry. *Geochemistry, Geophysics, Geosystems 14* (4): 1053–1068.

Cook, P.J. (1977). Loss of boron from shells during weathering and possible implications for the determination of paleosalinity. *Nature 268*: 426–427.

Cusack, M., Kamenos, N.A., Rollion-Bard, C., and Tricot, G. (2015). Red coralline algae assessed as marine pH proxies using ^{11}B MAS NMR. *Scientific Reports 5*: 1–3.

Dai, Y., Yu, J., and Johnstone, H.J.H. (2016). Distinct responses of planktonic foraminiferal B/Ca to dissolution on seafloor. *Geochemistry, Geophysics, Geosystems 17* (4): 1339–1348.

Darling, K.F. and Wade, C.M. (2008). The genetic diversity of planktic foraminifera and the global distribution of ribosomal RNA genotypes. *Marine Micropaleontology 67*: 216–238.

De Yoreo, J.J. and Sommerdijk, N.A.J.M. (2016). Investigating materials formation with liquid-phase and cryogenic TEM. *Nature Reviews Materials 1*: 16035.

DeConto, R.M. and Pollard, D. (2003). Rapid Cenozoic glaciation of Antarctica induced by declining atmospheric CO_2. *Nature 421*: 245–249.

DeConto, R.M., Pollard, D., Wilson, P.A. et al. (2008). Thresholds for Cenozoic bipolar glaciation. *Nature 455* (7213): 652–656.

Dekens, P.S., Lea, D.W., Pak, D.K., and Spero, H.J. (2002). Core top calibration of Mg/Ca in tropical foraminifera: refining paleotemperature estimation. *Geochemistry, Geophysics, Geosystems 3* (4): doi: 10.1029/2001GC000200.

DePaolo, D.J. (2011). Surface kinetic model for isotopic and trace element fractionation during precipitation of calcite from aqueous solutions. *Geochimica et Cosmochimica Acta 75* (4): 1039–1056.

Dickson, A.G. (1990a). Standard potential of the reaction: $AgCl(s)+1/2H_2(g)=Ag(s)+HCl(aq)$, and the standard acidity constant of the ion HSO_4^- in synthetic seawater from 273.15 to 318.15K. *The Journal of Chemical Thermodynamics 22*: 113–127.

Dickson, A.G. (1990b). Thermodynamics of the dissociation of boric acid in synthetic seawater from 273.15 to 318.15 K. *Deep Sea Research 37*: 755–766.

Dissard, D., Douville, E., Reynaud, S. et al. (2012). Light and temperature effects on $\delta^{11}B$ and B/Ca ratios of the zooxanthellate coral *Acropora* sp.: results from culturing experiments. *Biogeosciences 9* (11): 4589–4605.

Dyez, K.A., Hönisch, B., and Schmidt, G.A. (under review) Early Pleistocene obliquity-scale pCO2 variability at ~1.5 million years ago, Paleoceanography and Paleoclimatology.

Edgar, K.M., Anagnostou, E., Pearson, P.N., and Foster, G.L. (2015). Assessing the impact of diagenesis on $\delta^{11}B$, $\delta^{13}C$, $\delta^{18}O$, Sr/Ca and B/Ca values in fossil planktic foraminiferal calcite. *Geochimica et Cosmochimica Acta 166*: 189–209.

Elderfield, H., Bertram, C.J., and Erez, J. (1996). A biomineralization model for the incorporation of trace elements into foraminiferal calcium carbonate. *Earth and Planetary Science Letters 142* (3–4): 409–423.

Erez, J. (2003). The source of ions for biomineralization in foraminifera and their implications for paleoceanographic proxies. *Reviews in Mineralogy and Geochemistry 54*: 115–149.

Fairbanks, R.G., Wiebe, P.H., and Be, A.W.H. (1980). Vertical distribution and isotopic composition of living planktonic foraminifera in the western North Atlantic. *Science 207* (4426): 61–63.

Fairbanks, R.G., Sverdlove, M., Free, R. et al. (1982). Vertical distribution and isotopic fractionation of living planktonic foraminifera from the Panama Basin. *Nature 298* (5877): 841–844.

Fallon, S.J., McCulloch, M.T., van Woesik, R., and Sinclair, D.J. (1999). Corals at their latitudinal limits: laser ablation trace element systematics in Porites from Shirigai Bay, Japan. *Earth and Planetary Science Letters 172* (3–4): 221–238.

Farmer, E.C., Kaplan, A., de Menocal, P.B., and Lynch-Stieglitz, J. (2007). Corroborating ecological depth preferences of planktonic foraminifera in the tropical Atlantic with the stable oxygen isotope ratios of core top specimens. *Paleoceanography 22* (3): 1–14.

Farmer, J.R., Hönisch, B., and Uchikawa, J. (2016). Single laboratory comparison of MC-ICP-MS and N-TIMS boron isotope analyses in marine carbonates. *Chemical Geology 447*: 173–182.

Farmer, J.R., Hönisch, B., Robinson, L.F., and Hill, T.M. (2015). Effects of seawater-pH and biomineralization on the boron isotopic composition of deep-sea bamboo corals. *Geochimica et Cosmochimica Acta 155*: 86–106.

Fietzke, J., Ragazzola, F., Halfar, J. et al. (2015). Century-scale trends and seasonality in pH and temperature for shallow zones of the Bering Sea. *Proceedings of the National Academy of Sciences* doi: 10.1073/pnas.1419216112.

Fleet, M.E.L. (1965). Preliminary investigations into the sorption of boron by clay minerals. *Clay Minerals* 6: 3–16.

Foster, G.L. (2008). Seawater pH, pCO_2 and $[CO_3^{2-}]$ variations in the Caribbean Sea over the last 130kyr: a boron isotope and B/Ca study of planktic foraminifera. *Earth and Planetary Science Letters 271* (1–4): 254–266.

Foster, G.L. and Sexton, P.F. (2014). Enhanced carbon dioxide outgassing from the eastern equatorial Atlantic during the last glacial. *Geology 42* (11): 1003–1006.

Foster, G.L., Pogge von Strandmann, P.A.E., and Rae, J.W.B. (2010). Boron and magnesium isotopic composition of seawater. *Geochemistry, Geophysics, Geosystems 11* (8): Q08015.

Foster, G.L., Lear, C.H., and Rae, J.W.B. (2012). The evolution of pCO_2, ice volume and climate during the middle Miocene. *Earth and Planetary Science Letters 341–344*: 243–254.

Foster, G.L., Hönisch, B., Paris, G. et al. (2013). Interlaboratory comparison of boron isotope analyses of boric acid, seawater and marine $CaCO_3$ by MC-ICPMS and NTIMS. *Chemical Geology 358*: 1–14.

Gabitov, R.I., Rollion-Bard, C., Tripati, A., and Sadekov, A. (2014). In situ study of boron partitioning between calcite and fluid at different crystal growth rates. *Geochimica et Cosmochimica Acta 137*: 81–92.

Gaillardet, J. and Allègre, C.J. (1995). Boron isotopic compositions of corals: seawater or diagenesis record? *Earth and Planetary Science Letters 136*: 665–676.

Glas, M.S., Fabricius, K.E., de Beer, D., and Uthicke, S. (2012). The O2, pH and Ca2+ microenvironment of benthic foraminifera in a high CO2 world. *PLoS One 7* (11): e50010.

Goldberg, E.D. and Arrhenius, G.O.S. (1958). Chemistry of Pacific pelagic sediments. *Geochimica et Cosmochimica Acta 13* (2–3): 153–212.

Greenop, R., Foster, G.L., Wilson, P.A., and Lear, C.H. (2014). Middle Miocene climate instability associated with high-amplitude CO_2 variability. *Paleoceanography 29* (9): 845–853.

Greenop, R., Hain, M.P., Sosdian, S.M. et al. (2017). A record of Neogene seawater $\delta^{11}B$ reconstructed from paired $\delta^{11}B$ analyses on benthic and planktic foraminifera. *Climate of the Past 13* (2): 149–170.

Grottoli, A.G. (1999). Variability of stable isotopes and maximum linear extension in reef-coral skeletons at Kaneohe Bay, Hawaii. *Marine Biology 135*: 437–449.

Grottoli, A.G. (2002). Effect of light and brine shrimp on skeletal $\delta^{13}C$ in the Hawaiian coral *Porites compressa*: a tank experiment. *Geochimica et Cosmochimica Acta 66* (11): 1955–1967.

Gussone, N., Eisenhauer, A., Heuser, A. et al. (2003). Model for kinetic effects on calcium isotope fractionation ($\delta^{44}Ca$) in inorganic aragonite and cultured planktonic foraminifera. *Geochimica et Cosmochimica Acta 67* (7): 1375–1382.

Hain, M.P., Sigman, D.M., Higgins, J.A., and Haug, G.H. (2015). The effects of secular calcium and magnesium concentration changes on the thermodynamics of seawater acid/base chemistry: implications for Eocene and cretaceous ocean carbon chemistry and buffering. *Global Biogeochemical Cycles 29* (5): 517–533.

Harder, H. (1959). Beitrag zur Geochemie des Bors. I. Bor in Mineralen und magmatischen Gesteinen. *Nachrichten der Akademie Wissenschaften Göttingen, II. Mathematisch- Physikalische Klasse 5*: 67–122.

Harder, H. (1961). Einbau von Bor in detritische Tonminerale. Experimente zur Erklärung des Borgehaltes toniger Sedimente. *Geochimica et Cosmochima Acta 21*: 284–294.

Harder, H. (1970). Boron content of sediments as a tool in facies analysis. *Sedimentary Geology 4* (1–2): 153–175.

Harriss, R.C. (1969). Boron regulation in the oceans. *Nature 223* (5203): 290–291.

Haynes, L.L., Hönisch, B., Dyez, K.A. et al. (2017). Calibration of the B/Ca proxy in the planktic foraminifer *Orbulina universa* to Paleocene seawater conditions. *Paleoceanography 32* (6): 580–599.

He, M., Xiao, Y., Jin, Z. et al. (2013). Quantification of boron incorporation into synthetic calcite under controlled pH and temperature conditions using a differential solubility technique. *Chemical Geology 337–338*: 67–74.

Hemleben, C. and Bijma, J. (1994). Foraminiferal population dynamics and stable carbon isotopes. In: *Carbon Cycling in the Glacial Ocean: Constraints on the Ocean's Role in Climate Change* (ed. R. Zahn), 145–166. Springer-Verlag, Berlin Heidelberg.

Hemming, N.G. and Hanson, G.N. (1992). Boron isotopic composition and concentration in modern marine carbonates. *Geochimica et Cosmochimica Acta 56*: 537–543.

Hemming, N.G., Reeder, R.J., and Hanson, G.N. (1995). Mineral-fluid partitioning and isotopic fractionation of boron in synthetic calcium carbonate. *Geochimica et Cosmochima Acta 59* (2): 371–379.

Hemming, N.G., Reeder, R.J., and Hart, S.R. (1998a). Growth-step-selective incorporation of boron on the calcite surface. *Geochimica et Cosmochimica Acta 62* (17): 2915–2922.

Hemming, N.G., Guilderson, T.P., and Fairbanks, R.G. (1998b). Seasonal variations in the boron isotopic composition of a coral: a productivity signal? *Global Biogeochem Cycles 12* (4): 581–586.

Hendry, K.R., Rickaby, R.E.M., Meredith, M.P., and Elderfield, H. (2009). Controls on stable isotope and trace metal uptake in *Neogloboquadrina pachyderma* (sinistral) from an Antarctic sea-ice environment. *Earth and Planetary Science Letters 278* (1–2): 67–77.

Henehan, M.J., Foster, G.L., Bostock, H.C. et al. (2016). A new boron isotope-pH calibration for *Orbulina universa*, with implications for understanding and accounting for 'vital effects'. *Earth and Planetary Science Letters 454*: 282–292.

Henehan, M.J., Foster, G.L., Rae, J.W.B. et al. (2015). Evaluating the utility of B/Ca ratios in planktic foraminifera as a proxy for the carbonate system: a case study of *Globigerinoides ruber*. *Geochemistry, Geophysics, Geosystems 16*: 1–18.

Henehan, M.J., Rae, J.W.B., Foster, G.L. et al. (2013). Calibration of the boron isotope proxy in the planktonic foraminifera *Globigerinoides ruber* for use in palaeo-CO_2 reconstruction. *Earth and Planetary Science Letters 364*: 111–122.

Hobbs, M.Y. and Reardon, E.J. (1999). Effect of *pH* on boron coprecipitation by calcite: further evidence for nonequilibrium partitioning of trace elements. *Geochimica et Cosmochimica Acta 63* (7/8): 1013–1021.

Holcomb, M., DeCarlo, T.M., Gaetani, G.A., and McCulloch, M. (2016). Factors affecting B/Ca ratios in synthetic aragonite. *Chemical Geology 437* (Supplement C): 67–76.

Holcomb, M., Venn, A.A., Tambutte, E. et al. (2014). Coral calcifying fluid pH dictates response to ocean acidification. *Scientific Reports 4*: 1–4.

Holland, K., Eggins, S.M., Hönisch, B. et al. (2017). Calcification rate and shell chemistry response of the planktic foraminifer *Orbulina universa* to changes in microenvironment seawater carbonate chemistry. *Earth and Planetary Science Letters 464*: 124–134.

Hönisch, B. and Hemming, N.G. (2004). Ground-truthing the boron isotope paleo-pH proxy in planktonic foraminifera shells: partial dissolution and shell size effects. *Paleoceanography 19*: doi: 10.1029/2004PA001026.

Hönisch, B. and Hemming, N.G. (2005). Surface Ocean pH response to variations in pCO$_2$ through two full glacial cycles. *Earth and Planetary Science Letters 236* (1–2): 305–314.

Hönisch, B., Hemming, N.G., and Loose, B. (2007). Comment on "a critical evaluation of the boron isotope-pH proxy: the accuracy of ancient ocean pH estimates" by M. Pagani, D. Lemarchand, A. Spivack and J. Gaillardet. *Geochimica et Cosmochimica Acta 71* (6): 1636–1641.

Hönisch, B., Bickert, T., and Hemming, N.G. (2008). Modern and Pleistocene boron isotope composition of the benthic foraminifer *Cibicidoides wuellerstorfi*. *Earth and Planetary Science Letters 272* (1–2): 309–318.

Hönisch, B., Hemming, N.G., Archer, D. et al. (2009). Atmospheric carbon dioxide concentration across the mid-Pleistocene transition. *Science 324* (5934): 1551–1554.

Hönisch, B., Hemming, N.G., Grottoli, A.G. et al. (2004). Assessing scleractinian corals as recorders for paleo-pH: empirical calibration and vital effects. *Geochimica et Cosmochimica Acta 68* (18): 3675–3685.

Hönisch, B., Bijma, J., Russell, A.D. et al. (2003). The influence of symbiont photosynthesis on the boron isotopic composition of foraminifera shells. *Marine Micropaleontology 49*: 87–96.

Howes, E.L., Kaczmarek, K., Raitzsch, M. et al. (2017). Decoupled carbonate chemistry controls on the incorporation of boron into *Orbulina universa*. *Biogeosciences 14* (2): 415–430.

Ichikuni, M. and Kikuchi, K. (1972). Retention of boron by travertines. *Chemical Geology 9* (1–4): 13–21.

Ingri, N. (1963). Equilibrium studies of polyanions containing B^{III}, Si^{IV}, Ge^{IV} and V^{V}. *Svensk Kemisk Tidskrift 75* (4): 199–230.

Ingri, N., Lagerström, G., Frydman, M., and Sillén, L.G. (1957). Equilibrium studies of polyanions II Polyborates in NaClO$_4$ medium. *Acta Chemica Scandinavica 11*: 1034–1058.

Ishikawa, T. and Nakamura, E. (1993). Boron isotope systematics of marine sediments. *Earth and Planetary Science Letters 117* (3–4): 567–580.

Jørgensen, B.B., Erez, J., Revsbech, N.P., and Cohen, Y. (1985). Symbiotic photosynthesis in a planktonic foraminiferan, *Globigerinoides sacculifer* (Brady), studied with microelectrodes. *Limnology and Oceanography 30* (6): 1253–1267.

Kaczmarek, K., Nehrke, G., Misra, S. et al. (2016). Investigating the effects of growth rate and temperature on the B/Ca ratio and δ^{11}B during inorganic calcite formation. *Chemical Geology 421*: 81–92.

Kaczmarek, K., Langer, G., Nehrke, G. et al. (2015). Boron incorporation in the foraminifer *Amphistegina lessonii* under a decoupled carbonate chemistry. *Biogeosciences 12* (6): 1753–1763.

Kakihana, H. and Kotaka, M. (1977). Equilibrium constants for boron isotope-exchange reactions. *Bulletin of the Research Laboratory for Nuclear Reactors 2*: 1–12.

Kakihana, H., Kotaka, M., Satoh, S. et al. (1977). Fundamental studies on the ion-exchange of boron isotopes. *Bulletin of the Chemical Society of Japan 50*: 158–163.

Kasemann, S.A., Schmidt, D.N., Bijma, J., and Foster, G.L. (2009). In situ boron isotope analysis in marine carbonates and its application for foraminifera and palaeo-pH. *Chemical Geology 260* (1–2): 138–147.

Key, R.M., Kozyr, A., Sabine, C.L. et al. (2004). A global ocean carbon climatology: results from GLODAP. *Global Biogeochem Cycles 18*: doi: 10.1029/2004GB002247.

Kısakürek, B., Eisenhauer, A., Böhm, F. et al. (2011). Controls on calcium isotope fractionation in cultured planktic foraminifera, *Globigerinoides ruber* and *Globigerinella siphonifera*. *Geochimica et Cosmochimica Acta 75* (2): 427–443.

Kitano, Y., Okomura, M., and Idogaki, M. (1979). Influence of borate-boron on crystal from of calcium carbonate. *Geochemical Journal 13*: 223–224.

Klochko, K., Kaufman, A.J., Yao, W. et al. (2006). Experimental measurement of boron isotope fractionation in seawater. *Earth and Planetary Science Letters 248* (1–2): 261–270.

Klochko, K., Cody, G.D., Tossell, J.A. et al. (2009). Re-evaluating boron speciation in biogenic calcite and aragonite using ^{11}B MAS NMR. *Geochimica et Cosmochimica Acta 73* (7): 1890–1900.

Köhler-Rink, S. and Kühl, M. (2000). Microsensor studies of photosynthesis and respiration in larger symbiotic foraminifera. I – the physico-chemical microenvironment of *Marginopora vertebralis, Amphistegina lobifera* and *Amphisorus hemprichii*. *Marine Biology 137* (3): 473–486.

Köhler-Rink, S. and Kühl, M. (2005). The chemical microenvironment of the symbiotic planktonic foraminifer *Orbulina universa*. *Marine Biology Research 1* (1): 68–78.

Kotaka, M. and Kakihana, H. (1977). Thermodynamic isotope effect of trigonal planar and tetrahedral molecules. *Bulletin of the Research Laboratory for Nuclear Reactors 2*: 13–29.

Krief, S., Hendy, E.J., Fine, M. et al. (2010). Physiological and isotopic responses of scleractinian corals to ocean acidification. *Geochimica et Cosmochimica Acta 74* (17): 4988–5001.

Kühl, M., Cohen, Y., Dalsgaard, T. et al. (1995). Microenvironment and photosynthesis of zooxanthellae in scleractinian corals studied with microsensors for O_2, pH and light. *Marine Ecology Progress Series 117*: 159–172.

Landergren, S. (1945). Contribution to the geochemistry of boron II' the distribution of boron in some Swedish sediments, rocks and iron ores; the boron cycle in the upper lithosphere, Arkiv för Kemi. *Mineralogi och Geologi 19a*: 26–36.

Langdon, C., Takahashi, T., Sweeney, C. et al. (2000). Effect of calcium carbonate saturation state on the calcification rate of an experimental coral reef. *Global Biogeochemical Cycles 14* (2): 639–654.

Lazareth, C.E., Soares-Pereira, C., Douville, E. et al. (2016). Intra-skeletal calcite in a live-collected *Porites* sp.: impact on environmental proxies and potential formation process. *Geochimica et Cosmochimica Acta 176*: 279–294.

Lea, D.W. (2014). 8.14 – elemental and isotopic proxies of Past Ocean temperatures. In: *Treatise on Geochemistry*, 2e (ed. H.D. Holland and K.K. Turekian), 373–397. Oxford: Elsevier.

Lea, D.W., Martin, P.A., Chan, D.A., and Spero, H.J. (1995). Calcium uptake and calcification rate in the planktonic foraminifer Orbulina universa. *Journal of Foraminiferal Research 25* (1): 14–23.

Lécuyer, C., Grandjean, P., Reynard, B. et al. (2002). 11B/10B analysis of geological materials by ICP-MS plasma 54: application to the boron fractionation between brachiopod calcite and seawater. *Chemical Geology 186* (1–2): 45–55.

Lee, K., Kim, T.-W., Byrne, R.H. et al. (2010). The universal ratio of boron to chlorinity for the North Pacific and North Atlantic oceans. *Geochimica et Cosmochimica Acta 74* (6): 1801–1811.

Lemarchand, D., Gaillardet, J., Lewin, É., and Allègre, C.J. (2000). The influence of rivers on marine boron isotopes and implications for reconstructing past ocean pH. *Nature 408*: 951–954.

Lemarchand, D., Gaillardet, J., Lewin, E., and Allegre, C.J. (2002). Boron isotope systematics in large rivers: implications for the marine boron budget and paleo-pH reconstruction over the Cenozoic. *Chemical Geology 190* (1–4): 123–140.

Levin, L.A., Hönisch, B., and Frieder, C.A. (2015). Geochemical proxies for estimating faunal exposure to ocean acidification. *Oceanography 28* (2): 62–73.

Lindsay, C.G. and Jackson, M.D. (1994). Modified electron gas modelings of calcite and aragonite; comparison of polarizable anion and fully ionic methods. *American Mineralogist 79*: 215–220.

Lisiecki, L.E. and Raymo, M.E. (2005). A Pliocene-Pleistocene stack of 57 globally distributed benthic $\delta^{18}O$ records. *Paleoceanography 20*: doi: 10.1029/2004PA001071.

Liu, W.G., Xiao, Y.K., Peng, Z.C. et al. (2000). Boron concentration and isotopic composition of halite from experiments and salt lakes in the Qaidam Basin. *Geochimica et Cosmochimica Acta 64* (13): 2177–2183.

Liu, Y. and Tossell, J.A. (2005). Ab initio molecular orbital calculations for boron isotope fractionations on boric acids and borates. *Geochimica et Cosmochimica Acta 69* (16): 3995–4006.

Lombard, F., Erez, J., Michel, E., and Labeyrie, L. (2009). Temperature effect on respiration and photosynthesis of the symbiont-bearing planktonic foraminifera *Globigerinoides ruber*, *Orbulina universa*, and *Globigerinella siphonifera*. *Limnology and Oceanography 54* (1): 210–218.

Lowenstein, T.K., Kendall, B., and Anbar, A.D. (2014). 8.21 – the geologic history of seawater. In: *Treatise on Geochemistry*, 2e (ed. H.D. Holland and K.K. Turekian), 569–622. Oxford: Elsevier.

Lueker, T.J., Dickson, A.G., and Keeling, C.D. (2000). Ocean pCO_2 calculated from dissolved inorganic carbon, alkalinity, and equations for K1 and K2: validation based on laboratory measurements of CO_2 in gas and seawater at equilibrium. *Marine Chemistry 70* (1–3): 105–119.

Lüthi, D., Le Floch, M., Bereiter, B. et al. (2008). High-resolution carbon dioxide concentration record 650,000–800,000 years before present. *Nature 453* (7193): 379–382.

Marshall, B.J., Thunell, R.C., Spero, H.J. et al. (2015). Morphometric and stable isotopic differentiation of Orbulina universa morphotypes from the Cariaco Basin, Venezuela. *Marine Micropaleontology 120*: 46–64.

Martínez-Botí, M.A., Marino, G., Foster, G.L. et al. (2015a). Boron isotope evidence for oceanic carbon dioxide leakage during the last deglaciation. *Nature 518* (7538): 219–222.

Martínez-Botí, M.A., Foster, G.L., Chalk, T.B. et al. (2015b). Plio-Pleistocene climate sensitivity evaluated using high-resolution CO_2 records. *Nature 518* (7537): 49–54.

Mavromatis, V., Montouillout, V., Noireaux, J. et al. (2015). Characterization of boron incorporation and speciation in calcite and aragonite from co-precipitation experiments under controlled pH, temperature and precipitation rate. *Geochimica et Cosmochimica Acta 150*: 299–313.

Mayer, L., Pisias, N. and Janecek, T. (1991) Leg 138 Preliminary Report in: Program, O.D. (Ed.)

McConnaughey, T. (1989). ^{13}C and ^{18}O isotopic disequilibrium in biological carbonates: I. Patterns. *Geochimica et Cosmochima Acta 53*: 151–162.

McCulloch, M., Falter, J., Trotter, J., and Montagna, P. (2012a). Coral resilience to ocean acidification and global warming through pH up-regulation. *Nature Climate Change 2* (8): 623–627.

McCulloch, M., Trotter, J., Montagna, P. et al. (2012b). Resilience of cold-water scleractinian corals to ocean acidification: boron isotopic systematics of pH and saturation state up-regulation. *Geochimica et Cosmochimica Acta 87*: 21–34.

McElligott, S. and Byrne, R.H. (1997). Interaction of $B(OH)_3^0$ and HCO_3^- in Seawater: Formation of $B(OH)_2CO_3^-$. *Aquatic Geochemistry 3* (4): 345–356.

Millero, F.J. (1995). Thermodynamics of the carbon dioxide system in the oceans. *Geochimica et Cosmochimica Acta 59* (4): 661–667.

Millero, F.J., Ward, G.K., Surdo, A.L., and Huang, F. (2012). Effect of pressure on the dissociation constant of boric acid in water and seawater. *Geochimica et Cosmochimica Acta 76*: 83–92.

Misra, S. and Froelich, P.N. (2012). Lithium isotope history of Cenozoic seawater: changes in silicate weathering and reverse weathering. *Science 335* (6070): 818–823.

Morse, J.W. (1986). The surface chemistry of calcium carbonate minerals in natural waters – an overview. *Marine Chemistry 20* (1): 91–112.

Naik, S.S. and Naidu, P.D. (2014). Boron/calcium ratios in *Globigerinoides ruber* from the Arabian Sea: implications for controls on boron incorporation. *Marine Micropaleontology 107*: 1–7.

Ni, Y., Foster, G.L., Bailey, T. et al. (2007). A core top assessment of proxies for the ocean carbonate system in surface-dwelling foraminifers. *Paleoceanography 22*: doi: 10.1029/2006PA001337.

Nir, O., Vengosh, A., Harkness, J.S. et al. (2015). Direct measurement of the boron isotope fractionation factor: reducing the uncertainty in reconstructing ocean paleo-pH. *Earth and Planetary Science Letters 414*: 1–5.

Noireaux, J., Mavromatis, V., Gaillardet, J. et al. (2015). Crystallographic control on the boron isotope paleo-pH proxy. *Earth and Planetary Science Letters 430*: 398–407.

Nomaki, H., Yamaoka, A., Shirayama, Y., and Kitazato, H. (2007). Deep-sea benthic foraminiferal respiration rates measured under laboratory conditions. *Journal of Foraminiferal Research 37* (4): 281–286.

de Nooijer, L.J., Spero, H.J., Erez, J. et al. (2014). Biomineralization in perforate foraminifera. *Earth-Science Reviews 135* (0): 48–58.

Oi, T. (2000). Calculations of reduced partition function ratios of monomeric and Dimeric boric acids and borates by the ab initio molecular orbital theory. *Journal of Nuclear Science and Technology 37* (2): 166–172.

Oi, T., Nomura, M., Musashi, M. et al. (1989). Boron isotopic compositions of some boron minerals. *Geochimica et Cosmochimica Acta 53* (12): 3189–3195.

Oppo, D.W. and Fairbanks, R.G. (1989). Carbon isotope composition of tropical surface water during the past 22,000 years. *Paleoceanography 4* (4): 333–351.

Pagani, M., Lemarchand, D., Spivack, A., and Gaillardet, J. (2005). A critical evaluation of the boron isotope-pH proxy: the accuracy of ancient ocean pH estimates. *Geochimica et Cosmochimica Acta 69* (4): 953–961.

Palmer, M.R., Spivack, A.J., and Edmond, J.M. (1987). Temperature and pH controls over isotopic fractionation during adsorption of boron on marine clay. *Geochimica et Cosmochimica Acta 51* (9): 2319–2323.

Palmer, M.R., Pearson, P.N., and Cobb, S.J. (1998). Reconstructing past ocean pH-depth profiles. *Science 282*: 1468–1471.

Paris, G., Gaillardet, J., and Louvat, P. (2010a). Geological evolution of seawater boron isotopic composition recorded in evaporites. *Geology 38* (11): 1035–1038.

Paris, G., Bartolini, A., Donnadieu, Y. et al. (2010b). Investigating boron isotopes in a middle Jurassic micritic sequence: primary vs. diagenetic signal. *Chemical Geology 275* (3–4): 117–126.

Pearson, P.N. and Palmer, M.R. (1999). Middle Eocene seawater pH and atmospheric carbon dioxide concentrations. *Science 284*: 1824–1826.

Pearson, P.N. and Palmer, M.R. (2000). Atmospheric carbon dioxide concentrations over the past 60 million years. *Nature 406*: 695–699.

Pearson, P.N., Foster, G.L., and Wade, B.S. (2009). Atmospheric carbon dioxide through the Eocene-Oligocene climate transition. *Nature 461* (7267): 1110–U204.

Penman, D.E., Hönisch, B., Rasbury, E.T. et al. (2013). Boron, carbon, and oxygen isotopic composition of brachiopod shells: intra-shell variability, controls, and potential as a paleo-pH recorder. *Chemical Geology 340*: 32–39.

Penman, D.E., Hönisch, B., Zeebe, R.E. et al. (2014). Rapid and sustained surface ocean acidification during the Paleocene-Eocene thermal maximum. *Paleoceanography 29* (5): doi: 10.1002/2014PA002621.

Petit, J.R., Jouzel, J., Raynaud, D. et al. (1999). Climate and atmospheric history of the past 420,000 years from the Vostok ice core. *Antarctica, Nature 399*: 429–436.

Pierrot, D., Lewis, E., and Wallace, D.W.R. (2006) MS Excel Program Developed for CO_2 System Calculations, ORNL/CDIAC-105a.

Quintana Krupinski, N.B., Russell, A.D., Pak, D.K., and Paytan, A. (2017). Core-top calibration of B/Ca in Pacific Ocean *Neogloboquadrina incompta* and *Globigerina bulloides* as a surface water carbonate system proxy. *Earth and Planetary Science Letters 466*: 139–151.

Rae, J.W.B. (2018). Boron isotopes in foraminifera: systematics, Biomineralisation, and CO2 reconstruction. In: *Boron Isotopes: The Fifth Element* (ed. H. Marschall and G. Foster), 107–143. Cham: Springer International Publishing.

Rae, J.W.B., Foster, G.L., Schmidt, D.N., and Elliott, T. (2011). Boron isotopes and B/Ca in benthic foraminifera: proxies for the deep ocean carbonate system. *Earth and Planetary Science Letters 302* (3–4): 403–413.

Raitzsch, M. and Hönisch, B. (2013). Cenozoic boron isotope variations in benthic foraminifers. *Geology 41* (5): 591–594.

Raitzsch, M., Hathorne, E.C., Kuhnert, H. et al. (2011). Modern and late Pleistocene B/Ca ratios of the benthic foraminifer *Planulina wuellerstorfi* determined with laser ablation ICP-MS. *Geology 39* (11): 1039–1042.

Regenberg, M., Nürnberg, D., Steph, S. et al. (2006). Assessing the effect of dissolution on planktonic foraminiferal Mg/Ca ratios: evidence from Caribbean core tops. *Geochemistry, Geophysics, Geosystems 7* (7): 1–23.

Reynaud, S., Hemming, N.G., Juillet-Leclerc, A., and Gattuso, J.-P. (2004). Effect of pCO_2 and temperature on the boron isotopic composition of a zooxanthellate coral: *Acropora* sp. *Coral Reefs 23*: 539–546.

Ridgwell, A. and Schmidt, D.N. (2010). Past constraints on the vulnerability of marine calcifiers to massive carbon dioxide release. *Nature Geoscience 3* (3): 196–200.

Ries, J.B. (2011). A physicochemical framework for interpreting the biological calcification response to CO_2-induced ocean acidification. *Geochimica et Cosmochimica Acta 75* (14): 4053–4064.

Rink, S., Kühl, M., Bijma, J., and Spero, H.J. (1998). Microsensor studies of photosynthesis and respiration in the symbiotic foraminifer *Orbulina universa*. *Marine Biology 131* (4): 583–595.

Rollion-Bard, C. and Erez, J. (2010). Intra-shell boron isotope ratios in the symbiont-bearing benthic foraminiferan *Amphistegina lobifera*: implications for δ^{11}B vital effects and paleo-pH reconstructions. *Geochimica et Cosmochimica Acta 74* (5): 1530–1536.

Rollion-Bard, C., Chaussidon, M., and France-Lanord, C. (2003). pH control on oxygen isotopic composition of symbiotic corals. *Earth and Planetary Science Letters 215* (1–2): 275–288.

Rollion-Bard, C., Blamart, D., Trebosc, J. et al. (2011). Boron isotopes as pH proxy: a new look at boron speciation in deep-sea corals using ^{11}B MAS NMR and EELS. *Geochimica et Cosmochimica Acta 75* (4): 1003–1012.

Rosenthal, Y. and Lohmann, G.P. (2002). Accurate estimation of sea surface temperatures using dissolution-corrected calibrations for Mg/Ca paleothermometry. *Paleoceanography 17* (3): 16-1–16-6.

Ruiz-Agudo, E., Putnis, C.V., Kowacz, M. et al. (2012). Boron incorporation into calcite during growth: implications for the use of boron in carbonates as a pH proxy. *Earth and Planetary Science Letters 345–348*: 9–17.

Rustad, J.R. and Bylaska, E.J. (2007). Ab initio calculation of isotopic fractionation in B(OH)3(aq) and BOH4-(aq). *Journal of the American Chemical Society 129* (8): 2222–2223.

Rustad, J.R., Bylaska, E.J., Jackson, V.E., and Dixon, D.A. (2010). Calculation of boron-isotope fractionation between $B(OH)_{3(aq)}$ and $B(OH)_{4\ (aq)}^-$. *Geochimica et Cosmochimica Acta 74* (10): 2843–2850.

Sadekov, A., J. Kerr, G. Langer, et al. (2016) Understanding the mechanisms behind boron elemental and isotopic fractionation in the benthic foraminifera *Cibicidoides wuellerstorfi*, International Conference on Paleoceanography, Utrecht, The Netherlands.

Salmon, K.H., Anand, P., Sexton, P.F., and Conte, M. (2016). Calcification and growth processes in planktonic foraminifera complicate the use of B/Ca and U/Ca as carbonate chemistry proxies. *Earth and Planetary Science Letters 449*: 372–381.

Sanchez-Valle, C., Reynard, B., Daniel, I. et al. (2005). Boron isotopic fractionation between minerals and fluids: new insights from in situ high pressure-high temperature vibrational spectroscopic data. *Geochimica et Cosmochimica Acta 69* (17): 4301–4313.

Sanyal, A., Nugent, M., Reeder, R.J., and Bijma, J. (2000). Seawater pH control on the boron isotopic composition of calcite: evidence from inorganic calcite precipitation experiments. *Geochimica et Cosmochima Acta 64* (9): 1551–1555.

Sanyal, A., Bijma, J., Spero, H.J., and Lea, D.W. (2001). Empirical relationship between pH and the boron isotopic composition of *Globigerinoides sacculifer*: implications for the boron isotope paleo-pH proxy. *Paleoceanography 16* (5): 515–519.

Sanyal, A., Hemming, N.G., Broecker, W.S. et al. (1996). Oceanic pH control on the boron isotopic composition of foraminifera: evidence from culture experiments. *Paleoceanography 11* (5): 513–517.

Schlesinger, W.H. and Vengosh, A. (2016). Global boron cycle in the Anthropocene. *Global Biogeochemical Cycles 30* (2): 219–230.

Schlitzer, R. (2012) Ocean Data View, http://odv.awi.de.

Seki, O., Foster, G.L., Schmidt, D.N. et al. (2010). Alkenone and boron-based Pliocene pCO$_2$ records. *Earth and Planetary Science Letters 292* (1–2): 201–211.

Sen, S., Stebbins, J.F., Hemming, N.G., and Ghosh, B. (1994). Coordination environments of B impurities in calcite and aragonite polymorphs: a ^{11}B MAS NMR study. *American Mineralogist 79*: 819–825.

Siegenthaler, U., Stocker, T.F., Monnin, E. et al. (2005). Stable carbon cycle-climate relationship during the late Pleistocene. *Science 310* (5752): 1313–1317.

Simon, L., Lécuyer, C., Maréchal, C., and Coltice, N. (2006). Modelling the geochemical cycle of boron: implications for the long-term $\delta^{11}B$ evolution of seawater and oceanic crust. *Chemical Geology 225*: 61–76.

Sinclair, D.J., Kinsley, L.P.J., and McCulloch, M.T. (1998). High resolution analysis of trace elements in corals by laser ablation ICP-MS. *Geochimica et Cosmochimica Acta 62* (11): 1889–1901.

Skinner, A., LaFemina, J.P., and Jansen, H.J.F. (1994). Structure and bonding of calcite: a theoretical study. *American Mineralogist 79* (3–4): 205–214.

Spero, H.J. and Parker, S.L. (1985). Photosynthesis in the symbiotic planktonic foraminifer *Orbulina universa*, and its potential contribution to oceanic primary productivity. *Journal of Foraminiferal Research 15* (4): 273–281.

Spero, H.J. and Lea, D.W. (1993). Intraspecific stable isotope variability in the planktic foraminifera *Globigerinoides sacculifer*: results from laboratory experiments. *Marine Micropaleontology 22*: 221–234.

Spero, H.J., Mielke, K.M., Kalve, E.M. et al. (2003). Multispecies approach to reconstructing eastern equatorial Pacific thermocline hydrography during the past 360 kyr. *Paleoceanography 18* (1): 1–16.

Spivack, A.J. and Edmond, J.M. (1987). Boron isotope exchange between seawater and the oceanic crust. *Geochimica et Cosmochimica Acta 51* (5): 1033–1043.

Spivack, A.J. and You, C.-F. (1997). Boron isotopic geochemistry of carbonates and pore waters, ocean drilling program site 851. *Earth and Planetary Science Letters 152*: 113–122.

Spivack, A.J., You, C.-F., and Smith, H.J. (1993). Foraminiferal boron isotope ratios as a proxy for surface ocean pH over the past 21 Myr. *Nature 363*: 149–151.

Stewart, J.A., Gutjahr, M., Pearce, F. et al. (2015). Boron during meteoric diagenesis and its potential implications for Marinoan snowball Earth δ11B-pH excursions. *Geology 43* (7): 627–630.

Stoll, H., Langer, G., Shimizu, N., and Kanamaru, K. (2012). B/Ca in coccoliths and relationship to calcification vesicle pH and dissolved inorganic carbon concentrations. *Geochimica et Cosmochimica Acta 80* (0): 143–157.

Takagi, H., Kimoto, K., Fujiki, T. et al. (2016). Ontogenetic dynamics of photosymbiosis in cultured planktic foraminifers revealed by fast repetition rate fluorometry. *Marine Micropaleontology 122*: 44–52.

Takahashi, T., Sutherland, S.C., Chipman, D.W. et al. (2014). Climatological distributions of pH, pCO_2, total CO_2, alkalinity, and $CaCO_3$ saturation in the global surface ocean, and temporal changes at selected locations. *Marine Chemistry 164*: 95–125.

Tossell, J.A. (2006). Boric acid adsorption on humic acids: ab initio calculation of structures, stabilities, ^{11}B NMR and $^{11}B,^{10}B$ isotopic fractionations of surface complexes. *Geochimica et Cosmochima Acta 70*: 5089–5103.

Toyofuku, T., Matsuo, M.Y., de Nooijer, L.J. et al. (2017). Proton pumping accompanies calcification in foraminifera. *Nature Communications 8*: 14145.

Tripati, A.K., Roberts, C.D., and Eagle, R.A. (2009). Coupling of CO_2 and ice sheet stability over major climate transitions of the last 20 million years. *Science 326*: 1394.

Tripati, A.K., Roberts, C.D., Eagle, R.A., and Li, G. (2011). A 20 million year record of planktic foraminiferal B/Ca ratios: Systematics and uncertainties in pCO_2 reconstructions. *Geochimica et Cosmochimica Acta 75* (10): 2582–2610.

Trotter, J., Montagna, P., McCulloch, M. et al. (2011). Quantifying the pH 'vital effect' in the temperate zooxanthellate coral *Cladocora caespitosa*: validation of the boron seawater pH proxy. *Earth and Planetary Science Letters 303* (3–4): 163–173.

Tyrrell, T. and Zeebe, R.E. (2004). History of carbonate ion concentration over the last 100 million years. *Geochimica et Cosmochimica Acta 68* (17): 3521–3530.

Uchikawa, J., Penman, D.E., Zachos, J.C., and Zeebe, R.E. (2015). Experimental evidence for kinetic effects on B/Ca in synthetic calcite: implications for potential $B(OH)_4^-$ and $B(OH)_3$ incorporation. *Geochimica et Cosmochimica Acta 150*: 171–191.

Uchikawa, J., Harper, D.T., Penman, D.E. et al. (2017). Influence of solution chemistry on the boron content in inorganic calcite grown in artificial seawater. *Geochimica et Cosmochimica Acta 218* (Supplement C): 291–307.

Urey, H.C. (1947). The thermodynamic properties of isotopic substances. *Journal of the Chemical Society* 562–581.

Vengosh, A., Kolodny, Y., Starinsky, A. et al. (1991). Coprecipitation and isotopic fractionation of boron in modern biogenic carbonates. *Geochimica et Cosmochimica Acta 55*: 2901–2910.

Vengosh, A., Starinsky, A., Kolodny, Y. et al. (1992). Boron isotope variations during fractional evaporation of sea water: new constraints on the marine vs. nonmarine debate. *Geology 20*: 799–802.

Venn, A., Tambutté, E., Holcomb, M. et al. (2011). Live tissue imaging shows reef corals elevate pH under their calcifying tissue relative to seawater. *PLoS One 6* (5): e20013.

Venn, A.A., Tambutté, E., Holcomb, M. et al. (2013). Impact of seawater acidification on pH at the tissue–skeleton interface and calcification in reef corals. *Proceedings of the National Academy of Sciences 110* (5): 1634–1639.

de Villiers, J.P.R. (1971). Crystal structures of Aragonite, Strontianite, and Witherite. *American Mineralogist 56*: 758–767.

Walker, C.T. and Price, N.B. (1963). Departure curves for computing palaeosalinity from boron in illites and shales. *American Association of Petroleum Geologists 47*: 833–841.

Wall, M., Ragazzola, F., Foster, L.C. et al. (2015). pH up-regulation as a potential mechanism for the cold-water coral *Lophelia pertusa* to sustain growth in aragonite undersaturated conditions. *Biogeosciences 12*: 6869–6880.

Wara, M.W., Delaney, M.L., Bullen, T.D., and Ravelo, A.C. (2003). Possible roles of *p*H, temperature, and partial dissolution in determining boron concentration and isotopic composition in planktonic foraminifera. *Paleoceanography 18* (4): 1–9.

Wei, H., Jiang, S., Xiao, Y., and Hemming, N.G. (2014). Boron isotopic fractionation and trace element incorporation in various species of modern corals in Sanya bay, South China Sea. *Journal of Earth Science 25* (3): 431–444.

Wolf-Gladrow, D., Bijma, J., and Zeebe, R.E. (1999). Model simulation of the carbonate chemistry in the microenvironment of symbiont bearing foraminifera. *Marine Chemistry 64*: 181–198.

York, D., Evensen, N., Martinez, M., and Delgado, J. (2004). Unified equations for the slope, intercept, and standard errors of the best straight line. *American Journal of Physics* 367–375.

Yu, J. and Elderfield, H. (2007). Benthic foraminiferal B/Ca ratios reflect deep water carbonate saturation state. *Earth and Planetary Science Letters 258* (1–2): 73–86.

Yu, J., Elderfield, H., and Hönisch, B. (2007). B/Ca in planktonic foraminifera as a proxy for surface seawater pH. *Paleoceanography 22*: doi: 10.1029/2006PA001347.

Yu, J., Thornalley, D.J.R., Rae, J.W.B., and McCave, N.I. (2013a). Calibration and application of B/Ca, Cd/Ca, and δ^{11}B in *Neogloboquadrina pachyderma* (sinistral) to constrain CO_2 uptake in the subpolar North Atlantic during the last deglaciation. *Paleoceanography* doi: 10.1002/palo.20024.

Yu, J., Anderson, R.F., Jin, Z. et al. (2013b). Responses of the deep ocean carbonate system to carbon reorganization during the last glacial–interglacial cycle. *Quaternary Science Reviews 76*: 39–52.

Zeebe, R.E. (2005). Stable boron isotope fractionation between dissolved $B(OH)_3$ and $B(OH)_4^-$. *Geochimica et Cosmochimica Acta 69* (11): 2753–2766.

Zeebe, R.E. (2009). Hydration in solution is critical for stable oxygen isotope fractionation between carbonate ion and water. *Geochimica et Cosmochimica Acta 73* (18): 5283–5291.

Zeebe, R.E., Bijma, J., and Wolf-Gladrow, D. (1999). A diffusion-reaction model of carbon isotope fractionation in foraminifera. *Marine Chemistry 64*: 199–227.

Zeebe, R.E., Sanyal, A., Ortiz, J.D., and Wolf-Gladrow, D.A. (2001). A theoretical study of the kinetics of the boric acid-borate equilibrium in seawater. *Marine Chemistry 1814*: 113–124.

Zeebe, R.E. and Wolf-Gladrow, D.A. (2001). CO_2 *in seawater: Equilibrium, kinetics, isotopes*, vol. *65*, 346. Elsevier.

Zeebe, R.E., Wolf-Gladrow, D.A., Bijma, J., and Hönisch, B. (2003). Vital effects in foraminifera do not compromise the use of δ^{11}B as a paleo-pH indicator: evidence from modeling. *Paleoceanography 18* (2, 1043): 1–9.

Reconstructing Paleo-Acidity, pCO₂ and Deep-Ocean [CO₃²⁻]

Abstract

This chapter explains the translation of $\delta^{11}B$ and B/Ca to carbonate system parameters. In addition to explaining the equations behind this translation process, we also provide a series of sensitivity experiments to demonstrate the pH and pCO_2 uncertainties introduced by individual uncertainties of $\delta^{11}B_{foram}$, temperature, salinity, alkalinity, $\delta^{11}B_{sw}$, and varying seawater elemental composition over long geological time scales, including recommendations for propagating errors. The similarities between boron proxy calibrations published to date are further examined to explore potential systematics for proxy sensitivities in marine calcifiers that have not yet or can no longer be calibrated over a wide range of known environmental conditions. Finally, guidelines are provided for selecting sediment core sites and sample material for successful paleoreconstructions.

Keywords: boron isotope proxy sensitivity; paleocalibration; temperature effect; salinity effect; alkalinity effect; sensitivity study; error propagation

3.1 Introduction

The boron proxy systematics outlined in Chapter 2 have been applied by a number of studies to tropical corals and subarctic coralline algae on last glacial maximum (LGM) to Holocene timescales (Fietzke et al. 2015; Kubota et al. 2014; McCulloch et al. 2012; Pelejero et al. 2005; Wei et al. 2009), to planktic and benthic foraminifers from the Pleistocene-Pliocene (Bartoli et al. 2011; Ezat et al. 2017; Henehan et al. 2013; Hönisch and Hemming 2005; Hönisch et al. 2009; Martínez-Botí et al. 2015a; Martínez-Botí et al.

Boron Proxies in Paleoceanography and Paleoclimatology, First Edition. Bärbel Hönisch, Stephen M. Eggins, Laura L. Haynes, Katherine A. Allen, Katherine D. Holland, and Katja Lorbacher.
© 2019 John Wiley & Sons Ltd. Published 2019 by John Wiley & Sons Ltd.
Companion website: www.wiley.com/go/Hönisch/Boron_Paleoceanography

2015b; Palmer and Pearson 2003; Rae et al. 2014; Raitzsch et al. 2018; Seki et al. 2010; Tripati et al. 2009; Yu et al. 2010b; Yu et al. 2013b; Yu et al. 2014), the Miocene (Badger et al. 2013; Foster et al. 2012; Greenop et al. 2014), and Oligocene-Paleocene (Anagnostou et al. 2016; Gutjahr et al. 2017; Pearson and Palmer 1999; Pearson et al. 2009; Penman et al. 2014). Planktic foraminifers did not yet exist in the Paleozoic and reconstructions during that era have so far been based on brachiopods and bulk carbonates (Clarkson et al. 2015; Joachimski et al. 2005). Similarly, bulk carbonates have been investigated for the Neoproterozoic (Kasemann et al. 2005; Ohnemueller et al. 2014), but the interested reader should also review a recent modern calibration study that raises some concerns about the reliability of δ^{11}B in shallow water carbonate ooids and cements (Zhang et al. 2017).

The reader is referred to the above specific references for detailed discussions of the specific proxy records and paleoceanographic implications. Here we will use some of those published records to evaluate the pH, [CO$_3^{2-}$] and/or pCO$_2$ uncertainty introduced by the choice of (i) species-specific calibration, (ii) temperature and (iii) salinity corrections (iv), assumptions of a second parameter of the carbonate system, and (v) assumptions of δ^{11}B$_{sw}$. The reason we need to consider the effects of T, S, and p on boron proxy reconstructions of pH and pCO$_2$ is that they affect both boron and carbon equilibrium reactions in seawater (see Chapter 1 and Section 2.2.1), and possibly also α_{B3-B4} (Section 2.4.8). While temperature proxies have been established for foraminifers (e.g. Anand et al. 2003; Lea et al. 1999) and corals (e.g. Fallon et al. 2003; Wei et al. 2000), and pressure can be approximated reasonably well from typical growth habitats of specific organisms, there is no direct salinity proxy (although salinity is sometimes approximated by removing the temperature signal from δ^{18}O records in marine carbonates, e.g. Linsley et al. 2006; Schmidt et al. 2004). Furthermore, a second parameter of the marine carbonate system is required to translate pH estimates to pCO$_2$, but we do not yet have a proxy for such a parameter in the surface ocean, and pairing benthic foraminiferal δ^{11}B and B/Ca to reconstruct bottom water pH and [CO$_3^{2-}$] leads to large uncertainties in derived carbonate chemistry parameters (Rae et al. 2011; Yu et al. 2010a). Reconstructing atmospheric pCO$_2$ from planktic foraminiferal δ^{11}B therefore typically involves estimates of total alkalinity (TA), which is independent of variations in T, p, and pCO$_2$, but may change on timescales of tens of thousands of years and longer due to addition and removal by terrestrial weathering, calcification and CaCO$_3$ dissolution processes in the ocean (e.g. Clark et al. 2006; Kerr et al. 2017). Because alkalinity typically scales with salinity in the open ocean (Lee et al. 2006), and region-specific TA vs. S relationships are sometimes used to approximate paleo-alkalinity (e.g. Hönisch et al. 2009), any uncertainty of the salinity estimate (along with uncertainties in T) will need to be propagated into pCO$_2$ estimates from paired pH and alkalinity estimates. Tripati et al. (2011) performed a comprehensive sensitivity study for the effects of T, S, [B$_T$], and alkalinity on pCO$_2$ estimates from B/Ca in planktic foraminifers, and similar evaluations have been carried out in several boron isotope studies (e.g. Bartoli et al. 2011; Hönisch and Hemming 2005; Hönisch et al. 2009; Martínez-Botí et al. 2015b; Pearson et al. 2009). However, the individual effects of individual parameter uncertainties on boron isotope reconstructions have not yet been presented systematically, and this Chapter attempts to fill this gap.

Finally, deep-time (i.e. >10–20 Ma) reconstructions of absolute carbonate system parameter values are hampered by largely unknown vital effects in now extinct species, uncertain $\delta^{11}B_{sw}$, and [B$_T$] composition of paleoseawater. Such reconstructions typically revert to interpreting their data qualitatively, that is, as relative shifts in carbonate system parameter values across specific climate transitions. In addition, the major ion composition of seawater ([Mg] and [Ca]) affects the dissociation constants of carbon and boron in seawater (Hain et al. 2015; Nir et al. 2015; Tyrrell and Zeebe 2004), and thereby pH and pCO$_2$ reconstructions. As more boron proxy calibrations and records are generated, opportunities will arise to overcome the current limitations and uncertainties, and carbonate system parameters will be estimated more quantitatively for periods deeper in time. The question is, how can we do this based on our current understanding of the boron proxies? This chapter will start by introducing the basic concept of translating $\delta^{11}B_{CaCO3}$ to pH, followed by a discussion of empirical calibrations and the respective effects of T, S, alkalinity, and $\delta^{11}B_{sw}$ on pH and pCO$_2$ estimates.

3.2 Estimating Paleoseawater pH from Boron Isotopes

The first studies reconstructing surface ocean pH and atmospheric pCO$_2$ from boron isotopes in planktic foraminifers (Hönisch and Hemming 2005; Palmer and Pearson 2003; Palmer et al. 1998; Pearson and Palmer 1999, 2000; Sanyal et al. 1995) were under the impression that $\alpha_{B3-B4} = 1.0194$ (Kakihana and Kotaka 1977; Kakihana et al. 1977) describes the boron isotope fractionation in seawater reasonably well. This was borne out by laboratory calibrations of $\delta^{11}B$ in planktic foraminiferal and synthetic calcite (Sanyal et al. 1996, 2000, 2001), which match the shape and inflection point of $\delta^{11}B_{borate}$ using $\alpha_{B3-B4} = 1.0194$ and pK$^*_B$ after Dickson (Figures 2.4 and 2.10a, 1990). To infer seawater-pH from $\delta^{11}B_{CaCO3}$, these studies used the following equation:

$$pH = pK^*_B - \log\left(-\left(\delta^{11}B_{sw} - \delta^{11}B_{CaCO3}\right)/ \left(\delta^{11}B_{sw} - \alpha_{B3-B4}{}^*\delta^{11}B_{CaCO3} - \varepsilon_{B3-B4}\right)\right) \tag{3.1}$$

where pK$^*_B$ is the dissociation constant of boric acid in seawater (see Section 2.2.1), $\delta^{11}B_{sw}$ is the boron isotopic composition of seawater, α_{B3-B4} is the theoretical aqueous boron isotope fractionation factor after Kakihana et al. (1977) and ε_{B3-B4} the corresponding aqueous boron isotope fractionation (see Section 2.2.2). However, because δ^{11}B of marine carbonates is offset from δ^{11}B$_{borate}$, Hönisch and Hemming (2004) introduced a correction for a constant offset "a," so that absolute pH values could be calculated as:

$$pH = pK^*_B - \log\left(-\left(\delta^{11}B_{sw} - \delta^{11}B_{CaCO3} + a\right)\middle/\right.$$
$$\left.\left(\delta^{11}B_{sw} - \alpha_{B3-B4}^* \left(\delta^{11}B_{CaCO3} - a\right) - \varepsilon_{B3-B4}\right)\right) \tag{3.2}$$

The value of the constant offset is marine calcifier species-specific and in some instances also analytical technique-specific (see e.g. Figure 2.12). Having analyzed boron isotopes on several TIMS instruments, Hönisch and co-authors have cross-calibrated and documented the respective offsets over the years (Hönisch and Hemming 2004; Hönisch et al. 2009; Hönisch et al. 2003), so that the inferred pH from different sediment core sites and time periods can be compared directly. In contrast, early studies by Palmer and coauthors (Palmer and Pearson 2003; Palmer et al. 1998; Pearson and Palmer 1999, 2000) assumed that different foraminifer species record the same δ^{11}B$_{CaCO3}$ under similar environmental conditions. Because evidence for vital effect offsets between different marine calcifiers has now been firmly established (Figure 2.14), recent studies account for species offsets, albeit in a slightly different way than described by Eq. 3.2.

With the subsequent empirical determination of α_{B3-B4}~1.027 (Klochko et al. 2006; Nir et al. 2015), δ^{11}B$_{CaCO3}$ calibrations are now expressed as functions of δ^{11}B$_{borate}$, which allows direct comparison of proxy sensitivities across a range of pH, temperature, salinity and pressure conditions (Figures 2.11–2.14). δ^{11}B$_{borate}$ is thereby inferred from δ^{11}B$_{CaCO3}$ by linear regression of the form

$$\delta^{11}B_{borate} = \left(\delta^{11}B_{CaCO3} - c\right)/m \tag{3.3}$$

where "c" is the intercept and "m" the slope of the regression (Foster et al. 2012; Henehan et al. 2013). pH can then be calculated as follows:

$$pH = pK^*_B - \log\left(-\left(\delta^{11}B_{sw} - \delta^{11}B_{borate}\right)\middle/\right.$$
$$\left.\left(\delta^{11}B_{sw} - \alpha_{B3-B4}^* \delta^{11}B_{borate} - \varepsilon_{B3-B4}\right)\right) \tag{3.4}$$

Because the empirical δ^{11}B$_{borate}$ calibration equations (Eq. 3.3) include the species offset (but see Box 3.1 for a small offset correction between planktic foraminifers calibrated in laboratory culture and coretop sediments) and sensitivity deviations from aqueous δ^{11}B$_{borate}$ after Klochko et al. (2006), Eq. 3.4 consequently applies $\alpha_{B3-B4} = 1.0272$ and $\varepsilon_{B3-B4} = 27.2$ ‰ (Klochko et al. 2006), pK$^*_B$ after Dickson (1990) and Millero (1995), and δ^{11}B$_{sw} = 39.61$ ‰ in the modern ocean (Foster et al. 2010). Using this approach, consistent pH variations can be reconstructed from oligotrophic sediment core sites in the tropical ocean, and from a range of foraminifer species and analytical techniques. This can be seen in Figure 3.1, which displays pH reconstructions

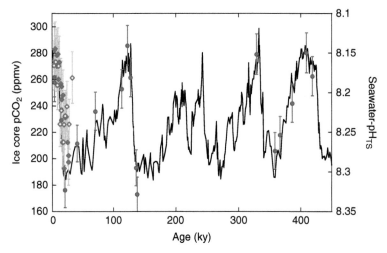

Figure 3.1 Comparison of Pleistocene paleoseawater-pH estimates from δ¹¹B in two species of planktic foraminifers – *G. sacculifer* (blue circles, technique: N-TIMS, Hönisch and Hemming 2005) and *Globigerinoides ruber* (red symbols, technique: MC-ICP-MS, Henehan et al. 2013) – and three core sites in the eastern (blue symbols) and western (red open and closed symbols) equatorial Atlantic Ocean. Reconstructions are based on culture calibrations corrected for size and gametogenic calcite effects (cf. Figure 3.2) and coherently suggest preindustrial and Pleistocene interglacial pH$_{TS}$~8.15 and a glacial/interglacial pH amplitude of ~0.15 units. Also shown are the pCO₂ estimates from ice cores (black line, Lüthi et al. 2008; Petit et al. 1999; Siegenthaler et al. 2005), with glacial pCO₂ ~100 ppmv lower than interglacial pCO₂.

Box 3.1 δ¹¹B$_{foram}$ Offsets Between Culture and Coretop Calibrations

The pH-reconstructions shown in Figure 3.1 are based on laboratory culture calibrations, but comparison of δ¹¹B$_{foram}$ from coretop sediments and cultured specimens has revealed small offsets (Figure 3.2) that need to be taken into account for accurate reconstruction of preindustrial surface ocean pH. In *G. ruber* the offset has been explained by the smaller shell size used for these paleorecon-structions (i.e. 300–355 μm) compared to the larger average shell size grown in the laboratory, which results in lower δ¹¹B$_{foram}$ (Henehan et al. 2013), see also Section 2.4.3. The *G. sacculifer* reconstruction is based on the shell size class 515–865 μm, which is comparable to shell sizes of gametogenic specimens in lab-oratory culture (Figure 3.2); however, coretop analyses of that shell size still reflect lower δ¹¹B$_{foram}$ than predicted from preindustrial δ¹¹B$_{borate}$. This difference is unlikely due to differences in light intensity between the sea surface and the labo-ratory, or to the use of 10x elevated B concentrations in the experiments of Sanyal et al. (2001), as the additional *G. sacculifer* calibration samples of Dyez et al. (sub-mitted) were grown in seawater with natural B concentrations (Table A2.2 (see online version), Figure 2.10) and the data agree within error with those of Sanyal et al. (2001). Instead, the offset between laboratory and field data is most likely related to the secretion of gametogenic calcite by *G. sacculifer* at greater water depth (Bé 1980), where the water column is dimmer and more acidic. Given the offsets from the culture equations are consistent across a range of sediment

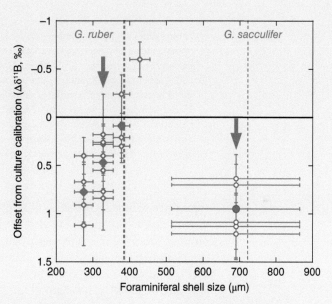

Figure 3.2 Difference ($\Delta\delta^{11}$B) between δ^{11}B$_{foram}$ predicted by laboratory calibrations and observations on naturally grown specimens from seafloor sediments. Preindustrial pH was estimated from GLODAP (Key et al. 2004) and World Ocean Atlas 2009 hydrographic data plotted in ODV (Schlitzer 2012). $\Delta\delta^{11}$B is shown versus average foraminiferal shell size and highlights the difference between typical shell size reached in laboratory culture (dashed lines) and the shell size fraction selected for the paleoreconstructions shown in Figure 3.1 (vertical arrows). Open symbols reflect data from spatially distant core sites, closed symbols depict the averages of individual core sites within the same shell size class. The original presentation by Henehan et al. (red symbols, 2013) included data from sediment traps and plankton tows, which are not shown here. *Globigerinoides sacculifer* data are from Hönisch and Hemming (2004, 2005) and Dyez et al. (submitted).

coretops from spatially distant locations, the magnitude of the offset can be removed by adjusting the intercept value ("c") of the δ^{11}B$_{foram}$ vs. δ^{11}B$_{borate}$ regression. The MC-ICP-MS *G. ruber* calibration for shell sizes 300–355 μm (applying updated York fit (Table A2.2) to *G. ruber* data of Henehan et al. 2013) thus becomes

$$\delta^{11}\text{B}_{borate} = \left(\delta^{11}\text{B}_{G.ruber} - 8.88\right)/0.60 \qquad (3.5)$$

and the N-TIMS *G. sacculifer* calibration for shell sizes >500 μm becomes

$$\delta^{11}\text{B}_{borate} = \left(\delta^{11}\text{B}_{G.sacculifer} - 6.42\right)/0.73 \qquad (3.6)$$

This correction is equivalent to the approach applied by Hönisch and Hemming (2005) and Hönisch et al. (2009), where the offset was expressed within the constant "a" in Eq. 3.2.

It is important to determine and verify such calibration offsets between culture and coretop studies because (i) laboratory cultures do not mimic the physico-chemical water column variations that wild specimens experience, and (ii) paleo-pH reconstructions (e.g. Figure 3.1) will yield pH estimates that are too low and consequently pCO$_2$ estimates that are too high if those offsets are unaccounted for. However, this caveat regarding calibrations really only applies to culture versus coretop comparisons, where cultured specimens may grow under slightly different conditions than wild specimens. If a calibration is based on coretop specimens of a given size class, any such offset is already built into the calibration and no further correction is required when generating downcore pH records.

from three sediment core sites (ODP 668B on the Sierra Leone Rise in the Eastern Equatorial Atlantic, ODP 999A in the Caribbean, and GeoB 1523-1 on the Ceara Rise in the Western Equatorial Atlantic) and two species of symbiont-bearing planktic foraminifers, *Globigerinoides sacculifer* and *Globigerinoides ruber*, analyzed by N-TIMS (Hönisch and Hemming 2005) and MC-ICP-MS (Henehan et al. 2013), respectively. Despite the geographical and analytical differences between studies, the inferred pH variations are similar in absolute and relative (i.e. glacial/interglacial) terms, which underscores the high degree of coherency between analytical techniques, δ^{11}B variations in different foraminifer species from oligotrophic core sites, and the technique- and species-specific calibrations applied.

3.2.1 Calibrating δ^{11}B$_{CaCO3}$ from Wild Specimens and for Extinct Species

The previous section has demonstrated how coherent pH reconstructions can be achieved when species- and technique-specific calibrations are available. Does that mean all laboratories now have to establish their own calibrations and perform their own culture experiments? The answer is both yes and no. Cumulative evidence from the suite of calibrations shown in Figure 2.14 and Table A2.2 (online) suggests that the sensitivity (i.e. the slope) of δ^{11}B recorded in many biogenic marine carbonates to seawater-pH and δ^{11}B$_{borate}$ is less than that predicted from the aqueous boron isotope fractionation (Klochko et al. 2006; Nir et al. 2015), but similar among different calcifiers. Studies indicating a more sensitive relationship are restricted to calibrations based on natural samples, which cover only a limited pH range and often a large temperature range. To explain this difference, Rae (2018) hypothesized that the lesser pH- and δ^{11}B$_{borate}$- sensitivity in laboratory cultures may be due to the large variation in Ω$_{calcite}$ and an associated growth rate effect. However, although calcite growth rate dependence on Ω$_{calcite}$ has been observed in inorganic precipitation experiments (Kaczmarek et al. 2016; Uchikawa et al. 2015), planktic foraminifera do not show a clear dependence on Ω$_{calcite}$ and appear to calcify ~100x more slowly at similar saturation states (e.g. Allen et al. 2016; Holland et al. 2017). Instantaneous precipitation rates could be much faster than the linear rates approximated by Allen et al. (2016) and Holland et al. (2017), and more research is needed to better constrain growth rates in marine calcifiers across laboratory conditions. Nonetheless, current data availability does not support the precipitation rate explanation for the differential sensitivity of the $\delta^{11}B$ proxy in field and culture studies. As discussed in Section 2.4.8, an alternative explanation for the more sensitive field calibrations can be based on the theoretical temperature effect on α_{B3-B4} (Figure 2.21), where, for example, the *Orbulina universa* field calibration of Henehan et al. (2016) can be aligned with the sensitivity of the *O. universa* laboratory calibrations (Hönisch et al. 2009; Sanyal et al. 1996) once the approximated temperature effect on α_{B3-B4} is applied (Figure 2.21d).

Applying the boron isotope proxy to extinct species requires the assumption of a pH-, or better, $\delta^{11}B_{borate}$- sensitivity, but because downcore conditions are highly uncertain they cannot be used to reliably calibrate a proxy. In a simple attempt to resolve this conundrum, we will now investigate whether the empirical calibrations shown in Figure 2.14 share a common sensitivity (i.e. slope). These calibrations often either cover a wide pH range but consist of only a few data points, or they include many data points, but their respective pH range is limited. As such each calibration may be associated with relatively large uncertainties in slope and intercept (Table A2.2), but a combination of calibrations with many more data points may allow us to reduce that uncertainty, and even investigate whether accounting for the theoretical temperature effect on the aqueous boron isotope fractionation constrains the slope further. If this is possible, studies that cannot establish species-specific calibrations may adopt the common slope and simply calibrate the intercept "c" by analyzing coretop samples, or by comparing extinct species to coexisting species with known intercept.

To ascertain whether calcifiers with different biomineralization pathways and vital effects might produce distinctive calibrations, we have plotted the slopes and uncertainties of the York fits of empirical calibrations (online Table A2.2 online) in Figure 3.3a, and clustered symbiont-bearing foraminifera, symbiont-barren foraminifera and scleractinian corals to compare their respective slopes. This figure shows that scleractinian coral calibrations cluster narrowly around a slope of m~0.73, symbiont-barren foraminifera are much more scattered with a slope value near 1, and laboratory cultures of symbiont-bearing foraminifera have slopes near m~0.7. Figure 3.3b shows the same calibrations after application of the estimated temperature effect on α_{B3-B4}. Calibrations whose slopes overlap within 95% confidence limits are plotted with filled symbols, but some calibration slopes do not agree with their cluster, and those data are shown with open symbols. We will now use calibrations that overlap within 95% confidence limits to create multi-species fits. However, this includes only calibrations shown with closed symbols in Figure. 3.3a and b, as some calibrations are excluded for a range of reasons: (i) The calibrations of bamboo corals and brachiopods are excluded

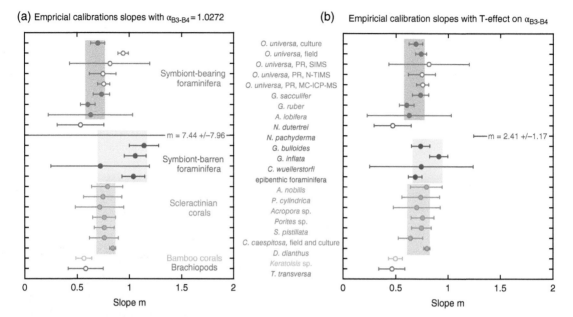

Figure 3.3 (a and b) $\delta^{11}B_{CaCO3}$ vs. $\delta^{11}B_{borate}$ slopes (m) of York fits of empirical calibrations listed in online Table A2.2. Slopes in (a) use $\alpha_{B3-B4} = 1.0272$ (Klochko et al. 2006) and slopes in (b) apply the estimated temperature effect on α_{B3-B4} (Section 2.4.8, Figure 2.21). Red symbols display symbiont-bearing foraminifera, dark blue symbols symbiont-barren foraminifera and orange symbols scleractinian corals. References for each calibration can be found in online Table A2.2. Solid symbols indicate individual calibrations whose slopes overlap within uncertainty of other calibrations within that calcifier group; shading has no quantitative significance but merely visualizes which calibrations are included in the normalizations shown in (c–h). The Puerto Rico (PR) calibrations of *O. universa* agree in slope with other symbiont-bearing foraminifera species, but these calibrations were omitted from the multispecies regressions in (c) and (d) to avoid overrepresentation of the same species. (c–h) York linear regressions of normalized empirical $\delta^{11}B_{CaCO3}$ calibrations versus $\delta^{11}B_{borate}$ calculated from (c, e, g) temperature-independent and (d, f, h) temperature-dependent α_{B3-B4} (see text for details of the normalization). Symbiont-bearing foraminifera (c, d) are normalized to the *Orbulina universa* laboratory calibration of Sanyal et al. (1996) and Hönisch et al. (2009), symbiont-barren foraminifera (e, f) to the *Globigerina bulloides* field calibration of Martínez-Botí et al. (2015a), and scleractinian corals (g, h) to the *Acropora nobilis* calibration of Hönisch et al. (2004). Regressions and 2σ standard errors on slope and intercept were calculated with the York fit method in Matlab (York et al. 2004), using the 2σ uncertainties in $\delta^{11}B_{CaCO3}$ and $\delta^{11}B_{borate}$. Except for the temperature-independent field calibrations of *O. universa* and symbiont-barren foraminifera, the slopes of these regressions support the lesser $\delta^{11}B_{borate}$ sensitivity observed in laboratory culture experiments (cf. Figure 2.13a and b). Data are from Sanyal et al. (1996, 2001), Hönisch et al. (2004, 2008, 2009), Henehan et al. (2013, 2016), Yu et al. (2013a), Martínez-Botí et al. (2015a), Dyez et al. (submitted), Krief et al. (2010), Trotter et al. (2011), Rae et al. (2011), Rollion-Bard and Erez (2010), Anagnostou et al. (2012) and Farmer et al. (2015), in addition to unpublished data from Hönisch and Rae (online Table A2.2).

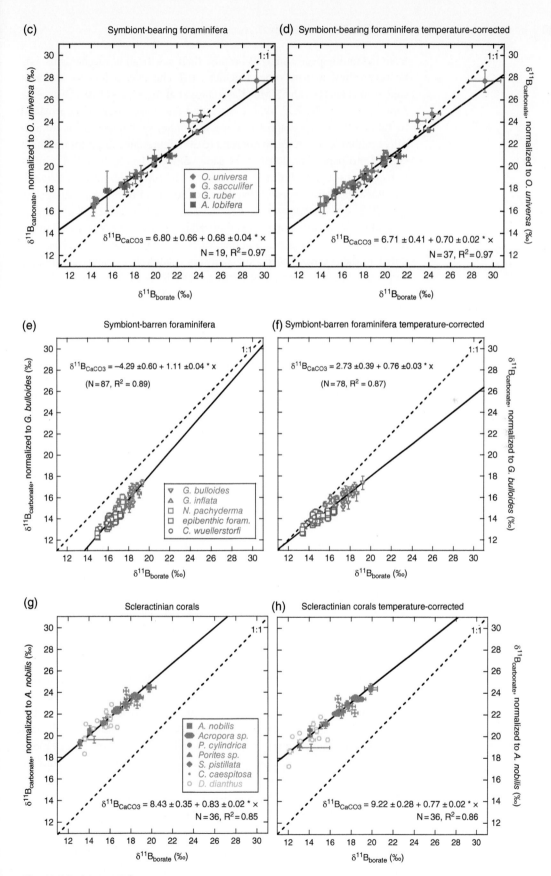

Figure 3.3 (*Continued*)

from the multi-species analysis because their calcification mechanisms may be very different from other calcifiers. (ii) The Puerto Rico *O. universa* calibrations analyzed by SIMS (Kasemann et al. 2009), N-TIMS (Dyez et al. submitted) and MC-ICP-MS (Hönisch and Rae, unpublished data provided in Table A2.2 (online)) agree in slope with other symbiont-bearing foraminifera calibrations, but we do not want to over-represent a single foraminifera species and parallel analyses of the same samples (i.e. the N-TIMS calibrations of Dyez et al. (submitted) and the MC-ICP-MS calibration of Hönisch and Rae (unpublished)), and therefore exclude them from the multi-species fits. (iii) The *Neogloboquadrina dutertrei* calibration of Foster (2008) displays a significantly smaller slope than other calibrations and is not included in the multi-species fits, further noting that the *N. dutertrei* calibration is based on only four data points that encompass a very small δ^{11}B$_{borate}$ range and include large uncertainties in the actual depth habitat. (iv) Because the slope of the *O. universa* field calibration of Henehan et al. (2016) agrees with other symbiont-bearing foraminifera calibrations only after application of the temperature effect on α$_{B3-B4}$, this calibration is only included in the multi-species fit of the temperature corrected slopes. (v) The large uncertainty of the *Neogloboquadrina pachyderma* calibration (Yu et al. 2013a) is much reduced after application of the estimated temperature effect but no longer overlaps with other symbiont-barren foraminifera calibration slopes. This calibration is therefore only included in the multi-species fit of the original calibration slopes.

Using the York fits of the empirical calibrations highlighted by solid symbols in Figure 3.3a and b, we determined the least squares y-axis offset of individual (i) symbiont-bearing foraminifera, (ii) symbiont-barren foraminifera and (iii) scleractinian coral calibrations. To do so, we calculated the York fit δ^{11}B$_{CaCO3}$ value for each original δ^{11}B$_{borate}$ data point of a given calibration and then used R (R-Core-Team 2015) to determine the species-specific offsets between individual York fits and a reference York fit. Any calibration could serve as the reference York fit without affecting the outcome of this analysis and we selected the laboratory *O. universa* calibration of Sanyal et al. (1996) and Hönisch et al. (2009) for symbiont-bearing foraminifera, the *Globigerina bulloides* calibration of Martínez-Botí et al. (2015a) for symbiont-barren foraminifera, and the *Acropora nobilis* calibration of Hönisch et al. (2004) for scleractinian corals. The species-specific offset between different calibrations is thereby determined as the least squares offset for the entire δ^{11}B$_{borate}$ range of a given calibration, so the data are not biased by any individual calibration data points or δ^{11}B$_{borate}$ values. After determination of the species-specific offsets we then applied these offsets to the original δ^{11}B$_{CaCO3}$ data points (i.e. the same offset is applied to all data points in each calibration) to normalize (i.e. shift) each group of calibrations onto the reference calibration. We then calculated new York fits and 2σ standard errors on slope and intercept for the combined calibration data with the York fit method in Matlab (York et al. 2004). For each group of calcifiers we considered two scenarios, one with δ^{11}B$_{borate}$ calculated using temperature-independent α_{B3-B4} and the other with temperature-dependent α_{B3-B4} (see also Section 2.4.8 and Klochko et al. 2006). The 2σ uncertainties in $\delta^{11}B_{CaCO3}$ are the same as reported in the original publications and 2σ uncertainties in $\delta^{11}B_{borate}$ are the recalculated values based on the revised boron isotope mass balance equations after Rae (2018), either using constant α_{B3-B4} = 1.0272 (Klochko et al. 2006) or using the estimated temperature dependent α_{B3-B4} described in Section 2.4.8, Figure 2.21 and (online Table A2.1).

In summary, this species normalization does not change the slope of any individual calibration because the offset for all data points in each individual calibration is the same. Application of the species-specific offset to each data point merely shifts the data onto the reference calibration and allows fitting a new regression through the combined multi-species data set. The specific choice of reference calibration thereby makes no difference to the sensitivity of the multi-species fit, and only determines the intercept. The intercept values of the multi-species fits therefore apply strictly only to each reference species (i.e. *O. universa*, *G. bulloides*, *A. nobilis*), whereas the slope of the multi-species fits applies to all species within that group of calcifiers.

The multi-species equation for symbiont-bearing foraminifera (Figure 3.3c) includes N = 19 data and reads

$$\delta^{11}B_{borate} = \left(\delta^{11}B_{symb.foram.} - c\right)/0.68 \pm 0.04 \tag{3.7}$$

for $\delta^{11}B_{borate}$ based on a constant α_{B3-B4} = 1.0272 (Klochko et al. 2006), and

$$\delta^{11}B_{borate} = \left(\delta^{11}B_{symb.foram.T} - c\right)/0.70 \pm 0.02 \tag{3.8}$$

for $\delta^{11}B_{borate}$ based on a temperature-dependent α_{B3-B4} (Figure 3.3d). Equation 3.8 includes N = 37 data points because it also includes the *O. universa* field calibration of Henehan et al. (cf. Figure 3.3b, 2016). In these equations $\delta^{11}B_{symb\ foram}$ is the boron isotopic composition of a given foraminifer species and "c" is the species and analytical technique-specific intercept.

Figure 3.3e and f include only symbiont-barren foraminifera (N = 87 and N = 78, respectively, with *G. bulloides* as the reference calibration), and Figure 3.3g and h only scleractinian corals (N = 38, with *A. nobilis* as the reference calibration). The equations are

$$\delta^{11}B_{borate} = \left(\delta^{11}B_{asymb.foram.} - c\right)/1.11 \pm 0.04 \tag{3.9}$$

$$\delta^{11}B_{borate} = \left(\delta^{11}B_{asymb.foram.T} - c\right)/0.76 \pm 0.03 \tag{3.10}$$

$$\delta^{11}B_{borate} = \left(\delta^{11}B_{coral.} - c\right)/0.83 \pm 0.02 \tag{3.11}$$

$$\delta^{11}B_{borate} = \left(\delta^{11}B_{coralT} - c\right)/0.77 \pm 0.02 \tag{3.12}$$

Of all biogenic carbonates calibrated to date, only symbiont-barren planktic and benthic foraminifera and the field calibration of *O. universa* display a similar sensitivity to the δ^{11}B of aqueous borate (m = 1, Figure 3.3a), and this is only true if α_{B3-B4} = 1.0272 (Klochko et al. 2006) is applied. Because the temperature correction increases α_{B3-B4} at lower temperatures and therefore shifts δ^{11}B$_{borate}$ towards lower values (cf. Figure 2.21), application of the estimated temperature dependency on α_{B3-B4} to these and the *O. universa* field calibration creates less sensitive slopes (m < 1) than predicted from the aqueous boron isotope fractionation (Figure 3.3b, Klochko et al. 2006; Nir et al. 2015). Importantly, the slopes of the temperature-corrected calibrations are relatively uniform and range between 0.70 ± 0.02 (symbiont-bearing foraminifera), 0.76 ± 0.03 (symbiont-barren foraminifera) and 0.77 ± 0.02 (scleractinian corals), suggesting boron isotope fractionation into marine carbonates may generally be less sensitive to pH and δ^{11}B$_{borate}$ than predicted by the fractionation between aqueous boron species (cf. Section 2.4.1). However, although the uncertainties of the intercept and slope generally decrease between the temperature-uncorrected (Figure 3.3c, e, and g) and temperature-corrected regressions (Figure 3.3d, f, and h), the goodness of fit is essentially indistinguishable between temperature-corrected and -uncorrected regressions. Therefore, and until the temperature effect on aqueous boron isotope fractionation has been determined experimentally and unequivocally, it is not reasonable to favor or reject the temperature correction. Until laboratory calibrations with symbiont-barren calcifiers have been established, it is instead advisable to factor the underlying uncertainty in proxy sensitivity due to temperature into the uncertainty estimate of any pH reconstruction. The question is, how large is that additional uncertainty?

Some studies that used sediment coretop samples for δ^{11}B$_{CaCO3}$ calibrations have argued that the general similarity of δ^{11}B$_{CaCO3}$ to δ^{11}B$_{borate}$ (Figure 2.11) allows application of a calibration slope m = 1 for paleo-reconstructions (e.g. Foster 2008; Rae et al. 2011). Henehan et al. (2013) compared pCO$_2$ reconstructions using the original reconstruction of Foster (2008), which assumed m = 1, with the same δ^{11}B$_{G. ruber}$ data set but using their laboratory *G. ruber* calibration (with slope m = 0.60). They found relatively small differences between the two approaches, but a slightly better match between their revised reconstruction (using m = 0.60) and atmospheric pCO$_2$ amplitudes observed in ice cores. For Pleistocene reconstructions, the difference in the calibration sensitivity thus has little consequence. However, we echo the concern raised by Henehan et al. (2013) that the reconstructed pH difference will be larger millions of years back in time, when seawater-pH was lower and the inferred difference in pH estimates from m = 1 and m = 0.60 was consequently larger. Verifying the pH-sensitivity in different calcifiers, and the temperature sensitivity of boron isotope fractionation in general, are therefore essential questions that need to be addressed.

In the meantime, the multi-species calibration fits (Figure 3.3c–h) provide an opportunity to quantify δ^{11}B$_{borate}$ from calcifiers that have not yet been (or can no longer be) calibrated across a wide range of seawater-pH.

By adopting these fits, calibration efforts could be limited to determining the species-specific intercept, for example, by analyzing modern or sub-recent specimens for which $\delta^{11}B_{borate}$ can be reasonably approximated, or by cross-calibrating coexisting calibrated and non-calibrated (e.g. extinct) species. An example of the latter is given in Raitzsch and Hönisch (2013), who cross-calibrated $\delta^{11}B$ in extinct benthic foraminifer species versus calibrated modern species. The multi-species fits provide a significant step forward because the slopes of empirical calibrations using natural (e.g. sediment coretop) samples have much larger uncertainties than the slopes of the multi-species fits (see online Table A2.2 for comparison).

3.2.2   Calibrating $\delta^{11}B_{CaCO3}$ in Complex Skeletal Structures

While Eqs. 3.7 through 3.12 provide an important opportunity to proceed with paleoreconstructions, a special word of caution is required with regard to the use of long-living calcifiers such as corals, brachiopods and mollusks for pH reconstructions and selecting calibration equations. In contrast to most foraminifera, which typically encompass only a short life span, and which are therefore typically analyzed as whole shells, long-living organisms provide an opportunity to study pH variations with annual or even seasonal resolution. However, the variety of biomineralization processes and the structural and geochemical complexity of skeletal hard parts of these different calcifiers complicate analyses, and analysis of random fragments can lead to large deviations from inferred seawater properties. For instance, Rollion-Bard et al. (2003), Blamart et al. (2007), and Wall et al. (2015) have observed large $\delta^{11}B$-differences between different structural components in aragonitic coral skeletons; Penman et al. (2013) found that the fibrous secondary layer of brachiopod shells records seawater-pH, whereas the primary exterior shell layer is more negative in $\delta^{11}B$; Heinemann et al. (2012) observed large intrashell $\delta^{11}B$ variations in shells of the mussel *Mytilus edulis* and B/Ca of the periostracum is approximately one order of magnitude higher compared to the carbonate; Liu et al. (2015) observed a 5‰ seasonal $\delta^{11}B$ range in shells of the clam *Arctica islandica*, and McCoy et al. (2011) found variable seasonal B/Ca resolution between the inner prismatic layer of the mussel *Mytilus californianus* and its beak region. It is beyond the scope of this book to describe the boron systematics in these specific organisms and provide detailed guidelines for sampling them. We therefore refer readers with specific interest in such organisms to the respective publications and encourage validation of boron proxy systematics within the shells and skeletons of calcifiers that have not yet been studied for boron proxies.

3.2.3 What Is the Consequence of Choosing a Temperature-Adjusted Calibration over a Calibration Based on Constant α_{B3-B4}?

As already noted, the effects of T, S and p on the equilibrium constants (i.e. pK values) for borate and carbonate dissociation in seawater are well understood and are systematically taken into account when calculating borate and carbonate equilibria (online Table A2.1 and carbonate system calculations as e.g. in CO2SYS, Pierrot et al. 2006). Here we explore the pH and pCO$_2$ uncertainty introduced by constant and temperature-dependent α_{B3-B4} using the Pleistocene *G. bulloides* dataset of Martínez-Botí et al. (2015a) as a case study. The empirical calibration for *G. bulloides* is based on a coretop sample set that encompasses a small surface ocean pH range of only 0.14 units and 1.8‰ in $\delta^{11}B_{borate}$, but a temperature range of 15 °C (online Table A2.2). Calibrations covering such a small pH range typically display a high degree of uncertainty in their $\delta^{11}B_{foram}$-sensitivity to $\delta^{11}B_{borate}$ (see York fits in online Table A2.2), and large variations in companion environmental parameters (in this case temperature) can potentially exert confounding influence on the calibration. Using the reported $\delta^{11}B_{G.\,bulloides}$ data and pH uncertainties to calculate the $\delta^{11}B_{borate}$ regression, the calibration is

$$\delta^{11}B_{borate} = \left(\delta^{11}B_{G.bulloides} + 4.74 \pm 2.51\right)/1.14 \pm 0.14 \qquad (3.13)$$

Note that these values differ slightly from the originally published calibration (Martínez-Botí et al. 2015a) because we recalculated their $\delta^{11}B_{borate}$ using the boron isotope mass balance equation of Rae (2018) and restricted the uncertainty calculations to the reported environmental pH-uncertainty (uncertainties in T, S and depth habitat were not reported in the original study).

Including the theoretical temperature effect on α_{B3-B4} (Section 2.4.8), the equation becomes

$$\delta^{11}B_{borate} = \left(\delta^{11}B_{G.bulloides} - 3.06 \pm 1.53\right)/0.74 \pm 0.09 \qquad (3.14)$$

As discussed in Section 2.4.8, the slope of the temperature-adjusted calibration is more similar to that of laboratory calibrations which are established across a wide range of pH and $\delta^{11}B_{borate}$ (see also Section 3.2.1 and Eq. 3.10 for the similarity to the multi-species $\delta^{11}B_{borate}$ sensitivity), and the respective uncertainties of both the intercept "c" and slope "m" are smaller compared to Eq. 3.13. Figure 3.4 displays the difference in pH and pCO$_2$ inferred from these two calibrations and using the downcore Mg/Ca temperature estimates of Martínez-Botí et al. (2015a) to calculate pH from reconstructed $\delta^{11}B_{borate}$. The first observation to be made from Figure 3.4 is that the amplitudes of both pH and pCO$_2$ increase when the temperature effect on α_{B3-B4} is included. This is partly because the slope of Eq. 3.14 is shallower than in Eq. 3.13, but also because α_{B3-B4} varies downcore with the Mg/Ca temperature estimate. However, as discussed earlier, temperature

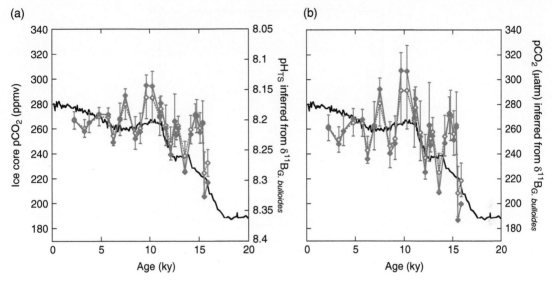

Figure 3.4 Comparison of deglacial pH changes (a) and pCO$_2$ disequilibrium (b) in the South Atlantic, reconstructed from δ^{11}B in shells of the planktic foraminifer *Globigerina bulloides* (δ^{11}B$_{foram}$ data are from Martínez-Botí et al. 2015a). Open symbols reflect the reconstruction based on constant α$_{B3-B4}$ = 1.0272 and corresponding Eq. 3.13, closed symbols the same reconstruction using calibration Eq. 3.14 and assuming α$_{B3-B4}$ varies with temperature. Also shown are atmospheric pCO$_2$ estimates from the Vostok ice core (black line, Petit et al. 1999). Note that the amplitude of the pH and pCO$_2$ estimates increases when the temperature adjustment is included, but most data fall within the uncertainty estimate of the unadjusted calibration.

variations across more recent climate events (e.g. glacial-interglacial cycles) are typically relatively small and therefore the total temperature range of <3 °C in this specific data set results in relatively small contributions to pH uncertainties. Deviations from the reconstructed pH using Eq. 3.13 are therefore no larger than $^{+0.02}_{-0.04}$ units, with larger deviations in samples with greater differences from the data set's average temperature. Translated to pCO$_2$, values can be lower by as much as −16 µatm, or higher by up to +22 µatm. Most of these temperature-adjusted estimates fall within the uncertainty of the constant α$_{B3-B4}$ reconstruction (i.e. ±19 µatm), indicating that the difference between the constant and temperature-adjusted α$_{B3-B4}$ and associated calibration fits translates to relatively minor differences in estimated pH and pCO$_2$ using this particular data set.

To test how large a temperature shift must be for the pH and pCO$_2$ differences to become significant, we perform an additional sensitivity study. In this case, δ^{11}B$_{G. bulloides}$, p, S, and alkalinity are kept constant, but pH and pCO$_2$ vary as a function of the non-adjusted or temperature-adjusted δ^{11}B$_{G. bulloides}$ calibration equation (i.e. Eqs. 3.13 and 3.14), and α$_{B3-B4}$ varies for the pH and pCO$_2$ calculations from a reference temperature of 14 °C by ±10 °C. The result of this test is displayed in Figure 3.5, where inferred pH deviates to higher values at temperature increases above the reference temperature of

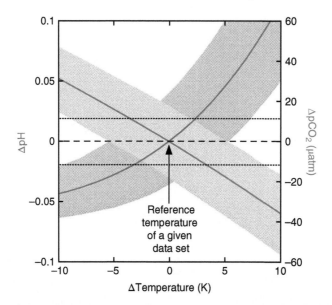

Figure 3.5 Deviation of pH (red line) and pCO$_2$ (blue line) between the temperature-adjusted (Eq. 3.14) and the unadjusted (Eq. 3.13) δ^{11}B$_{G.\ bulloides}$ calibration across a range of temperatures. The dashed black line indicates zero deviation, the dotted black lines indicate average pH and pCO$_2$ uncertainties estimated by Martínez-Botí et al. (2015a) for their deglacial reconstruction. Shaded pink and light blue areas reflect the respective pH and pCO$_2$ uncertainty from the analytical δ^{11}B$_{foram}$ uncertainty (±0.25 ‰) for the temperature-adjusted calibration at a given temperature. All data are calculated for constant δ^{11}B$_{G.\ bulloides}$, S, p, and TA; the only variable is the effect of temperature on the δ^{11}B$_{foram}$ vs. δ^{11}B$_{borate}$ calibration and α_{B3-B4} for the pH calculation from reconstructed δ^{11}B$_{borate}$.

the hypothetical data set (i.e. ΔT > 0 °C), and to lower pH values at ΔT < 0 °C. In terms of pCO$_2$, estimates diverge in the opposite direction to pH, i.e. to lower pCO$_2$ at ΔT > 0 °C and higher pCO$_2$ at ΔT < 0 °C. These temperature-adjusted inferred pH and pCO$_2$ deviations exceed the δ^{11}B$_{foram}$ based analytical uncertainty (e.g. ±0.25 ‰, Martínez-Botí et al. 2015a) on the pH and pCO$_2$ estimates when downcore temperature differs by more than ~±3 °C from the reference temperature of the data set. Consequently, future pH and pCO$_2$ reconstructions across climate events where temperature changes exceed ±3 °C should estimate the potential contribution of the temperature effect on resultant uncertainties. This could be done as a separate calculation analogous to studies that have estimated pH and pCO$_2$ based on different δ^{11}B$_{sw}$ scenarios (e.g. Bartoli et al. 2011; Pearson et al. 2009), and would be open to further evaluation and refinement as empirical studies confirm the existence and magnitude of the temperature effect on α_{B3-B4}. Although this provides an additional burden on the generators of boron isotope-based paleo-reconstructions, there currently exists no sound basis on which to dismiss the temperature effect. Furthermore, even though the temperature effect on α_{B3-B4} used herein is approximated based on theoretical studies (Section 2.4.8 and Figure. 2.21), including it as an uncertainty is important for comparing time-equivalent paleo-pCO_2 reconstructions using different proxy approaches. Clearly, the existence and magnitude of a temperature effect on α_{B3-B4} needs to be determined experimentally to further constrain this uncertainty. Fortunately, large climate shifts are relatively rare in Earth's recent history, and planktic foraminifers, which are used for open ocean pH and pCO_2 reconstructions, have only inhabited the surface ocean for the past ~100 million years. The Paleocene-Eocene Thermal Maximum (~56 million years ago) provides one example of an acidification event with a temperature shift of ~5 °C, where the temperature effect on α_{B3-B4} has already been taken into account (Penman et al. 2014). Seasonal temperature variations in near shore environments can also be quite large, such that pH estimates from $\delta^{11}B$ in corals or other coastal calcifiers should always evaluate the uncertainty introduced to their reconstruction using the theoretical temperature effect on α_{B3-B4}.

3.2.4   *How Do Past Salinity Variations Affect pH and pCO₂ Estimates from Boron Isotopes?*

Salinity is one of the most fundamental properties of seawater but, unfortunately, we still lack proxies that can reconstruct it with sufficiently high accuracy and precision. Na/Ca in foraminifera shells is loosely correlated with salinity (Allen et al. 2016; Mezger et al. 2016; Wit et al. 2013), and it has been suggested that analyzing a sufficiently large population of shells may adequately reduce sample uncertainties due to the large inter-individual variability (Mezger et al. 2016; Wit et al. 2013). However, discrepancies between field and culture calibrations are large (Allen et al. 2016; Mezger et al. 2016), and convincing reconstructions of, for example, glacial-interglacial salinity variations have yet to be generated. Similarly, the deconvolution of oxygen isotope records into temperature, sea level, and local salinity variations is fraught with uncertainty (e.g. Arbuszewski et al. 2010; Hönisch et al. 2013). However, paleosalinity can be reasonably approximated by scaling salinity to global sea level changes. Assuming that the balance of local evaporation, precipitation, and river runoff at a given core site did not change in the past, salinity can be estimated as

$$S = S_{modern}/3800\,m^{*}\left(3800\,m + RSL\right)$$

where S_{modern} is the modern salinity at the sampling location and habitat depth of the studied calcifier, 3800 m is the average depth of the modern ocean, and RSL is the relative global sea level difference between a given time in the past and the modern.

Cenozoic sea level estimates have been published by de Boer et al. (2010, 2012) and salinity can be assumed constant from the Eocene to the mid Mesozoic, when no major ice sheets existed. For the last glacial maximum (LGM) we can use $S_{modern} = 34.8$ and last glacial RSL ~ −130 m (Fairbanks 1989;

Siddall et al. 2003; Waelbroeck et al. 2002), to estimate $S_{LGM} = 36.0$, that is, 1.2‰ higher than the modern. At typical surface ocean conditions of T = 25 °C and p = 1 bar this 1.2‰ salinity difference causes pK_B^* to decrease from 8.599 (modern) to 8.592 (LGM) (online Table A2.1).

In addition to the well quantified effect of salinity on pK_B^* (Dickson 1990), Klochko et al. (2006) also observed a salinity effect on α_{B3-B4}. In artificial seawater with S = 35 they determined $\alpha_{B3-B4} = 1.0272 \pm 0.0006$, whereas pure water yielded $\alpha_{B3-B4} = 1.0308 \pm 0.0023$. The uncertainty of the pure water estimates is particularly large, but for the sake of our sensitivity studies we use both constraints to determine LGM α_{B3-B4} was 0.00012 units higher at $S_{LGM} = 36.0$ than modern α_{B3-B4} (at S = 34.8). This translates to a difference in ε_{B3-B4} of 0.12‰.

To evaluate the effect of salinity on the pH and pCO$_2$ reconstructions, we can use any published dataset (e.g. Hönisch and Hemming 2005) and calculate the pH and pCO$_2$ difference between (i) constant S, (ii) pK_B^* varying as a function of S and (iii) pK_B^* and α_{B3-B4} varying as a function of S. The result of this exercise is shown in Figure 3.6, where the difference in pH is <0.01 units and in pCO$_2$ < 10 µatm in all variants. Using constant modern S for all samples shifts glacial pH estimates towards slightly lower values, whereas adding the observed salinity effect on α_{B3-B4} shifts glacial pH to slightly higher values (Figure 3.6a). pCO$_2$ shifts in the opposite direction,

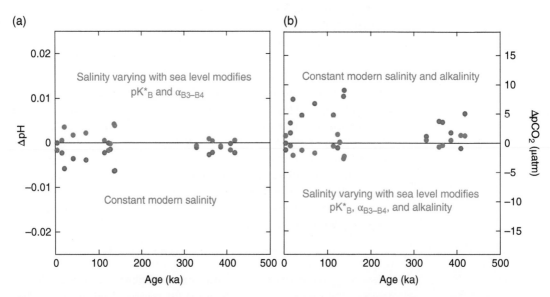

Figure 3.6 Deviation of (a) pH and (b) pCO$_2$ estimates from the same estimates using the commonly applied pK_B^* variation with salinity (solid black lines in (a) and (b)). Blue symbols apply constant modern salinity (and constant modern alkalinity in (b)) to all data, red symbols vary pK_B^* and α_{B3-B4} (and alkalinity in (b)) with salinity. The y-axes limits indicate average pH and pCO$_2$ uncertainties for the intermediate pH and pCO$_2$ estimates (i.e. pK_B^* and alkalinity vary with salinity but α_{B3-B4} is constant), which are based on the analytical uncertainty of $\delta^{11}B_{foram}$. The $\delta^{11}B_{foram}$, T, S, p and alkalinity data used for this figure are from Hönisch and Hemming (2005).

that is, higher pCO_2 corresponds to lower pH (Figure 3.6b). It should be noted that the pCO_2 estimate using constant salinity also uses constant alkalinity, whereas alkalinity scales with salinity in all other scenarios and therefore adds to the salinity effect on pCO_2. However, when compared to the pH and pCO_2 uncertainties arising from analytical uncertainty on $\delta^{11}B_{CaCO3}$ (e.g. for the Pleistocene: ±0.025 pH units and ±19 µatm, respectively), the uncertainty introduced by salinity variations is well within analytical uncertainty estimates of paleo-pH and pCO_2 reconstructions from $\delta^{11}B_{foram}$. Published uncertainties typically propagate the individual uncertainties of $\delta^{11}B_{foram}$, T, S, and alkalinity on the pCO_2 estimate into a combined pCO_2 uncertainty. This will be addressed in Section 3.2.7.

3.2.5 Does Variable Seawater Elemental Composition on Geological Time Scales Affect Boron Isotope Reconstructions?

The previous section considered the effect of seawater salinity on boron isotope reconstructions, but on geological times we also need to consider how the relative elemental composition of seawater (i.e. [Mg], [Ca], $[B_T]$) has differed from the modern ocean (Figure 3.7). As hinted at in Section 2.2.1,

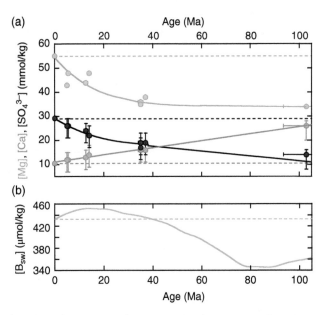

Figure 3.7 The elemental composition of seawater has changed through time and it has been argued that this affects the dissociation of carbonic and boric acid, and therefore pCO_2 estimates based on boron isotope reconstructions. [Mg], [Ca] and [SO₄³⁻] data in (a) are based on fluid inclusions in halites (Brennan et al. 2013; Horita et al. 2002; Timofeeff et al. 2006), the boron concentration in (b) is based on the model of Lemarchand et al. (2000, 2002), assuming constant river flux (see also Figure 2.24). The dashed lines indicate modern [Mg], [Ca], [SO₄³⁻] and [B] concentrations in seawater with S = 35.

strong anion-cation interactions affect acid/base chemistry in seawater by changing the activity of dissolved B(OH)$_4^-$ and carbonate ions, and thus the dissociation constants of boric acid, pK$^*_B$, and carbonic acid, pK$^*_1$ and pK$^*_2$ (Hain et al. 2015; Nir et al. 2015; Tyrrell and Zeebe 2004). While the effect of ion pairs on pK$^*_B$ is generally considered to be small (Byrne and Kester 1974; Hain et al. 2015; Hershey et al. 1986), their potential effect on boron proxy studies on geological time scales has been assessed recently.

Using seawater ion composition estimates from Ligi et al. (2013) and Holt et al. (2014), as well as Pitzer ion interaction constraints in PHREEQC (wwwbrr.cr.usgs.gov/projects/GWC_coupled/phreeqc), Nir et al. (2015) compute variations in boron speciation and isotopic composition for Neogene (Ca^{2+} and Mg^{2+}-poor) and middle Cretaceous (Ca^{2+}-rich and SO$_4^{2-}$-poor) times. Their calculations suggest pK$^*_B$ values that translate into pH uncertainties up to ±0.07 units (Figure 3.8a). However, estimates of past seawater ion composition are fraught with uncertainty, and Nir et al. (2015) applied extreme [Mg] and [Ca] values to illustrate their case (compare their values specified on Figure 3.8a to Figure 3.7a).

Hain et al. (2015) followed the same line of argument but made the case that using Pitzer constants in PHREEQC does not accurately represent modern chemical equilibrium constants in solutions with ionic strength as high as seawater. To overcome the PHREEQC limitations, Hain et al. (2015) developed an implementation (MyAMI) of the Ionic Interaction Model MIAMI (Millero and Pierrot 1998), which predicts the effect of changing seawater elemental composition on seawater acid/base chemistry. Using the same [Ca^{2+}], [Mg^{2+}] and [SO$_4^{2-}$] concentrations as Nir et al. (2015), MyAMI predicts a much smaller range in pK$^*_B$, and pH changes no larger than ±0.04 units (Figure 3.8a). Using [Ca^{2+}], [Mg^{2+}] and [SO$_4^{2-}$] concentrations that more accurately reflect the current data consensus of generally decreasing [Ca^{2+}], increasing [Mg^{2+}] and increasing [SO$_4^{2-}$] since the Cretaceous (Figure 3.7a, Brennan et al. 2013; Horita et al. 2002; Timofeeff et al. 2006), MyAMI predicts pK$^*_B$ generally decreased since the Cretaceous. Following those seawater compositional constraints, MyAMI predicts Cretaceous seawater pH may be underestimated by ~0.04 units, and Neogene seawater by ~0.02 pH units (Figure 3.8b).

These two studies (Hain et al. 2015; Nir et al. 2015) indicate somewhat inaccurate pH estimates will result if the chemical evolution of seawater is not taken into account and modern pK$^*_B$ is instead applied to all time periods. However, the difference in pK$^*_B$ is relatively small. The magnitude and direction of estimated change in pK$^*_B$ is subject to the model (PHREEQC or MyAMI) used to estimate pK$^*_B$, and the seawater elemental composition used to parameterize the model. The Python source code of MyAMI is available at https://github.com/MathisHain/MyAMI.

Estimates of past seawater ionic composition are constantly being improved and the most comprehensive compilation should be applied as new paleo datasets are generated. To date, deep-time paleo-pH reconstructions typically regard ion pairing effects on pK$^*_B$ as a less important and

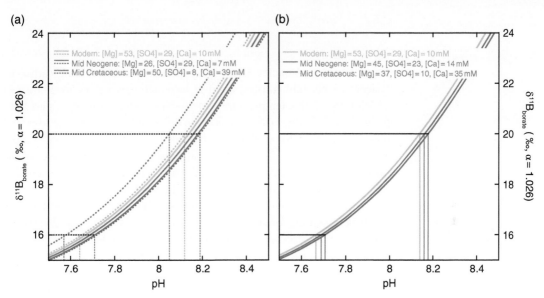

Figure 3.8 $\delta^{11}B_{borate}$ predicted from $\alpha_{B3-B4} = 1.026$ after Nir et al. (applying Pitzer constants in PHREEQC, dashed lines, 2015) and Hain et al. (applying Pitzer constants in MyAMI model, solid lines, 2015), using T = 25 °C, S = 35, p = 1 bar, $\delta^{11}B_{sw} = 39.61$ ‰ (Foster et al. 2010) and $[B_T] = 432.6\,\mu mol\,kg^{-1}$ (Lee et al. 2010), but varying pK^*_B as a function of $[Ca^{2+}]$, $[Mg^{2+}]$ and $[SO_4^{2-}]$. (a) The pK^*_B values determined by Nir et al. (2015) were originally published on the Macinnes pH scale, and we approximated these values for the total pH scale by subtracting −0.19 (see Table 1.1 for common pH scale definitions). The pK^*_B values after Nir et al. (2015) thus approximate to $pK^*_B = 8.58$ for the modern (green dashed line), $pK^*_B = 8.65$ for the mid Neogene (blue dashed line) and $pK^*_B = 8.51$ for the mid Cretaceous (red dashed

line). For comparison, using the same $[Ca^{2+}]$, $[SO_4^{2-}]$ and $[Mg^{2+}]$ as Nir et al. (2015), Hain et al. (2015) predict a much smaller range: $pK^*_B = 8.60$ for the modern (green solid line), $pK^*_B = 8.64$ for the mid Neogene (blue solid line) and $pK^*_B = 8.62$ for the mid Cretaceous (red solid line). (b) Using $[Ca^{2+}]$, $[SO_4^{2-}]$ and $[Mg^{2+}]$ as summarized by Hain et al. (2015), pK^*_B values range from $pK^*_B = 8.60$ in the modern ocean, $pK^*_B = 8.62$ in the mid Neogene and $pK^*_B = 8.64$ in the mid Cretaceous. The actual $[Ca^{2+}]$, $[SO_4^{2-}]$ and $[Mg^{2+}]$ values used for each estimate are shown in the legend. Following these predictions and assuming that $\delta^{11}B_{carbonate} = \delta^{11}B_{borate}$, the same $\delta^{11}B$ value (two examples shown as stippled (a) and solid (b) black lines) translates to small pH-deviations compared to modern conditions.

somewhat uncertain constraint (e.g. Anagnostou et al. 2016; Hain et al. 2015). In contrast, the ion pairing effects on the carbonate dissociation constants pK^*_1 and pK^*_2 may be more significant (Hain et al. 2015) and a few recent studies have included corresponding MyAMI estimates in their deep time carbonate chemistry reconstructions (e.g. Anagnostou et al. 2016; Greenop et al. 2017). However, Zeebe and Tyrrell (2018) recently argued that the implementation of the Ca concentration on pK^*_1 and pK^*_2 (but not on pK^*_B) in MyAMI is erroneous. Any further assessment of the significance of variable pK^*_1 and pK^*_2 on paleo-pH and -pCO₂ reconstructions will have to await revision of the underlying constraints, and we cannot discuss this issue further at this time.

3.2.6   How Do Past Alkalinity Variations Affect pCO$_2$ Estimates from Boron Isotopes?

As explained in Chapter 1, the six parameters of the marine carbonate system can be fully constrained if at least two parameters are known in addition to salinity, temperature and pressure. While estimating past temperature and salinity variations has been described in previous sections, and pressure can be inferred from the habitat depth of a given organism, pH inferred from $\delta^{11}B_{CaCO3}$ needs to be paired with a second parameter of the carbon system to calculate, for example, surface ocean pCO$_2$. Unfortunately, we do not have any precise proxies for dissolved inorganic carbon (DIC), alkalinity, or the individual concentrations of [HCO$_3{}^-$] and [CO$_3{}^{2-}$], except for B/Ca in benthic foraminifers, which is related to [CO$_3{}^{2-}$], and based on which we will discuss the pairing of reconstructed pH and [CO$_3{}^{2-}$] in Section 3.3.2. However, bottom water [CO$_3{}^{2-}$] cannot be paired with surface ocean pH to reconstruct pCO$_2$. Studies attempting to infer pCO$_2$ from boron isotopes therefore typically try to estimate regional surface ocean alkalinity. Alkalinity is the preferred quantity here because DIC, in addition to weathering and CaCO$_3$ deposition at the seafloor, is also modified by regional variations in the biological pump. Any inaccuracy in the DIC estimate therefore amplifies the pCO$_2$ uncertainty compared to that introduced by an inaccurate alkalinity estimate.

Because alkalinity varies with salinity in the ocean (Broecker and Peng 1982), several studies have scaled Pleistocene alkalinity to estimates of salinity, where salinity is assumed to vary with past sea level (see Section 3.2.4). The modern regional alkalinity versus salinity relationship can thereby be determined from regional hydrographic data (e.g. Foster 2008; Hönisch and Hemming 2005). Although this approach may estimate Pleistocene alkalinity reasonably well, regional alkalinity/salinity relationships could vary on time scales of tens to hundreds of thousands of years, if terrestrial weathering and the ocean calcium carbonate pump varied in strength (e.g. Clark et al. 2006; Fry et al. 2015; Kerr et al. 2017). Paleo-pCO$_2$ reconstructions using this approach therefore often place an additional uncertainty on their alkalinity estimate. For instance, Hönisch and Hemming (2005) scaled alkalinity to sea level-modulated salinity changes, and estimated last glacial alkalinity was ~70 µmol kg^{-1} (i.e. 3%) higher than interglacial (modern) alkalinity. The authors further factored the uncertainty of the modern alkalinity/salinity relationship, i.e. ±27 µmol kg^{-1} into their pCO$_2$ error calculation, which contributes ±3.1 µatm to the pCO$_2$ uncertainty estimate (Hönisch and Hemming 2005).

In comparison, deep time pCO$_2$ reconstructions often estimate alkalinity or [CO$_3{}^{2-}$] as a function of the carbonate compensation depth (CCD, e.g. Foster et al. 2012; Pearson and Palmer 2000; Pearson et al. 2009), as alkalinity sources (i.e. weathering and CaCO$_3$ dissolution) and sinks (i.e. calcification and CaCO$_3$ burial in the sediment) need to be balanced in the ocean, and this balance is regulated via CaCO$_3$ dissolution at the seafloor. The idea behind this approach is that the CCD provides an estimate of [CO$_3^{2-}$] via the saturation state $\Omega_{calcite}$ (as long as [Ca^{2+}] and [Mg^{2+}], temperature and pressure are known), and, if paired with a boron isotope estimate of deep ocean pH, deep ocean alkalinity can be calculated. By assuming a regional surface to deep ocean alkalinity gradient during the past is similar to the modern day, or attempting to estimate this gradient for the time period and region of interest, surface ocean alkalinity can be inferred. This approach is associated with a number of uncertainties, one of them being the strength of the biological pump and its effects on respiratory calcite dissolution at the seafloor (Archer and Maier-Reimer 1994; Tyrrell and Zeebe 2004), and another being the vertical alkalinity gradient (Foster et al. 2012). Recent modeling estimates suggest the CCD, alkalinity, and DIC can be largely decoupled, and that the lysocline, or better the transition between the lysocline and the CCD may place stronger constraints on alkalinity and DIC. Depending on how well CCD variations are known in the past, some studies may assume constant alkalinity across their study interval and apply a large uncertainty estimate of up to $\pm200\,\mu\mathrm{mol\,kg}^{-1}$ (e.g. Foster et al. 2012). The reader is referred to the respective studies for details of their alkalinity estimate and resulting pCO$_2$ uncertainty estimates.

Here we perform a sensitivity study using the original alkalinity estimate of each study and modify alkalinity by $\pm200\,\mu\mathrm{mol\,kg}^{-1}$ to demonstrate the effect of alkalinity uncertainties on the pCO$_2$ estimate (Figure. 3.9). Where original studies paired their pH estimates with [CO$_3^{2-}$] estimates from model studies (e.g. Bartoli et al. 2011; Pearson et al. 2009), we calculated alkalinity (using CO2SYS) from the originally published pH-[CO$_3^{2-}$] pairs (Pearson et al. 2009), respectively applied the Caribbean alkalinity/salinity relationship of Foster (2008) to all reconstructions based on ODP 999A (i.e. Bartoli et al. 2011; Seki et al. 2010). The studies of Pearson and Palmer (2000) and Foster et al. (2012) provide additional opportunities for testing the effect of extreme alkalinity variations on pCO$_2$ estimates, as both studies used CCD estimates to model relatively large variations in alkalinity for their respective records. Foster et al. (2012) estimated Miocene alkalinity as low as $1292\,\mu\mathrm{mol\,kg}^{-1}$, and Pearson and Palmer's (2000) Paleocene alkalinity estimates are as high as $4000\,\mu\mathrm{mol\,kg}^{-1}$. In comparison, alternative model estimates (using constraints from GEOCARB III, Berner and Kothavala 2001) by Tyrrell and Zeebe (2004) and Ridgwell (2005) suggest Cenozoic alkalinity may have fluctuated within the bounds of $\sim2000\pm200\,\mu\mathrm{mol\,kg}^{-1}$. The studies of Foster et al. (2012) and Pearson and Palmer (2000) are therefore represented by two scenarios, (i) their original alkalinity estimates and (ii) a constant alkalinity of $2000\,\mu\mathrm{mol\,kg}^{-1}$. Furthermore, if not already applied by the original studies, all pH estimates from $\delta^{11}\mathrm{B}_{foram}$ are based on York fits as listed in online Table A2.2. All other environmental parameters (i.e. $\delta^{11}\mathrm{B}_{sw}$, T, S, and p) were taken directly from the original studies.

Except for the Pearson and Palmer (2000) estimates, the $\pm200\,\mu\mathrm{mol\,kg}^{-1}$ uncertainty assumed here translates to pCO$_2$ differences on the order of ±20–$100\,\mu$atm (dashed lines in Figure 3.9), which is similar to or slightly

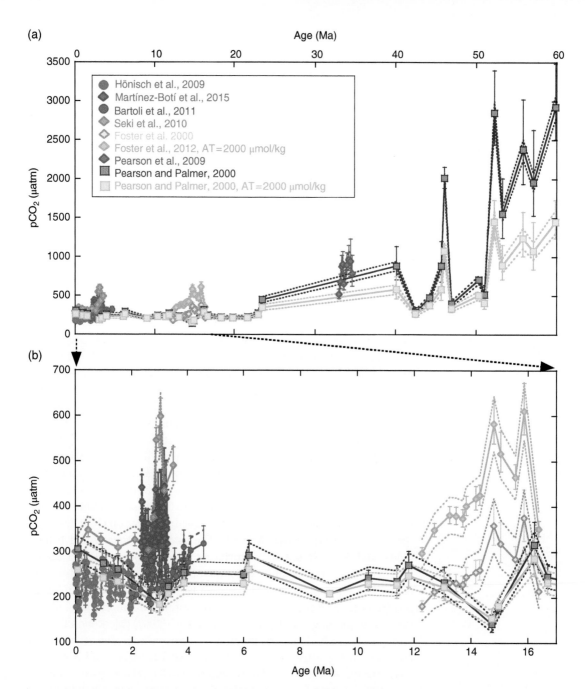

Figure 3.9 Sensitivity of surface ocean pCO$_2$ estimates from δ^{11}B$_{foram}$ to different estimates of paleoseawater alkalinity. All calculations have been done using alkalinity estimates applied by original studies (or with only minor modifications thereof, see text for details); the studies of Foster et al. (2012) and Pearson and Palmer (2000) have additionally been calculated with a constant alkalinity of 2000 μmol kg^{-1} (see legend for identification). Dashed lines reflect pCO$_2$ uncertainties only from alkalinity uncertainties of ±200 μmol kg^{-1}, whereas error bars indicate pCO$_2$ uncertainty based on the analytical uncertainty of δ^{11}B$_{foram}$ only. (b) shows the same data as (a) but magnifies the interval 0–17 Ma. During this interval, the pCO$_2$ uncertainty introduced by an alkalinity uncertainty of ±200 μmol kg^{-1} often exceeds the analytical uncertainty of δ^{11}B$_{foram}$, but such a large uncertainty is unlikely appropriate for the most recent time scales. In contrast, the original alkalinity approximations of Pearson and Palmer (2000) (gray symbols in (a)) exceed newer estimates (Ridgwell and Zeebe 2005; Tyrrell and Zeebe 2004) up to twofold and pCO$_2$ estimates using constant TA = 2000 μmol kg^{-1} (light green symbols) fall much closer to independent pCO$_2$ estimates from leaf stomata (Figure 1.4). Miocene pCO$_2$ estimates from Foster et al. (2012) were calculated using both their original alkalinity estimate based on the depth of the CCD (i.e. 1292 μmol kg^{-1}, gray diamonds) and using TA = 2000 ± 200 μmol kg^{-1} (turquoise diamonds, see panel (b) for a magnification). This comparison demonstrates that large alkalinity uncertainties (>500 μmol kg^{-1}) can affect pCO$_2$ estimates from δ^{11}B quite significantly.

greater than the pCO$_2$ uncertainty introduced by the analytical uncertainty of $\delta^{11}B_{foram}$ (error bars in Figure 3.9b). As described above, the $\pm200\,\mu mol\,kg^{-1}$ uncertainty is particularly conservative for late Pleistocene studies, where glacial/interglacial whole ocean alkalinity changes were unlikely much larger than 3% (e.g. Hain et al. 2010) and uncertainties unlikely larger than $\pm50\,\mu mol\,kg^{-1}$. In contrast, Figure 3.9a highlights how large alkalinity uncertainties exceeding $\pm500\,\mu mol\,kg^{-1}$ contribute to the extreme high Paleogene pCO$_2$ estimates suggested by Pearson and Palmer (2000) (see also Figure 1.4). Furthermore, Miocene pCO$_2$ may have been higher than suggested by Foster et al. (2012) and Greenop et al. (2014), and potentially more in line with ongoing revisions of the alkenone-pCO$_2$ proxy (Bolton and Stoll 2013; Bolton et al. 2016). This sensitivity analysis therefore highlights the need for factoring the alkalinity variations and their uncertainty into the pCO$_2$ uncertainty when deep time estimates are attempted.

In summary, developing independent proxies for alkalinity and/or DIC are priority targets for improving estimates of surface ocean pCO$_2$ from boron isotope records. Until such proxies have been developed and validated, the CaCO$_3$ saturation state at the seafloor (constrained by lysocline and CCD) is likely the best approach to approximating seawater alkalinity. However, the uncertainties of this approximation need to be weighed carefully and included in the reported uncertainty of the pCO$_2$ estimate.

3.2.7   What Is the Effect of Variations in $\delta^{11}B_{sw}$ on pH and pCO$_2$ Estimates?

Despite the long residence time of boron in seawater, the boron isotopic composition of seawater has changed over the Cenozoic (Figure 2.24). Given the analytical uncertainty of boron isotope measurements is on the order of ±0.2–0.3‰, and $\delta^{11}B_{sw}$ changes by ~0.1‰ per million years (Lemarchand et al. (2000, 2002) and Figure 2.24), we can apply modern $\delta^{11}B_{sw}$ for no more than two million years before paleo-pH and -pCO$_2$ estimates are biased systematically; reconstructions of earlier periods need to take the secular evolution of $\delta^{11}B_{sw}$ into account. As discussed in Section 2.5, independent estimates of paleo-$\delta^{11}B_{sw}$ are converging to a relatively coherent evolution pattern (Figure 2.24), but some uncertainty remains, in particular for Mesozoic and older time periods (e.g. Clarkson et al. 2015; Joachimski et al. 2005; Kasemann et al. 2005; Ohnemueller et al. 2014). Here we apply the $\delta^{11}B_{sw}$ estimates of Lemarchand et al. (2000, 2002) and Raitzsch and Hönisch (2013) to published $\delta^{11}B_{foram}$ studies reconstructing Cenozoic pH and pCO$_2$. The respective pH and pCO$_2$ estimates using $\delta^{11}B_{sw}$ of the original studies are shown for comparison (Figure 3.10).

Despite widespread concern about the effect of uncertain $\delta^{11}B_{sw}$ on the accuracy of paleo-pH and pCO$_2$ reconstructions from boron isotopes (e.g. Pagani et al. 2005; Tripati et al. 2011), the different scenarios tested here

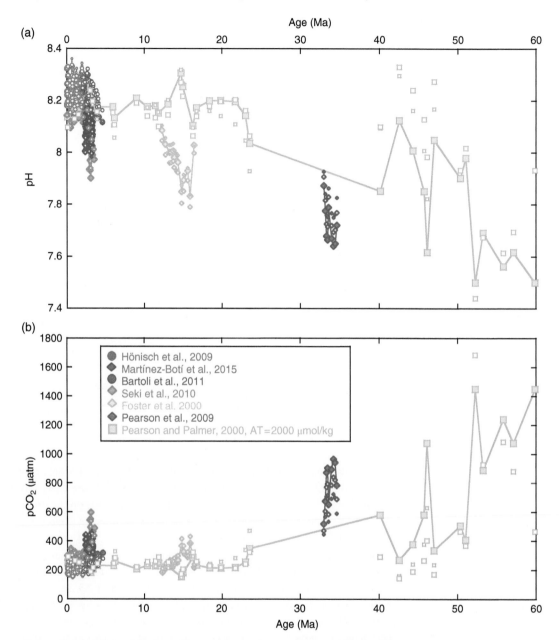

Figure 3.10 Sensitivity of surface ocean pH (a) and pCO$_2$ (b) estimates from $\delta^{11}B_{foram}$ to different estimates of $\delta^{11}B_{sw}$. Solid symbols and lines reflect pH and pCO$_2$ estimates using $\delta^{11}B_{sw}$ applied in original studies, intermediate size open symbols use $\delta^{11}B_{sw}$ after Lemarchand et al. (2000, 2002), and small open symbols use $\delta^{11}B_{sw}$ after Raitzsch and Hönisch (2013). See Figure 2.24 for estimates of $\delta^{11}B_{sw}$. The $\delta^{11}B_{sw}$ reconstruction of Raitzsch and Hönisch (2013) covers only the past 50 My and pH and pCO$_2$ estimates using that approach are consequently limited to the past 50 My. See Figure 3.9 for symbol legend and $\delta^{11}B_{foram}$ analytical uncertainties. Following Figure 3.9, pH and pCO$_2$ estimates after Pearson and Palmer (2000) use TA = 2000 µmol kg^{-1}.

yield remarkably similar results, with pH-differences smaller than −0.05 to +0.04, −0.03 to +0.11 and − 0.43 to +0.06 units for the Pleistocene-Pliocene, Miocene-Oligocene, and Eocene-Paleocene, respectively (Figure 3.10a). Translated to pCO$_2$, differences range from +40 to −25, +26 to −49 and + 984 to −235 µatm for the Pleistocene-Pliocene, Miocene-Oligocene, and Eocene-Paleocene, respectively (Figure 3.10b). While pH and pCO$_2$ differences generally increase further back in time, the largest pCO$_2$ spread is due to the original $\delta^{11}B_{sw}$ estimates of Pearson and Palmer (2000), which differ substantially from the $\delta^{11}B_{sw}$ estimates of Lemarchand et al. (2000, 2002) and Raitzsch and Hönisch (2013). Although the difference between these estimates is overall reassuringly small, the uncertainty in $\delta^{11}B_{sw}$ is often similar to the analytical uncertainty of $\delta^{11}B_{foram}$ (cf. Figure 3.9). This comparison demonstrates that refining and improving $\delta^{11}B_{sw}$ estimates through time is of primary importance for reconstructing accurate paleo-pH and pCO$_2$ from $\delta^{11}B_{foram}$. While the search for a pristine archive of $\delta^{11}B_{sw}$ continues, we will include the suggested variability in published $\delta^{11}B_{sw}$ estimates in the following error propagation for paleo-pH and pCO$_2$ uncertainty from $\delta^{11}B_{foram}$.

3.2.8   *What Is the Total Uncertainty of pH and pCO$_2$ Reconstructions from Boron Isotope Reconstructions?*

In Sections 3.2.3–3.2.6 we have explored the respective effects of temperature, salinity, alkalinity, and $\delta^{11}B_{sw}$ variations on paleo-pH and pCO$_2$ reconstructions from $\delta^{11}B_{foram}$. While temperature and salinity exert relatively small effects on pH and pCO$_2$ (Figures 3.4–3.6), propagated uncertainty in estimates of past alkalinity (Figure 3.9) and $\delta^{11}B_{sw}$ (Figure 3.10) can be as large as or larger than that of the analytical uncertainty of $\delta^{11}B_{foram}$. Paleoceanographic reconstructions often display uncertainty estimates based on the analytical uncertainty of a given proxy only, with calibration uncertainties rarely included, and error bars inconsistently shown as one or two standard deviations. Based on the data density, and how large the analytical uncertainty is relative to the calibration uncertainty, different approaches can be justified, but when it comes to such a politically charged field as estimating climate sensitivity from paleo-pCO$_2$ estimates, it is important to account for the uncertainties of individual components carefully, and to add them to the analytical uncertainty. Error bars should neither inflate nor diminish the true uncertainty of a pCO$_2$ estimate. To combine the respective uncertainties of individual environmental parameters (2σ), the individual uncertainties can be propagated in quadrature to give the combined uncertainty δpCO_2 by calculating the square root of the sum of the squared individual pCO$_2$ uncertainties:

$$\delta pCO_2 = \sqrt{\begin{array}{l}\left(\delta pCO_2\_\delta^{11}B_{\,foram}\right)^2 + \left(\delta pCO_2\_T\right)^2 \\ + \left(\delta pCO_2\_S\right)^2 + \left(\delta pCO_2\_alk\right)^2 + \left(\delta pCO_2\_\delta^{11}Bsw\right)^2\end{array}}$$

(3.15)

where $\delta pCO_2\text{--}\delta^{11}B_{foram}$ is the pCO$_2$ uncertainty stemming from the analytical uncertainty of $\delta^{11}B_{foram}$, and $\delta pCO_2\text{\_}T$, $\delta pCO_2\text{\_}S$, $\delta pCO_2\text{\_}alk$ and $\delta pCO_2\text{--}\delta^{11}B_{sw}$ are the respective pCO$_2$ uncertainties resulting from the individual uncertainties of the temperature, salinity, alkalinity and $\delta^{11}B_{sw}$ estimates. Where the calibration uncertainty is larger than the analytical uncertainty, it should also be included in the analysis.

As an alternative, more recent studies have applied a Monte Carlo approach for estimating pCO$_2$ uncertainties, generating ten thousand simulations of pH and pCO$_2$ by randomly sampling the relevant input parameters ($\delta^{11}B_{CaCO3}$, T, S, and alkalinity) within their given uncertainty bounds (2σ) (e.g. Anagnostou et al. 2016; Martínez-Botí et al. 2015a; Martínez-Botí et al. 2015b). Importantly, the $\delta^{11}B_{CaCO3}$ uncertainty estimates in Martínez-Botí et al. (2015a) include both the analytical uncertainty and the uncertainty of the calibration equation, an important addition because of the limited pH range of the *G. bulloides* coretop calibration (cf. Figures. 2.11 and 2.14). The reader is referred to the original publications for details of the Monte Carlo approach.

We have illustrated the respective magnitudes of the individual and propagated uncertainties in Figure. 3.11, which shows them both for the pH and pCO$_2$ uncertainty estimates across a range of $\delta^{11}B_{CaCO3}$. Both panels assume individual uncertainties for $\delta^{11}B_{CaCO3} = \pm 0.25$ ‰, T $= \pm 1$ °C, S $= \pm 1$ ‰, and $\delta^{11}B_{sw} = \pm 1$ ‰; the alkalinity uncertainty of $\pm 100\,\mu mol\,kg^{-1}$ applies only to the pCO$_2$ estimates and does not affect pH (Figure 3.11b). The respective pH estimates for each condition are also shown and highlight the greater uncertainty in both pH and pCO$_2$ at low pH, where the sensitivity of the boron isotope proxy decreases (cf. Figure 2.2). For the same reason, the lower (i.e. negative) uncertainty bound in the pH estimate is always larger than the upper (i.e. positive) uncertainty bound, and vice versa for the pCO$_2$ estimate. Because of this, it is important to calculate the positive and negative errors separately, and to calculate them with 2σ uncertainties on all parameters that are associated with an uncertainty. Doing the calculation with 1σ uncertainty inputs and doubling the resulting propagated error to obtain 2σ uncertainties gives uncertainty estimates that are erroneously too symmetrical, as 1σ uncertainties on the $\delta^{11}B$ estimate do not fully capture the non-linearity of the $\delta^{11}B_{borate}$ vs. pH relationship (cf. Figure 2.2). The sum of squared uncertainties (Eq. 3.15) and Monte Carlo error propagation methods thereby yield equivalent results, as shown in Figure 3.11c and d. The Matlab Monte Carlo script can be found in the online supplement.

As mentioned earlier, the largest uncertainty is due to the analytical uncertainty of $\delta^{11}B_{CaCO3}$, which captures almost the entire propagated error including the individual uncertainties of the T, S, $\delta^{11}B_{CaCO3}$ and alkalinity estimates. However, going back into the Pliocene and beyond, when the $\delta^{11}B_{sw}$ was different from the modern ocean, the uncertainty of the $\delta^{11}B_{sw}$ estimate has to be taken into account and can dominate the uncertainty estimates. For transient climate events with a duration of ~2 Myr or less, sensitivity studies using a range of possible $\delta^{11}B_{sw}$ values provide viable solutions, but records spanning several million years, and in particular those aiming to reconstruct absolute pH and pCO$_2$ values through time, must include a

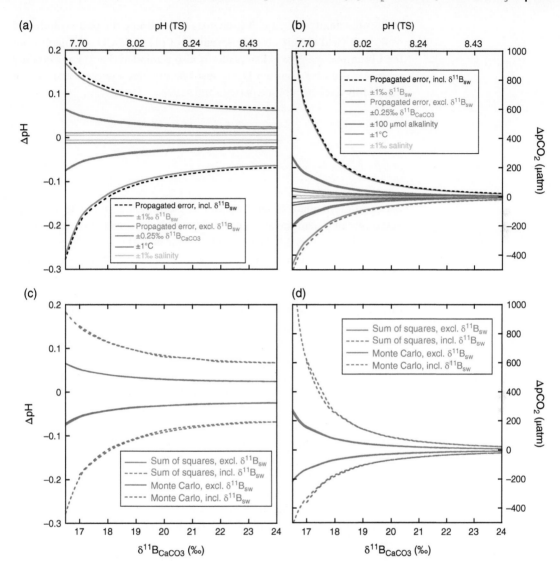

Figure 3.11 Illustration of individual T, S, $\delta^{11}B_{CaCO3}$, alkalinity and $\delta^{11}B_{sw}$ uncertainty estimates on (a, c) pH and (b, d) pCO$_2$ uncertainty estimates from $\delta^{11}B_{CaCO3}$. All calculations use the $\delta^{11}B_{CaCO3}$ vs. $\delta^{11}B_{borate}$ sensitivity from Eq. 3.7 and CO2SYS (Pierrot et al. 2006), as outlined in Chapter 1. Excluding any uncertainty in $\delta^{11}B_{sw}$, the largest contribution to the propagated pH and pCO$_2$ uncertainty (gray solid lines in (a) and (b)) is due to the analytical uncertainty on $\delta^{11}B_{CaCO3}$. Pliocene time scales and beyond also need to consider the uncertainty in $\delta^{11}B_{sw}$, which, depending on the magnitude of the uncertainty, can be much larger than all other uncertainties combined. The dashed black lines in (a) and (b) display the error propagation including the $\delta^{11}B_{sw}$ uncertainty. (c) and (d) highlight that propagated uncertainties from Eq. 3.11 (red lines) and the Monte Carlo approach (blue lines) are equivalent.

$\delta^{11}B_{sw}$ uncertainty estimate. It is obvious that refining the past evolution of $\delta^{11}B_{sw}$ will yield the greatest improvement of pH and pCO$_2$ estimates from the boron isotope proxy. The reader is also reminded that the uncertainty estimates used for Figure 3.11 are just one possible example, and do not necessarily apply to all time periods and scales.

3.3 Estimating Marine Carbonate Chemistry from B/Ca Ratios

B/Ca ratios in planktic and benthic foraminifers respond to different environmental parameters, and the respective sensitivities have been described in Section 2.6. Reconstructing marine carbonate chemistry from B/Ca generally requires similar constraints as reconstructions from $\delta^{11}B$ – the boron and carbon dissociation constants need to be constrained by auxiliary T, S, and p estimates, whereas [CO$_3^{2-}$]$_{sat}$ as the reference for benthic B/Ca (Figure 2.33) is assumed to be relatively constant through time, at least as long as the pressure and temperature conditions did not change beyond glacial/interglacial sea level and temperature variations (Yu et al. 2008). Furthermore, paleoreconstructions beyond the Pleistocene need to account for variations in the seawater boron concentration through time (Section 2.5, Figure 2.24, Lemarchand et al. 2000; Lemarchand et al. 2002; Simon et al. 2006). Although the model estimates of Lemarchand et al. (2000, 2002) suggest [B$_T$] was only ~5% lower 10 million years ago compared to the modern ocean (Figure 2.24), [Ca] was ~20% higher compared to the modern ocean (e.g. Lowenstein et al. 2014). Whereas the B/Ca ratio in planktic foraminifera is linearly correlated with the boron concentration of seawater (Allen et al. 2011; Haynes et al. 2017), [Ca] has no effect on B/Ca ratios at least in the planktic foraminifer species *O. universa* (Haynes et al. 2017). This is in contrast to inorganic precipitation experiments of Uchikawa et al. (2015), which suggest a strong positive correlation between B/Ca and [Ca]. Constraining these complexities is likely the biggest challenge in extending the B/Ca framework further back in time.

B/Ca calibrations in the epibenthic foraminifer species *Cibicidoides wuellerstorfi* and *Cibicidoides mundulus* have been confirmed in several studies and they are independent of location (Figure 2.33), such that direct comparisons can be made between spatially distant core sites (e.g. Yu et al. 2010b; Yu et al. 2014). Because *C. wuellerstorfi* and *C. mundulus*, and the precursor species *Cibicidoides praemundulus*, existed through much of the Cenozoic, reconstructions covering millions of years back in time are theoretically feasible. In addition, cross-calibration of coexisting *Oridorsalis umbonatus* and the now extinct *Nuttalides truempyi* revealed no offset between these two species, thus providing extended application potential of the B/Ca proxy to the Late Cretaceous, when *N. truempyi* inhabited the seafloor (Brown et al. 2011).

In contrast, B/Ca in planktic foraminifers varies greatly between shells recovered from seafloor sediments and shells grown in laboratory culture (Figure 2.27), and contemporaneous but spatially distant downcore records show large offsets as well (Figure 2.28). However, within individual sediment cores planktic B/Ca values tend to be quite consistent, which is promising for local reconstructions. The comparison of contemporaneous records from different regions is not advisable until the environmental controls on planktic foraminiferal B/Ca are unequivocally identified.

In comparison to boron isotopes, the B/Ca proxy requires further validation, nonetheless, more and more records are being generated around the world, with some extending into deeper Earth history. The following sections highlight the most important discoveries, and we will make some recommendations for reconstructions going beyond currently available calibrations.

3.3.1 Approximating B/Ca Proxy Sensitivity in Uncalibrated or Extinct Foraminifer Species

As described in Section 2.6, species specific calibrations are required for B/Ca in both planktic and benthic foraminifers. In comparison to $\delta^{11}B$, calibrating extinct species, or estimating the proxy sensitivity in uncalibrated species, requires not only determining the species-specific boron discrimination during shell precipitation (i.e. the y-intercept), but also the species-specific slope. An example of this challenge was published by Penman et al. (2014) who used the relative B/Ca proxy sensitivity of the modern planktic foraminifer species *O. universa*, *G. sacculifer*, and *G. ruber* to approximate the pH change at the PETM from the extinct species *Morozovella velascoensis* and *Acarinina soldadoensis*. The three modern calibration ranges approximate the range of measured paleo B/Ca-data values, and as such should give a reasonable range of pH estimates, albeit with large uncertainties.

An alternative approach might be to explore the general trend in planktic and benthic B/Ca calibrations, where the proxy sensitivity increases with greater species-specific B incorporation (Figures 2.27 and 2.33). We illustrate this pattern by plotting the average B/Ca ratio of each species versus the slope of each species-specific calibration (Figure 3.12). Because benthic foraminifers are sensitive to ΔCO_3^{2-} (Figure 2.33) and planktic foraminifers may instead be controlled by the aqueous $[B(OH)_4^-]/DIC$ ratio, both groups are shown relative to these respective controlling parameters, and both groups show a strong positive relationship with the species-specific magnitude of boron incorporation. However, symbiont-bearing and symbiont-barren planktic foraminifera seem to follow different sensitivities, and B/Ca in the benthic *Nuttalides umbonifera* (Brown et al. 2011) shows an unusually high sensitivity to ΔCO_3^{2-} (Figure 2.33). Further exploration of these relationships is advisable as more species are calibrated Figure 3.13.

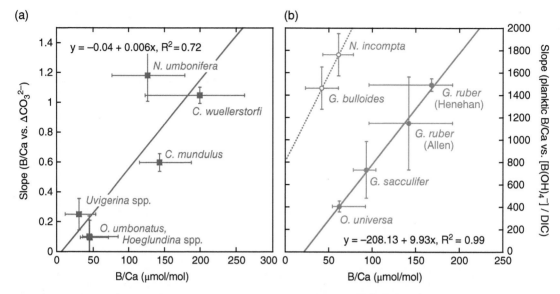

Figure 3.12 Correlation between the B/Ca incorporation in different benthic (a) and planktic (b) foraminifer species and their respective proxy sensitivities (i.e. calibration slopes). Calibration slopes in (a) are based on coretop observations by Yu and Elderfield (2007), Yu et al. (2013b), Rae et al. (2011), Brown et al. (2011) and Raitzsch et al. (2011); the species-specific B/Ca incorporation is reflected by the B/Ca median of each calibration data set and the x-"error" bars are the minimum and maximum B/Ca values in each calibration data set. Calibration slopes in (b) are based on laboratory culture experiments with symbiont-bearing planktic foraminifera (red symbols) by Allen et al. (*Orbulina universa*, 2011), Allen et al. (*Globigerinoides sacculifer* and *Globigerinoides ruber*, 2012), and Henehan et al.

(*G. ruber*, 2013), as well as coretop calibrations by Quintana Krupinski et al. (light blue symbols, 2017). Laboratory calibration slopes are plotted on the respective ambient [B(OH)$_4^-$]/DIC value, coretop calibration slopes on the median of the respective calibration data set. x-"error" bars are the minimum and maximum B/Ca values from species-specific coretop analyses shown in Figure 2.27. The y-error in both panels is given by the standard errors of each slope value. The equations describe linear regressions through these data for benthic and symbiont-bearing planktic foraminifera, the dashed line connecting the two symbiont-barren planktic species bears no mathematical significance and is just shown to describe a potentially different relationship in these species.

The strength of the multi-species regressions shown in Figure 3.12 indicates some potential for approximating calibrations of extinct or uncalibrated foraminifer species, where the species-specific incorporation may be estimated from a suite of contemporaneous B/Ca values in the same species – for instance by the median of a data population. However, in particular planktic foraminifera record highly variable B/Ca ratios in modern calibration studies (Figure 2.27), and Pleistocene downcore variations rarely agree with the expectation of higher B/Ca during glacial periods (Figure 2.28). Species-specific B incorporation in ancient foraminifera therefore needs to be carefully assessed over a wide range of environmental conditions (including T, S, pH, and p), so that the proxy sensitivity is neither underestimated (i.e. the calibration data set is biased towards low pH or

ΔCO_3^{2-} conditions) or overestimated (i.e. the calibration data set is biased towards high pH or ΔCO_3^{2-} conditions). The reader is further reminded that this approach may work better for benthic foraminifers, where B/Ca appears to be exclusively related to ΔCO_3^{2-}. In contrast, the B/Ca controls in planktic foraminifers from coretop sediments seem to vary widely (Figure 2.27), and the excellent correlation shown in Figure 3.12b may only be true for laboratory cultured specimens.

3.3.2 Constraining the Full Carbonate System from Paired $\delta^{11}B$ and B/Ca Analyses

As described in Chapter 1, all parameters of the marine carbon system can be calculated if two of its six components (i.e. pH, pCO$_2$, DIC, alkalinity, or the individual concentrations of [CO$_2$], [HCO$_3^-$] and [CO$_3^{2-}$]) are known in addition to temperature, salinity, and pressure. If temperature and salinity can be constrained from independent proxies, pressure can be assumed based on the habitat of the boron proxy carrier organism, and pH can be inferred from $\delta^{11}B$, then we only need one additional carbon parameter to fully constrain the system. The development of the B/Ca proxy and its proposed relationship to [CO$_3^{2-}$] could theoretically provide the second parameter of the carbon system, in particular if B/Ca analyses are performed on remains of the same proxy carrier organism. The first attempt at exploring this approach was done by Foster (2008), who provided evidence that B/Ca in planktic foraminifera may be related to [CO$_3^{2-}$]. Pairing reconstructed [CO$_3^{2-}$] from B/Ca with pH reconstructed from $\delta^{11}B$, Foster estimated surface ocean pCO$_2$ that agreed well with ice core values over the past 130 kyr. Although a promising result at the time, the inverse B/Ca vs. [CO$_3^{2-}$] relationship observed by Foster (2008) has not been confirmed by subsequent studies (e.g. Figure 2.27), and the use of K$_D$-driven B/Ca reconstructions has been discouraged since Allen and Hönisch (2012) showed its confounding effect on pCO$_2$ reconstructions (see also Figure 2.29 and Section 2.6.1). Based on this line of evidence, the B/Ca relationship to [CO$_3^{2-}$] in planktic foraminifera is most likely an artifact, and we do not further explore this approach for this group of organisms.

In contrast, the B/Ca proxy in benthic foraminifers is directly and linearly related to ΔCO_3^{2-}, and assuming that [CO$_3^{2-}$]$_{sat}$ at a given location does not change over time, [CO$_3^{2-}$] can be estimated from B/Ca and paired with pH estimates from $\delta^{11}B$ to estimate, for example, alkalinity, and/or DIC, for which we do not currently have any geochemical proxies. Combining the B/Ca-[CO$_3^{2-}$] and $\delta^{11}B$-pH proxy estimates from *C. wuellerstorfi*, Yu et al. (2010a) and Rae et al. (2011) estimated alkalinity and DIC, and realized that the pair of pH and [CO$_3^{2-}$] is poorly suited for constraining other parameters of the carbonate system. This is because pH and [CO$_3^{2-}$] are highly correlated and using these two quantities to constrain other carbon system

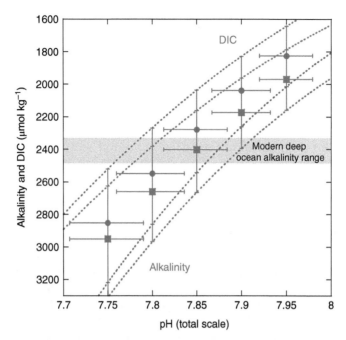

Figure 3.13 Presentation of the alkalinity (red) and DIC (blue) uncertainty resulting from pairing bottom water [CO$_3$$^{2-}$] and pH reconstructed from benthic foraminiferal B/Ca and δ^{11}B. The uncertainties are fully propagated errors based on the 2σ individual uncertainty of B/Ca (±2.6%, Yu et al. 2010a) and δ^{11}B analyses (±0.25 ‰, Rae et al. 2011). Because the δ^{11}B uncertainty translates to a larger pH uncertainty at lower pH, the DIC and alkalinity uncertainties decrease at higher pH. However, the absolute estimates and uncertainties by far exceed the natural range of modern alkalinity (shown as the gray bar) in the modern deep ocean. The modern DIC range is not shown but encompasses a similar range, albeit with ~200 μmol kg^{-1} lower absolute values compared to alkalinity. Assuming ~125 m lower glacial sea level (e.g. Fairbanks 1989; Siddall et al. 2003), glacial alkalinity and DIC values may have been only ~3% higher compared to the modern, highlighting that this paired approach to reconstruct alkalinity and DIC is unfortunately not useful.

parameters leads to amplification of uncertainties in DIC and alkalinity to a degree that renders the estimates of little use.

We illustrate this problem by calculating DIC and alkalinity across a range of deep-ocean pH at constant [CO$_3$$^{2-}$] = 90 μmol kg^{-1}, T = 2 °C, S = 35 and p = 401 bar (Figure 3.13). Because the B/Ca proxy in benthic foraminifers is described by a linear relationship, the uncertainty estimate on predicted [CO$_3$$^{2-}$] is relatively constant, but the curvature of the δ^{11}B vs. pH relationship (Figure 2.2) prescribes greater uncertainty at lower pH. Consequently, the uncertainties of the DIC and alkalinity estimates decrease at higher pH, but uncertainties ranging from 180 to 360 μmol kg^{-1} far exceed the natural range of these two parameters in the ocean. Because the applied analytical uncertainties for B/Ca (±2.6%, Yu et al. 2010a) and δ^{11}B (±0.25 ‰, Rae et al. 2011) are at the current limit of analytical precision, there is little room for improvement in these estimates, and the search for independent alkalinity and DIC proxies will need to continue. However, this does not diminish the value of the B/Ca proxy, as estimates of deep ocean [CO$_3^{2-}$] made in isolation and in conjunction with $\delta^{13}C$ have proven to be powerful indicators of carbon transfer between ocean water masses and the atmosphere (e.g. Allen et al. 2015; Yu et al. 2010b; Yu et al. 2013b; Yu et al. 2014).

3.4 Guidelines for Selecting Sediment Core Sites and Sample Sizes

As with all other proxies, the choice of sampling site depends on the research question, but when reconstructing carbonate chemistry from boron proxies, certain specific aspects are worth considering when selecting a core site.

1 Where estimating atmospheric pCO$_2$ from planktic foraminifers is the aim, an ocean location needs to be selected that is in CO$_2$ equilibrium with the atmosphere. Open ocean locations outside of upwelling areas and major current boundaries are most likely excellent candidates, and in the modern ocean their suitability can be confirmed with modern observations such as those compiled by Takahashi et al. (2002, 2009). Further back in time, and specifically at times when the position of the continents differed from today, core sites should be carefully assessed for potential upwelling, because the addition of respired CO$_2$ from greater water depths will elevate surface ocean pCO$_2$ above atmospheric levels. High productivity and cool temperatures are typically associated with upwelling areas and proxies specific to these parameters will provide valuable insight.

2 Where the strength of CO$_2$ sources to the atmosphere is of interest, upwelling regions are prime target locations, however, the high productivity of such environments may falsify B proxy records as organic matter degradation decreases bottom and pore water saturation, resulting in partial shell dissolution. This could bias B proxies towards lower-than-original values (see also Section 2.4.9) and lead to an overestimate of the CO$_2$ source strength. To avoid such bias, the preservation state of planktic foraminifers can be assessed using shell weight analyses and microstructural observations using scanning electron microscopy (e.g. Hertzberg and Schmidt 2013; Hönisch and Hemming 2004; Regenberg et al. 2013). This can also be qualitatively assessed by the crushing index, a subjective measure of the relative strength of foraminifera shells upon crushing them between glass slides during preparation for chemical cleaning.

3 Water depth is a critical issue with proxies that are prone to partial dissolution and any cores that are located near or below the regional lysocline should be carefully assessed for evidence of dissolution. Importantly, the depth of the bottom water saturation horizon alone

is not necessarily a reliable indicator for the preservation state of planktic foraminifer shells, as organic matter degradation under high productivity areas may cause partial shell dissolution, in particular in sediments accumulated at shallow water depths where the organic matter rain is high.

4 Partial shell dissolution is generally not considered to be a major problem with benthic foraminifers (e.g. Edgar et al. 2013), but degradation of organic matter can still modify porewater carbonate chemistry from ocean bottom water chemistry. Furthermore, paleoreconstructions using benthic foraminifers typically rely on few shells, and Yu et al. (2013b) provide some helpful guidelines for studying B/Ca (or B proxies in general): (i) study high sedimentation cores to minimize the effect of bioturbation; (ii) analyze multiple shells (preferably greater than eight) to reduce the effect of individual shell variability; (iii) select shells of consistent morphology (see also Rae et al. 2011) and document that morphology visually for comparison with studies at different core sites; and (iv) replicate the results to estimate overall uncertainty of the reconstruction.

References

Allen, K.A. and Hönisch, B. (2012). The planktic foraminiferal B/Ca proxy for seawater carbonate chemistry: a critical evaluation. *Earth and Planetary Science Letters 345–348*: 203–211.

Allen, K.A., Hönisch, B., Eggins, S.M., and Rosenthal, Y. (2012). Environmental controls on B/Ca in calcite tests of the tropical planktic foraminifer species *Globigerinoides ruber* and *Globigerinoides sacculifer*. *Earth and Planetary Science Letters 351–352*: 270–280.

Allen, K.A., Hönisch, B., Eggins, S.M. et al. (2011). Controls on boron incorporation in cultured tests of the planktic foraminifer *Orbulina universa*. *Earth and Planetary Science Letters 309* (3–4): 291–301.

Allen, K.A., Hönisch, B., Eggins, S.M. et al. (2016). Trace element proxies for surface ocean conditions: a synthesis of culture calibrations with planktic foraminifera. *Geochimica et Cosmochimica Acta 193*: 197–221.

Allen, K.A., Sikes, E.L., Hönisch, B. et al. (2015). Southwest Pacific deep water carbonate chemistry linked to high southern latitude climate and atmospheric CO$_2$ during the last glacial termination. *Quaternary Science Reviews 122*: 180–191.

Anagnostou, E., Huang, K.F., You, C.F. et al. (2012). Evaluation of boron isotope ratio as a pH proxy in the deep sea coral *Desmophyllum dianthus*: evidence of physiological pH adjustment. *Earth and Planetary Science Letters 349–350*: 251–260.

Anagnostou, E., John, E.H., Edgar, K.M. et al. (2016). Changing atmospheric CO$_2$ concentration was the primary driver of early Cenozoic climate. *Nature 533* (7603): 380–384.

Anand, P., Elderfield, H., and Conte, M.H. (2003). Calibration of Mg/Ca thermometry in planktonic foraminifera from a sediment trap time series. *Paleoceanography 18* (2): 15.

Arbuszewski, J., deMenocal, P., Kaplan, A., and Farmer, E.C. (2010). On the fidelity of shell-derived $\delta^{18}O_{seawater}$ estimates. *Earth and Planetary Science Letters 300* (3–4): 185–196.

Archer, D. and Maier-Reimer, E. (1994). Effect of deep-sea sedimentary calcite preservation on atmospheric CO$_2$ concentration. *Nature 367* (6460): 260–263.

Badger, M.P.S., Lear, C.H., Pancost, R.D. et al. (2013). CO$_2$ drawdown following the middle Miocene expansion of the Antarctic ice sheet. *Paleoceanography 28* (1): 42–53.

Bartoli, G., Hönisch, B., and Zeebe, R.E. (2011). Atmospheric CO$_2$ decline during the Pliocene intensification of Northern Hemisphere glaciations. *Paleoceanography 26* (4): PA4213.

Bé, A.W.H. (1980). Gametogenic calcification in a spinose planktonic foraminifer, *Globigerinoides sacculifer* (BRADY). *Marine Micropaleontology 5*: 283–310.

Berner, R.A. and Kothavala, Z. (2001). GEOCARB III: a revised model of atmospheric CO$_2$ over phanerozoic time. *American Journal of Science 301* (2): 182–204.

Blamart, D., Rollion-Bard, C., Meibom, A. et al. (2007). Correlation of boron isotopic composition with ultrastructure in the deep-sea coral *Lophelia pertusa*: implications for biomineralization and paleo-pH. *Geochemistry, Geophysics, Geosystems 8*: 11.

de Boer, B., van de Wal, R.S.W., Lourens, L.J., and Bintanja, R. (2012). Transient nature of the Earth's climate and the implications for the interpretation of benthic records. *Palaeogeography, Palaeoclimatology, Palaeoecology 335–336*: 4–11.

de Boer, B., van de Wal, R.S.W., Bintanja, R. et al. (2010). Cenozoic global ice-volume and temperature simulations with 1-D ice-sheet models forced by benthic $\delta^{18}O$ records. *Annals of Glaciology 51* (55): 23–33.

Bolton, C.T. and Stoll, H.M. (2013). Late Miocene threshold response of marine algae to carbon dioxide limitation. *Nature 500* (7464): 558–562.

Bolton, C.T., Hernandez-Sanchez, M.T., Fuertes, M.-A. et al. (2016). Decrease in coccolithophore calcification and CO$_2$ since the middle Miocene. *Nature Communications 7*: 10284.

Brennan, S.T., Lowenstein, T.K., and Cendón, D.I. (2013). The major ion composition of Cenozoic sewater: the past 36 million years from fluid inclusions in marine halite. *American Journal of Science 313*: 713–775.

Broecker, W.S. and Peng, T.-H. (1982). *Tracers in the Sea*. Lamont Doherty Earth Observatory, Columbia University: Palisades, New York.

Brown, R.E., Anderson, L.D., Thomas, E., and Zachos, J.C. (2011). A core-top calibration of B/Ca in the benthic foraminifers *Nuttallides umbonifera* and *Oridorsalis umbonatus*: a proxy for Cenozoic bottom water carbonate saturation. *Earth and Planetary Science Letters 310* (3–4): 360–368.

Byrne, R.H. and Kester, D.R. (1974). Inorganic speciation of boron in seawater. *Journal of Marine Research 32* (2): 119–127.

Clark, P.U., Archer, D., Pollard, D. et al. (2006). The middle Pleistocene transition: characteristics, mechanisms, and implications for long-term changes in atmospheric pCO$_2$. *Quaternary Science Reviews 25* (23–24): 3150–3184.

Clarkson, M.O., Kasemann, S.A., Wood, R.A. et al. (2015). Ocean acidification and the Permo-Triassic mass extinction. *Science 348*: 229–232.

Dickson, A.G. (1990). Thermodynamics of the Dissociation of Boric Acid in Synthetic Seawater from 273.15 to 318.15 K. *Deep Sea Research 37*: 755–766.

Dyez, K.A., Hönisch, B., and G.A. Schmidt (submitted) Early Pleistocene obliquity-scale pCO2 variability at ~1.5 million years ago, Paleoceanography and Paleoclimatology.

Edgar, K.M., Pälike, H., and Wilson, P.A. (2013). Testing the impact of diagenesis on the δ^{18}O and δ^{13}C of benthic foraminiferal calcite from a sediment burial depth transect in the equatorial Pacific. *Paleoceanography 28*: 468–480.

Ezat, M.M., Rasmussen, T.L., Hönisch, B. et al. (2017). Episodic release of CO2 from the high-latitude North Atlantic Ocean during the last 135 kyr. *Nature Communications 8*: 14498.

Fairbanks, R.G. (1989). A 17,000-year glacio-eustatic sea level record: influence of glacial melting rates at the Younger Dryas event and deep-ocean circulation. *Nature 342*: 637–642.

Fallon, S., McCulloch, M., and Alibert, C. (2003). Examining water temperature proxies in Porites corals from the great barrier reef: a cross-shelf comparison. *Coral Reefs 22*: 389–404.

Farmer, J.R., Hönisch, B., Robinson, L.F., and Hill, T.M. (2015). Effects of seawater-pH and biomineralization on the boron isotopic composition of deep-sea bamboo corals. *Geochimica et Cosmochimica Acta 155*: 86–106.

Fietzke, J., Ragazzola, F., Halfar, J. et al. (2015). Century-scale trends and seasonality in pH and temperature for shallow zones of the Bering Sea. *Proceedings of the National Academy of Sciences* doi: 10.1073/pnas.1419216112.

Foster, G.L. (2008). Seawater pH, pCO$_2$ and [CO$_3^{2-}$] variations in the Caribbean Sea over the last 130 kyr: a boron isotope and B/Ca study of planktic foraminifera. *Earth and Planetary Science Letters 271* (1–4): 254–266.

Foster, G.L., Pogge von Strandmann, P.A.E., and Rae, J.W.B. (2010). Boron and magnesium isotopic composition of seawater. *Geochemistry, Geophysics, Geosystems 11* (8): Q08015.

Foster, G.L., Lear, C.H., and Rae, J.W.B. (2012). The evolution of pCO$_2$, ice volume and climate during the middle Miocene. *Earth and Planetary Science Letters 341–344*: 243–254.

Fry, C.H., Tyrrell, T., Hain, M.P. et al. (2015). Analysis of global surface ocean alkalinity to determine controlling processes. *Marine Chemistry 174*: 46–57.

Greenop, R., Foster, G.L., Wilson, P.A., and Lear, C.H. (2014). Middle Miocene climate instability associated with high-amplitude CO$_2$ variability. *Paleoceanography 29* (9): doi: 10.1002/2014PA002653.

Greenop, R., Hain, M.P., Sosdian, S.M. et al. (2017). A record of Neogene seawater δ^{11}B reconstructed from paired δ^{11}B analyses on benthic and planktic foraminifera. *Climate of the Past 13* (2): 149–170.

Gutjahr, M., Ridgwell, A., Sexton, P.F. et al. (2017). Very large release of mostly volcanic carbon during the palaeocene–eocene thermal maximum. *Nature 548* (7669): 573–577.

Hain, M.P., Sigman, D.M., and Haug, G.H. (2010). Carbon dioxide effects of Antarctic stratification, North Atlantic intermediate water formation, and subantarctic nutrient drawdown during the last ice age: diagnosis and synthesis in a geochemical box model. *Global Biogeochemical Cycles 24* (4): 1–19.

Hain, M.P., Sigman, D.M., Higgins, J.A., and Haug, G.H. (2015). The effects of secular calcium and magnesium concentration changes on the thermodynamics of seawater acid/base chemistry: implications for Eocene and cretaceous ocean carbon chemistry and buffering. *Global Biogeochemical Cycles 29* (5): 517–533.

Haynes, L.L., Hönisch, B., Dyez, K.A. et al. (2017). Calibration of the B/Ca proxy in the planktic foraminifer *Orbulina universa* to Paleocene seawater conditions. *Paleoceanography 32* (6): 580–599.

Heinemann, A., Fietzke, J., Melzner, F. et al. (2012). Conditions of *Mytilus edulis* extracellular body fluids and shell composition in a pH-treatment experiment: acid-base status, trace elements and δ^{11}B. *Geochemistry, Geophysics, Geosystems 13* (1): Q01005.

Henehan, M.J., Foster, G.L., Bostock, H.C. et al. (2016). A new boron isotope-pH calibration for *Orbulina universa*, with implications for understanding and accounting for 'vital effects'. *Earth and Planetary Science Letters 454*: 282–292.

Henehan, M.J., Rae, J.W.B., Foster, G.L. et al. (2013). Calibration of the boron isotope proxy in the planktonic foraminifera *Globigerinoides ruber* for use in palaeo-CO$_2$ reconstruction. *Earth and Planetary Science Letters 364*: 111–122.

Hershey, J.P., Fernandez, M., Milne, P.J., and Millero, F.J. (1986). The ionization of boric acid in NaCl, Na-Ca-Cl and Na-Mg-Cl solutions at 25°C. *Geochimica et Cosmochimica Acta 50*: 143–148.

Hertzberg, J.E. and Schmidt, M.W. (2013). Refining Globigerinoides ruber Mg/Ca paleothermometry in the Atlantic Ocean. *Earth and Planetary Science Letters 383*: 123–133.

Holland, K., Eggins, S.M., Hönisch, B. et al. (2017). Calcification rate and shell chemistry response of the planktic foraminifer *Orbulina universa* to changes in microenvironment seawater carbonate chemistry. *Earth and Planetary Science Letters 464*: 124–134.

Holt, N.M., García-Veigas, J., Lowenstein, T.K. et al. (2014). The major-ion composition of carboniferous seawater. *Geochimica et Cosmochimica Acta 134*: 317–334.

Hönisch, B. and Hemming, N.G. (2004). Ground-truthing the boron isotope paleo-*p*H proxy in planktonic foraminifera shells: partial dissolution and shell size effects. *Paleoceanography 19*: doi: 10.1029/2004PA001026.

Hönisch, B. and Hemming, N.G. (2005). Surface Ocean pH response to variations in pCO$_2$ through two full glacial cycles. *Earth and Planetary Science Letters 236* (1–2): 305–314.

Hönisch, B., Bickert, T., and Hemming, N.G. (2008). Modern and Pleistocene boron isotope composition of the benthic foraminifer *Cibicidoides wuellerstorfi*. *Earth and Planetary Science Letters 272* (1–2): 309–318.

Hönisch, B., Hemming, N.G., Archer, D. et al. (2009). Atmospheric carbon dioxide concentration across the mid-Pleistocene transition. *Science 324* (5934): 1551–1554.

Hönisch, B., Hemming, N.G., Grottoli, A.G. et al. (2004). Assessing scleractinian corals as recorders for paleo-*p*H: empirical calibration and vital effects. *Geochimica et Cosmochimica Acta 68* (18): 3675–3685.

Hönisch, B., Bijma, J., Russell, A.D. et al. (2003). The influence of symbiont photosynthesis on the boron isotopic composition of foraminifera shells. *Marine Micropaleontology 49*: 87–96.

Hönisch, B., Allen, K.A., Lea, D.W. et al. (2013). The influence of salinity on Mg/Ca in planktic foraminifers – evidence from cultures, core-top sediments and complementary δ^{18}O. *Geochimica et Cosmochimica Acta 121*: 196–213.

Horita, J., Zimmermann, H., and Holland, H.D. (2002). Chemical evolution of seawater during the Phanerozoic: implications from the record of marine evaporites. *Geochimica et Cosmochimica Acta 66*: 3733–3756.

Joachimski, M.M., Simon, L., van Geldern, R., and Lecuyer, C. (2005). Boron isotope geochemistry of Paleozoic brachiopod calcite: implications for a secular change in the boron isotope geochemistry of seawater over the Phanerozoic. *Geochimica et Cosmochimica Acta 69* (16): 4035–4044.

Kaczmarek, K., Nehrke, G., Misra, S. et al. (2016). Investigating the effects of growth rate and temperature on the B/Ca ratio and δ^{11}B during inorganic calcite formation. *Chemical Geology 421*: 81–92.

Kakihana, H. and Kotaka, M. (1977). Equilibrium constants for boron isotope-exchange reactions. *Bulletin of the Research Laboratory for NuclearReactors 2*: 1–12.

Kakihana, H., Kotaka, M., Satoh, S. et al. (1977). Fundamental studies on the ion-exchange of boron isotopes. *Bulletin of the Chemical Society of Japan 50*: 158–163.

Kasemann, S.A., Schmidt, D.N., Bijma, J., and Foster, G.L. (2009). In situ boron isotope analysis in marine carbonates and its application for foraminifera and palaeo-pH. *Chemical Geology 260* (1–2): 138–147.

Kasemann, S.A., Hawkesworth, C.J., Prave, A.R. et al. (2005). Boron and calcium isotope composition in Neoproterozoic carbonate rocks from Namibia: evidence for extreme environmental change. *Earth and Planetary Science Letters 231* (1–2): 73–86.

Kerr, J., Rickaby, R., Yu, J. et al. (2017). The effect of ocean alkalinity and carbon transfer on deep-sea carbonate ion concentration during the past five glacial cycles. *Earth and Planetary Science Letters 471*: 42–53.

Key, R.M., Kozyr, A., Sabine, C.L. et al. (2004). A global ocean carbon climatology: results from GLODAP. *Global Biogeochemical Cycles 18*: doi:10.1029/2004GB002247.

Klochko, K., Kaufman, A.J., Yao, W. et al. (2006). Experimental measurement of boron isotope fractionation in seawater. *Earth and Planetary Science Letters 248* (1–2): 261–270.

Krief, S., Hendy, E.J., Fine, M. et al. (2010). Physiological and isotopic responses of scleractinian corals to ocean acidification. *Geochimica et Cosmochimica Acta 74* (17): 4988–5001.

Kubota, K., Yokoyama, Y., Ishikawa, T. et al. (2014). Larger CO2 source at the equatorial Pacific during the last deglaciation. *Scientific Reports 4*: 5261.

Lea, D.W., Mashiotta, T.A., and Spero, H.J. (1999). Controls on magnesium and strontium uptake in planktonic foraminifera determined by live culturing. *Geochimica et Cosmochimica Acta 63* (16): 2369–2379.

Lee, K., Kim, T.-W., Byrne, R.H. et al. (2010). The universal ratio of boron to chlorinity for the North Pacific and North Atlantic oceans. *Geochimica et Cosmochimica Acta 74* (6): 1801–1811.

Lee, K., Tong, L.T., Millero, F.J. et al. (2006). Global relationships of total alkalinity with salinity and temperature in surface waters of the world's oceans. *Geophysical Research Letters 33* (19).

Lemarchand, D., Gaillardet, J., Lewin, É., and Allègre, C.J. (2000). The influence of rivers on marine boron isotopes and implications for reconstructing past ocean pH. *Nature 408*: 951–954.

Lemarchand, D., Gaillardet, J., Lewin, E., and Allegre, C.J. (2002). Boron isotope systematics in large rivers: implications for the marine boron budget and paleo-pH reconstruction over the Cenozoic. *Chemical Geology 190* (1–4): 123–140.

Ligi, M., Bonatti, E., Cuffaro, M., and Brunelli, D. (2013). Post-Mesozoic rapid increase of seawater Mg/Ca due to enhanced mantle-seawater interaction. *Science Reporter 3*: 2752.

Linsley, B.K., Kaplan, A., Gouriou, Y. et al. (2006). Tracking the extent of the South Pacific convergence zone since the early 1600s. *Geochemistry, Geophysics, Geosystems 7* (5): n/a–n/a.

Liu, Y.W., Aciego, S.M., and Wanamaker, A.D. Jr. (2015). Environmental controls on the boron and strontium isotopic composition of aragonite shell material of cultured *Arctica islandica*. *Biogeosciences 12* (11): 3351–3368.

Lowenstein, T.K., Kendall, B., and Anbar, A.D. (2014). 8.21 – the geologic history of seawater. In: *Treatise on Geochemistry*, 2e (ed. H.D. Holland and K.K. Turekian), 569–622. Oxford: Elsevier.

Lüthi, D., Le Floch, M., Bereiter, B. et al. (2008). High-resolution carbon dioxide concentration record 650,000-800,000 years before present. *Nature 453* (7193): 379–382.

Martínez-Botí, M.A., Marino, G., Foster, G.L. et al. (2015a). Boron isotope evidence for oceanic carbon dioxide leakage during the last deglaciation. *Nature 518* (7538): 219–222.

Martínez-Botí, M.A., Foster, G.L., Chalk, T.B. et al. (2015b). Plio-Pleistocene climate sensitivity evaluated using high-resolution CO$_2$ records. *Nature 518* (7537): 49–54.

McCoy, S.J., Robinson, L., Pfister, C.A. et al. (2011). Exploring B/Ca as a pH proxy in bivalves: relationships between *Mytilus californianus* B/Ca and environmental data from the Northeast Pacific. *Biogeosciences 8* (9): 2567–2579.

McCulloch, M., Falter, J., Trotter, J., and Montagna, P. (2012). Coral resilience to ocean acidification and global warming through pH up-regulation. *Nature Climate Change 2* (8): 623–627.

Mezger, E.M., de Nooijer, L.J., Boer, W. et al. (2016). Salinity controls on Na incorporation in Red Sea planktonic foraminifera. *Paleoceanography 31* (12): 1562–1582.

Millero, F.J. (1995). Thermodynamics of the carbon dioxide system in the oceans. *Geochimica et Cosmochimica Acta 59* (4): 661–667.

Millero, F.J. and Pierrot, D. (1998). A chemical equilibrium model for natural waters. *Aquatic Geochemistry 4*: 153–199.

Nir, O., Vengosh, A., Harkness, J.S. et al. (2015). Direct measurement of the boron isotope fractionation factor: reducing the uncertainty in reconstructing ocean paleo-pH. *Earth and Planetary Science Letters 414*: 1–5.

Ohnemueller, F., Prave, A.R., Fallick, A.E., and Kasemann, S.A. (2014). Ocean acidification in the aftermath of the Marinoan glaciation. *Geology 42* (12): 1103–1106.

Pagani, M., Lemarchand, D., Spivack, A., and Gaillardet, J. (2005). A critical evaluation of the boron isotope-pH proxy: the accuracy of ancient ocean pH estimates. *Geochimica et Cosmochimica Acta 69* (4): 953–961.

Palmer, M.R. and Pearson, P.N. (2003). A 23,000-year record of surface water pH and PCO$_2$ in the western equatorial Pacific Ocean. *Science 300*: 480–482.

Palmer, M.R., Pearson, P.N., and Cobb, S.J. (1998). Reconstructing past ocean pH-depth profiles. *Science 282*: 1468–1471.

Pearson, P.N. and Palmer, M.R. (1999). Middle Eocene seawater pH and atmospheric carbon dioxide concentrations. *Science 284*: 1824–1826.

Pearson, P.N. and Palmer, M.R. (2000). Atmospheric carbon dioxide concentrations over the past 60 million years. *Nature 406*: 695–699.

Pearson, P.N., Foster, G.L., and Wade, B.S. (2009). Atmospheric carbon dioxide through the Eocene-Oligocene climate transition. *Nature 461* (7267): 1110–U204.

Pelejero, C., Calvo, E., McCulloch, M.T. et al. (2005). Preindustrial to modern Interdecadal variability in coral reef pH. *Science 309* (5744): 2204–2207.

Penman, D.E., Hönisch, B., Rasbury, E.T. et al. (2013). Boron, carbon, and oxygen isotopic composition of brachiopod shells: intra-shell variability, controls, and potential as a paleo-pH recorder. *Chemical Geology 340*: 32–39.

Penman, D.E., Hönisch, B., Zeebe, R.E. et al. (2014). Rapid and sustained surface ocean acidification during the Paleocene-Eocene thermal maximum. *Paleoceanography 29* (5): doi: 10.1002/2014PA002621.

Petit, J.R., Jouzel, J., Raynaud, D. et al. (1999). Climate and atmospheric history of the past 420,000 years from the Vostok ice core Antarctica. *Nature 399*: 429–436.

Pierrot, D., Lewis, E., and Wallace, D.W.R. (2006) MS Excel Program Developed for CO2 System Calculations, ORNL/CDIAC-105a. Carbon Dioxide Information Analysis Center, Oak Ridge National Laboratory, U.S. Department of Energy, Oak Ridge, Tennessee. doi: 10.3334/CDIAC/otg. CO2SYS_XLS_CDIAC105a.

Quintana Krupinski, N.B., Russell, A.D., Pak, D.K., and Paytan, A. (2017). Core-top calibration of B/Ca in Pacific Ocean *Neogloboquadrina incompta* and *Globigerina bulloides* as a surface water carbonate system proxy. *Earth and Planetary Science Letters 466*: 139–151.

Rae, J.W.B. (2018). Boron isotopes in foraminifera: systematics, biomineralisation, and CO2 reconstruction. In: *Boron Isotopes: The Fifth Element* (ed. H. Marschall and G. Foster), 107–143. Cham: Springer International Publishing.

Rae, J.W.B., Foster, G.L., Schmidt, D.N., and Elliott, T. (2011). Boron isotopes and B/Ca in benthic foraminifera: proxies for the deep ocean carbonate system. *Earth and Planetary Science Letters 302* (3–4): 403–413.

Rae, J.W.B., Sarnthein, M., Foster, G.L. et al. (2014). Deep water formation in the North Pacific and deglacial CO_2 rise. *Paleoceanography 29* (6): doi: 10.1002/2013PA002570.

Raitzsch, M. and Hönisch, B. (2013). Cenozoic boron isotope variations in benthic foraminifers. *Geology 41* (5): 591–594.

Raitzsch, M., Hathorne, E.C., Kuhnert, H. et al. (2011). Modern and late Pleistocene B/Ca ratios of the benthic foraminifer *Planulina wuellerstorfi* determined with laser ablation ICP-MS. *Geology 39* (11): 1039–1042.

Raitzsch, M., Bijma, J., Benthien, A. et al. (2018). Boron isotope-based seasonal paleo-pH reconstruction for the Southeast Atlantic – a multispecies approach using habitat preference of planktonic foraminifera. *Earth and Planetary Science Letters 487*: 138–150.

Regenberg, M., Schröder, J.F., Jonas, A.-S. et al. (2013). Weight loss and elimination of planktonic foraminiferal tests in a dissolution experiment. *The Journal of Foraminiferal Research 43* (4): 406–414.

Ridgwell, A. (2005). A mid Mesozoic revolution in the regulation of ocean chemistry. *Marine Geology 217* (3–4): 339–357.

Ridgwell, A. and Zeebe, R.E. (2005). The role of the global carbonate cycle in the regulation and evolution of the earth system. *Earth and Planetary Science Letters 234* (3–4): 299–315.

Rollion-Bard, C. and Erez, J. (2010). Intra-shell boron isotope ratios in the symbiont-bearing benthic foraminiferan *Amphistegina lobifera*: implications for $\delta^{11}B$ vital effects and paleo-pH reconstructions. *Geochimica et Cosmochimica Acta 74* (5): 1530–1536.

Rollion-Bard, C., Chaussidon, M., and France-Lanord, C. (2003). pH control on oxygen isotopic composition of symbiotic corals. *Earth and Planetary Science Letters 215* (1–2): 275–288.

Sanyal, A., Hemming, N.G., Hanson, G.N., and Broecker, W.S. (1995). Evidence for a higher pH in the glacial ocean from boron isotopes in foraminifera. *Nature 373*: 234–236.

Sanyal, A., Nugent, M., Reeder, R.J., and Bijma, J. (2000). Seawater pH control on the boron isotopic composition of calcite: evidence from inorganic calcite precipitation experiments. *Geochimica et Cosmochimica Acta 64* (9): 1551–1555.

Sanyal, A., Bijma, J., Spero, H.J., and Lea, D.W. (2001). Empirical relationship between *p*H and the boron isotopic composition of *Globigerinoides sacculifer*: implications for the boron isotope paleo-*p*H proxy. *Paleoceanography 16* (5): 515–519.

Sanyal, A., Hemming, N.G., Broecker, W.S. et al. (1996). Oceanic *p*H control on the boron isotopic composition of foraminifera: evidence from culture experiments. *Paleoceanography 11* (5): 513–517.

Schlitzer, R. (2012) Ocean Data View.

Schmidt, M.W., Spero, H.J., and Lea, D.W. (2004). Links between salinity variation in the Caribbean and North Atlantic thermohaline circulation. *Nature 428*: 160–163.

Seki, O., Foster, G.L., Schmidt, D.N. et al. (2010). Alkenone and boron-based Pliocene pCO$_2$ records. *Earth and Planetary Science Letters 292* (1–2): 201–211.

Siddall, M., Rohling, E.J., Almogi-Labin, A. et al. (2003). Sea-level fluctuations during the last glacial cycle. *Nature 423* (6942): 853–858.

Siegenthaler, U., Stocker, T.F., Monnin, E. et al. (2005). Stable carbon cycle-climate relationship during the late Pleistocene. *Science 310* (5752): 1313–1317.

Simon, L., Lécuyer, C., Maréchal, C., and Coltice, N. (2006). Modelling the geochemical cycle of boron: implications for the long-term δ^{11}B evolution of seawater and oceanic crust. *Chemical Geology 225*: 61–76.

Takahashi, T., Sutherland, S.C., Sweeney, C. et al. (2002). Global Sea-air CO$_2$ flux based on climatological surface ocean pCO$_2$, and seasonal biological and temperature effects. *Deep Sea Research Part II: Topical Studies in Oceanography 49* (9–10): 1601–1622.

Takahashi, T., Sutherland, S.C., Wanninkhof, R. et al. (2009). Climatological mean and decadal change in surface ocean pCO$_2$, and net sea-air CO$_2$ flux over the global oceans. *Deep Sea Research Part II: Topical Studies in Oceanography 56* (8–10): 554–577.

Timofeeff, M.N., Lowenstein, T.K., da Silva, M.A., and Harris, N.B. (2006). Secular variation in the major-ion chemistry of seawater: evidence from fluid inclusions in cretaceous halites. *Geochimica et Cosmochimica Acta 70*: 1977–1994.

Tripati, A.K., Roberts, C.D., and Eagle, R.A. (2009). Coupling of CO$_2$ and ice sheet stability over major climate transitions of the last 20 million years. *Science* 1178296. doi: 10.1126/science.1178296.

Tripati, A.K., Roberts, C.D., Eagle, R.A., and Li, G. (2011). A 20 million year record of planktic foraminiferal B/Ca ratios: systematics and uncertainties in pCO$_2$ reconstructions. *Geochimica et Cosmochimica Acta 75* (10): 2582–2610.

Trotter, J., Montagna, P., McCulloch, M. et al. (2011). Quantifying the pH 'vital effect' in the temperate zooxanthellate coral *Cladocora caespitosa*: validation of the boron seawater pH proxy. *Earth and Planetary Science Letters 303* (3–4): 163–173.

Tyrrell, T. and Zeebe, R.E. (2004). History of carbonate ion concentration over the last 100 million years. *Geochimica et Cosmochimica Acta 68* (17): 3521–3530.

Uchikawa, J., Penman, D.E., Zachos, J.C., and Zeebe, R.E. (2015). Experimental evidence for kinetic effects on B/Ca in synthetic calcite: implications for potential B(OH)$_4^-$ and B(OH)$_3$ incorporation. *Geochimica et Cosmochimica Acta 150*: 171–191.

Waelbroeck, C., Labeyrie, L., Michel, E. et al. (2002). Sea-level and deep water temperature changes derived from benthic foraminifera isotopic records. *Quaternary Science Reviews 21* (1–3): 295–305.

Wall, M., Ragazzola, F., Foster, L.C. et al. (2015). pH up-regulation as a potential mechanism for the cold-water coral *Lophelia pertusa* to sustain growth in aragonite undersaturated conditions. *Biogeosciences 12*: 6869–6880.

Wei, G., Sun, M., Li, X., and Nie, B. (2000). Mg/Ca, Sr/Ca and U/Ca ratios of a *Porites* coral from Sanya Bay, Hainan Island, South China Sea and their relationships to sea surface temperature. *Palaeogeography, Palaeoclimatology, Palaeoecology 162*: 59–74.

Wei, G., McCulloch, M.T., Mortimer, G. et al. (2009). Evidence for ocean acidification in the great barrier reef of Australia. *Geochimica et Cosmochimica Acta 73*: 2332–2346.

Wit, J.C., de Nooijer, L.J., Wolthers, M., and Reichart, G.J. (2013). A novel salinity proxy based on Na incorporation into foraminiferal calcite. *Biogeosciences 10* (10): 6375–6387.

York, D., Evensen, N., Martinez, M., and Delgado, J. (2004). Unified equations for the slope, intercept, and standard errors of the best straight line. *The American Journal of Physiology 72* (3): 367.

Yu, J. and Elderfield, H. (2007). Benthic foraminiferal B/Ca ratios reflect deep water carbonate saturation state. *Earth and Planetary Science Letters 258* (1–2): 73–86.

Yu, J., Elderfield, H., and Piotrowski, A.M. (2008). Seawater carbonate ion-$\delta^{13}C$ systematics and application to glacial-interglacial North Atlantic Ocean circulation. *Earth and Planetary Science Letters 271* (1–4): 209–220.

Yu, J., Thornalley, D.J.R., Rae, J.W.B., and McCave, N.I. (2013a). Calibration and application of B/Ca, Cd/Ca, and $\delta^{11}B$ in *Neogloboquadrina pachyderma* (sinistral) to constrain CO₂ uptake in the subpolar North Atlantic during the last deglaciation. *Paleoceanography* doi: 10.1002/palo.20024.

Yu, J., Foster, G.L., Elderfield, H. et al. (2010a). An evaluation of benthic foraminiferal B/Ca and $\delta^{11}B$ for deep ocean carbonate ion and pH reconstructions. *Earth and Planetary Science Letters 293* (1–2): 114–120.

Yu, J., Broecker, W.S., Elderfield, H. et al. (2010b). Loss of carbon from the Deep Sea since the last glacial maximum. *Science 330* (6007): 1084–1087.

Yu, J., Anderson, R.F., Jin, Z. et al. (2013b). Responses of the deep ocean carbonate system to carbon reorganization during the last glacial–interglacial cycle. *Quaternary Science Reviews 76*: 39–52.

Yu, J., Anderson, R.F., Jin, Z. et al. (2014). Deep South Atlantic carbonate chemistry and increased interocean deep water exchange during last deglaciation. *Quaternary Science Reviews 90*: 80–89.

Zeebe, R.E. and Tyrrell, T. (2018) Comment on "The effects of secular calcium and magnesium concentration changes on the thermodynamics of seawater acid/base chemistry: Implications for Eocene and Cretaceous ocean carbon chemistry and buffering" by Hain et al. (2015). Global Biogeochemical Cycles, 32, 895-897.

Zhang, S., Henehan, M.J., Hull, P.M. et al. (2017). Investigating controls on boron isotope ratios in shallow marine carbonates. *Earth and Planetary Science Letters 458*: 380–393.

4 Boron Concentration and Isotope Ratio Analysis

Abstract

This chapter reviews methods for analysis of boron concentrations and isotope ratios in marine carbonates. Emphasis is placed on describing sample preparation procedures, in particular the diverse approaches currently used for matrix separation and boron purification, and the sampling cleaning and handling requirements that are needed to avoid sample contamination from the environment and during laboratory processing. Detailed descriptions of instrumental methods are provided that represent current state-of-the-art, in particular inductively coupled plasma mass spectrometry (ICP-MS) methods used for elemental boron analysis and both ICPMS and thermal ionization mass spectrometry (TIMS) methods for boron isotope ratios analysis. An overview of the capabilities of established and emerging microanalysis techniques to map the distribution of boron concentrations and isotopic ratios at nanometer and micrometer-scale resolution is also provided, as well as a brief outline of some other more specialized techniques that are providing insights into the mode of incorporation of boron and the distributions of boron and boron isotope ratios within marine carbonates.

Keywords: Boron; boron isotopes; boron analysis; boron isotope analysis; instrumental techniques

4.1 Introduction

Our capacity to understand and constrain changes in seawater pH, calcium carbonate saturation state, and pCO_2 levels in the atmosphere relies on being able to measure boron elemental concentrations and boron isotopic compositions of various marine carbonate archives with high levels of accuracy and

Boron Proxies in Paleoceanography and Paleoclimatology, First Edition. Bärbel Hönisch, Stephen M. Eggins, Laura L. Haynes, Katherine A. Allen, Katherine D. Holland, and Katja Lorbacher.
© 2019 John Wiley & Sons Ltd. Published 2019 by John Wiley & Sons Ltd.
Companion website: www.wiley.com/go/Hönisch/Boron_Paleoceanography

precision (Section 3.2). It also demands understanding how boron and boron isotopes are incorporated into and recorded at varying spatial and temporal scales within the different marine carbonate proxy archives.

Our ability to make precise and accurate boron concentration and boron isotope ratio analyses has benefitted from the recent development of more sensitive and robust analytical instruments, as well as improved procedures and methods that facilitate more reliable and efficient analysis. Nonetheless, despite these advances, the analysis of boron concentrations and isotope ratios in marine carbonates remains challenging. This is manifest in the relatively small number of laboratories that consistently report high quality results, the often different and at times contested analytical methods employed by different laboratories, and the ongoing effort to improve existing or develop new methods of analysis (Foster et al. 2013).

An important question that confronts any new and even existing boron analysts is what analytical technique(s) and method(s) should I invest my time and effort in developing and implementing? This equates to the fundamental question in analytical chemistry of what technique is most "fit for purpose." In other words, what technique is best able to measure the required number of samples with the necessary accuracy and precision to evaluate my scientific question(s). More than one technique may be "fit for purpose" in which case the most time and cost efficient would logically be chosen, but such choices are also subject to considerations of ease of use, relevant in-house expertise, and existing instrumentation, and laboratory facilities. If in a position to acquire new instruments and laboratory facilities, other considerations might include what other research questions could be addressed using the same instruments and facilities.

The choice of analytical technique for measuring boron concentrations and isotope ratios in marine carbonates can be restricted by the amount of boron available for analysis. This limitation is encountered with the analysis of boron and boron isotopes in foraminifer shells, for which low boron concentrations and availability of only milligram or often lesser amounts of shell calcite, require the use of high sensitivity mass spectrometry techniques, such as SF-ICP-MS, negative ion – thermal ionization mass spectrometry (N-TIMS), and multiple collector – inductively coupled plasma mass spectrometry (MC-ICP-MS) (cf. Table 4.1 for a Glossary of analysis techniques and acronyms).

By way of example, to reconstruct past changes in deep sea carbonate ion concentration ($[CO_3^{2-}]$) and pH across the last deglaciation a researcher might choose to measure the B/Ca ratio and $\delta^{11}B$ shell compositions of the benthic foraminifera *Cibicidoides wuellerstorfi*. The accuracy and precision of B/Ca analyses needed to obtain a target uncertainty for $[CO_3^{2-}]$ can be obtained by differentiating the empirical B/Ca-$[CO_3^{2-}]$ relationship for *C. wuellerstorfi* (Yu and Elderfield 2007; see Eq. 2.23) to give the sensitivity of reconstructed $[CO_3^{2-}]$ to measured B/Ca as:

$$d\left[CO_3^{2-}\right]/d(B/Ca) = 1.14 \pm 0.048 \quad (\mu mol/kg)/(\mu mol/mol) \quad (4.1)$$

Table 4.1 Glossary of analysis techniques and acronyms.

APT	Atom Probe Tomography
ICP-OES	Inductively Coupled Plasma Optical Emission Spectroscopy
LA-ICP-MS	Laser Ablation – Inductively Coupled Plasma Mass Spectrometry
MC-ICP-MS	Multiple Collector – Inductively Coupled Plasma Mass Spectrometry
N-TIMS	Negative ion – Thermal Ionization Mass Spectrometry
NMR	Nuclear Magnetic Resonance
NanoSIMS	Nanoscale Secondary Ion Mass Spectrometry
P-TIMS	Positive ion – Thermal Ionization Mass Spectrometry
Q-ICP-MS	Quadrupole – Inductively Coupled plasma Mass Spectrometry
SF-ICP-MS	Sector Field – Inductively Coupled Plasma Mass Spectrometry
SIMS	Secondary Ion Mass Spectrometry
STXM	Scanning Transmission X-ray Microscopy
TOF-SIMS	Time of Flight – Secondary Ion Mass Spectrometry
TE-N-TIMS	Total Evaporation – Negative ion – Thermal Ionization Mass Spectrometry

Given the $[CO_3^{2-}]$ of the deep North Atlantic is likely to have increased by $\sim35\,\mu mol\,kg^{-1}$ over the last deglaciation (Yu et al. 2010) a target resolution better than 10% or $\pm3\,\mu mol\,kg^{-1}$ (2σ) might be set for reconstructed $[CO_3^{2-}]$ values which, from Eq. 4.1, would require measuring B/Ca to better than $\pm3.5\,\mu mol\,mol^{-1}$ (2σ).

In the same way, the empirical relationship between seawater pH and $\delta^{11}B$ shell compositions of *C. wuellerstorfi* could be differentiated to give the sensitivity of reconstructed seawater pH to measured differences in $\delta^{11}B_{foram}$ ($\Delta pH/\Delta\delta^{11}B_{foram}$). However, in the absence of any documented calibration for $\delta^{11}B_{C.\,wuellerstorfi}$ over a wide seawater pH range it could be assumed that $\delta^{11}B_{C.\,wuellerstorfi}$, like other marine carbonates, will be related to seawater pH via an empirical isotopic fractionation factor ($\alpha_{(B3-B4)}$; Hönisch et al. 2008; Rae et al. 2011; Chapter 2.1.2):

$$pH_{sw} = pK_B - \log_{10}(-\left(\delta^{11}B_{sw} - \delta^{11}B_{C.wuellerstorfi}\right)/ \\ \left(\delta^{11}B_{sw} - \alpha_{B3-B4} {}^*\delta^{11}B_{C.wuellerstorfi} - 1000^*\left(\alpha_{B3-B4} - 1\right)\right) \tag{4.2}$$

In this case the resolvable seawater pH difference (ΔpH) increases as the seawater pH varies away from the pK_B value (see Figure 3.11a). The relative $\delta^{11}B$-pH sensitivity is also subject to the choice of $\alpha_{(B3-B4)}$ value, which is uncertain and might also be temperature dependent (Zeebe 2005; Chapter 2.3.8). Assuming for simplicity a constant $\alpha_{B3-B4} = 1.0272$ (Section 2.2.2) and that $\delta^{11}B_{C.\,wuellerstorfi}$ is unfractionated from seawater borate, to achieve a target resolution sufficient to resolve detail within a total deglacial deep ocean pH change of 0.1 pH units (for deep ocean pH around 7.8) would require measuring $\delta^{11}B$ to a precision better than ±0.6 ‰ (2σ). This is calculated by dividing 0.1 pH units by the $\Delta pH/\Delta\delta^{11}B$ sensitivity value obtained at pH 7.8 from Figure 4.1.

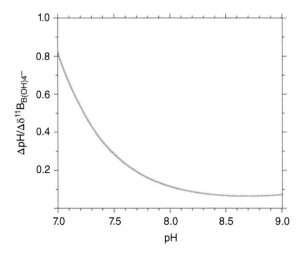

Figure 4.1 Variation in $\Delta pH_{sw}/\Delta\delta^{11}B_{B(OH)4^-}$ as a function of seawater pH derived from Eq. 4.2 assuming $\delta^{11}B_{C. wuellerstorfi}$ is equivalent to $\delta^{11}B$ of seawater borate (i.e. $\delta^{11}B_{B(OH)4^-}$). The analytical precision required to reconstruct pH_{sw} to a specified target pH uncertainty varies as a function of seawater pH and can be estimated by dividing the target pH uncertainty by the value of $\Delta pH_{sw}/\Delta\delta^{11}B_{B(OH)4^-}$ at the relevant seawater pH.

The above example ignores sampling and analytical errors, as well as uncertainties deriving from other factors that may influence the B/Ca and $\delta^{11}B$ seawater carbonate ion and pH proxy sensitivities (e.g. temperature, salinity; see Sections 3.2 and 3.3). Nonetheless, it provides a useful starting point for evaluating analytical techniques that could meet the required levels of accuracy and precision for one's research purpose. In the case of boron isotope analysis, the precision achievable using different analytical techniques varies widely and is a trade-off between their ion yield efficiencies and analyzable amounts of boron (sample size), and is fundamentally limited by counting statistics. The N-TIMS, MC-ICP-MS, and P-TIMS techniques are all capable of achieving precision better than 0.5 ‰ but have different optimum sample size requirements (Figure 4.2). The precision obtainable with any of these techniques is also a function of other limitations, including sampling errors due to sample heterogeneity, the reproducibility of sample preparation procedures, and the increasing contributions from sample contamination and blank subtraction errors when measuring small amounts of boron.

Returning to our example using *C. wuellerstorfi* to reconstruct deep ocean pH, the available sample is likely to be limited to <1 mg of calcium carbonate (ca. 10 shells or fewer) and therefore to analyzing <10 ng of boron (based on reported boron concentrations in epifaunal benthic foraminifers see Table 4.2). From Figure 4.2 it can be seen this would require analysis by N-TIMS or MC-ICP-MS to achieve the target $\delta^{11}B$ precision. However, not all published methods using these techniques have demonstrated the ability to achieve this level of precision with such small sample sizes. Accordingly, further careful evaluation of the analytical procedures used and basis for any analytical precision estimates reported in different studies would be required. Critical questions include whether the precision

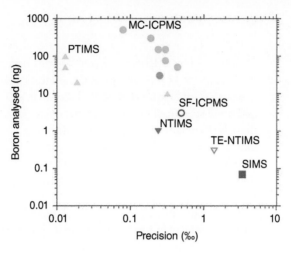

Figure 4.2 Comparison of reported analytical precision (±2σ) versus amount of boron analyzed for various boron isotope analysis methods, modified after Foster et al. (2006). Measurements by PTIMS are from He et al. (2013), by MC-ICP-MS from McCulloch et al. (2014; light green circles) and Foster (2008; dark green circle), by SF-ICPMS from Misra et al. (2014b), by NTIMS from Hönisch and Hemming (2005), by TE-NTIMS from Ni et al. (2010), and by SIMS from Kasemann et al. (2009).

Table 4.2 Boron concentrations reported for marine carbonates.

Proxy Archive	B (ppm)	B/Ca (μmol mol)⁻¹	Reference
Benthic foraminifera			
epifaunal	15–30	131–500	Vengosh et al. (1991), Yu and Elderfield (2007), Rae et al. (2011)
infaunal	1–9	12–86	Yu and Elderfield (2007), Rae et al. (2011)
Planktic foraminifera	4–22	37–207	Vengosh et al. (1991), Ni et al. (2007), Foster (2008), Allen and Hönisch (2012), Misra et al. (2014a), Henehan et al. (2015, 2016)
Scleractinian corals	39–123	361–1140	Vengosh et al. (1991), Hemming and Hanson (1992), Gaillardet and Allègre (1995), Sinclair et al. (1998), Hönisch et al. (2004), Montagna et al. (2007), Wei et al. (2009), Krief et al. (2010), Trotter et al. (2011), Anagnostou et al. (2012)
Deep Sea Corals	8–114	75–1056	Blamart et al. (2007), Rollion-Bard et al. (2011a), Farmer et al. (2015), Stewart et al. (2016)
Coccolithophorids	1–6	5–56	Stoll et al. (2012)
Brachiopods	16–104	148–963	Lécuyer et al. (2002), Joachimski et al. (2005), Penman et al. (2013)
Mollusks	1–60	28–120	Vengosh et al. (1991), McCoy et al. (2011), Heinemann et al. (2012)
Echinoderms	37–44	340–404	Hemming and Hanson (1992)
Coralline Algae	51–77	471–716	Hemming and Hanson (1992), Fietzke et al. (2015), Zhang et al. (2017)

estimates are based simply on analyses of pure boron solutions, or incorporate full method reproducibility using real calcium carbonate samples. This raises the importance of assessing and reporting analytical reproducibility and the need for consistent Quality Assessment and Quality Control measures when using any analytical technique.

4.2 Inter-Laboratory Comparison Studies

A number of inter-laboratory comparisons have been undertaken to assess the quality of boron isotope ratio analyses being reported in geological materials (Gonfiantini et al. 2003; Tonarini et al. 2003; Aggarwal et al. 2009a) and more specifically for marine carbonates (Foster et al. 2013). The first of these by Gonfiantini et al. (2003) distributed a range of materials, of which only a seawater and a limestone sample are relevant to analysis of marine carbonates. This study reported disturbingly disparate $\delta^{11}B$ compositions for these samples by the contributing laboratories, ranging over 4 ‰ for the measured $\delta^{11}B$ composition of seawater and over 10 ‰ for the limestone. A subsequent study by Aggarwal et al. (2009a) addressed the potential confounding issue of boron heterogeneity in powdered solid samples by distributing only boron containing synthetic solutions. This subsequent study included examples of now preferred techniques, specifically TIMS and MC-ICP-MS, but still found a wide range in reported $\delta^{11}B$ values (>5 ‰) by participating laboratories using either of these preferred instrumental techniques. Importantly, these inter-laboratory comparison studies revealed most laboratories were grossly underestimating the uncertainty of their $\delta^{11}B$ analyses. They also highlighted the inability of many geochemical laboratories with $\delta^{11}B$ analysis capability to make routine $\delta^{11}B$ measurements that could be applied to reconstruct past seawater pH values to an accuracy better than ~0.5 pH units.

A more recent inter-laboratory comparison reported by Foster et al. (2013) restricted its distribution of seawater and carbonate standards, all as solutions, to a small group of laboratories specializing in the analysis of $\delta^{11}B$ in marine carbonates. The participating laboratories are notable for employing different analytical techniques, two using MC-ICP-MS and two using N-TIMS. Excellent agreement between these laboratories was found for the analysis of seawater ($\delta^{11}B = 39.65 \pm 0.41$ ‰, 2σ) and no systematic bias between laboratories was found for the analysis of pure boron solutions. However, poorer agreement was reported for $\delta^{11}B$ analysis of calcium carbonate samples, for which the pooled variance of ±1.46 ‰ (2σ) significantly exceeds the uncertainty estimates provided by the individual laboratories. This indicated systematic bias between the laboratories, although the nature of the bias and which laboratories or techniques were biased could not be established (Foster et al. 2013). Nonetheless the four participating laboratories were generally able to reproduce relative differences in $\delta^{11}B$ composition between the analyzed carbonate samples, although planktic foraminifera samples, which have the lowest B concentration, were not reproduced as well by laboratories that had optimized their method to measuring larger B concentrations (e.g., in corals). This finding was of great practical significance as it demonstrated the different laboratories and techniques used were capable of making consistent seawater pH reconstructions subject to applied $\delta^{11}B_{cc}$ – pH calibrations being determined in (or corrected for bias between) the respective laboratories. Farmer et al. (2016) have since made MC-ICP-MS and N-TIMS analyses of the same samples in the same laboratory, and shown the biases between laboratories reported in Foster et al. (2013) are likely to originate from the different mass spectrometry techniques (Section 4.7) rather than procedural methods (Sections 4.5 and 4.6). Nonetheless, it is important to recognize technique-specific differences when translating $\delta^{11}B$ data to $\delta^{11}B_{borate}$ and pH, as the intercept of $\delta^{11}B_{CaCO3}$ vs. $\delta^{11}B_{borate}$ calibrations varies between techniques (Section 2.4.1, Figure 2.12). Using an inappropriate calibration, for example, an N-TIMS calibration with unknown sample data measured by MC-ICP-MS or vice versa, could lead to erroneous $\delta^{11}B_{borate}$ and seawater pH estimates. As reported in data comparisons from different TIMS (Hönisch et al. 2003; Hönisch and Hemming 2005; Hönisch et al. 2009; Bartoli et al. 2011; Foster et al. 2013) and different MC-ICP-MS instruments (Foster et al. 2013), care must also be taken to validate whether a given calibration is applicable to all instruments of the same type, or whether an additional intercept correction needs to be applied to yield unbiased results.

No inter-laboratory comparison study has yet focused on the analysis of boron concentrations or B/Ca molar ratios in calcium carbonates, although an inter-laboratory comparison involving 21 laboratories (Hathorne et al. 2013) has reported the analysis of trace metal concentrations (including B/Ca) for reference materials JCp-1 (a *Porites* coral) and JCt-1 (a *Tridacna gigas* giant clam). This comparison study incorporated a wide range of analytical techniques (including mass spectrometry) and focused on the analysis of Sr/Ca, Mg/Ca and other element/Ca ratios used for seawater temperature reconstruction, with only a small number of participating laboratories reporting B/Ca results. These include a pooled mean B/Ca molar ratio value of $460 \pm 45\,\mu mol\,mol^{-1}$ for JCp-1 ($2\,\sigma, n = 6$) and $191 \pm 19\,\mu mol\,mol^{-1}$ for JCt-1 ($2\,\sigma, n = 3$), which suggest an overall ability to measure B/Ca compositions in marine carbonate to only around ±10%. This would translate to an uncertainty in reconstructed $[CO_3^{2-}]$ of $10–12\,\mu mol\,kg^{-1}$, but as with the $\delta^{11}B$ inter-laboratory comparison studies, a study carried out only by laboratories specializing in B/Ca analysis could be expected to produce more consistent and precise results. Hathorne et al. (2013) made the important recommendation that all studies determining element/Ca ratios in marine carbonates should report determined values for a reference material such as JCp-1. Unfortunately, both the JCp-1 and JCt-1 reference materials are now difficult to obtain, and the paleoceanography community is in critical need of suitable reference materials for both trace element and stable isotope analysis of marine carbonates.

To summarize, the limited inter-laboratory comparison studies undertaken to date highlight the need for further assessment of both boron isotope and boron concentration data being produced for marine carbonates. The inter-laboratory comparison study conducted by Foster et al. (2013) found laboratories specializing in boron isotope analysis were able to produce accurate and precise data for pure boron solutions and seawater, and could reproduce isotopic differences between calcium carbonate samples. There remains a critical and unmet need for standard reference materials (SRMs) to be made widely available to the community of marine carbonate analysts without which the quality (accuracy and precision) of reported boron isotope ratios and boron concentrations cannot be adequately evaluated.

4.3 Standard Reference Materials and Data Quality Assurance

The analysis and reporting of boron isotope ratio and boron concentration data for SRMs should be considered essential to data quality assessment and assurance. In the case of boron isotopes, the National Institute of Standards and Technology (NIST) SRM 951 boric acid is the primary SRM against which all other reference materials and samples are measured. NIST SRM 951 has an assigned $^{11}B/^{10}B$ isotope ratio of 4.04362 ($\delta^{11}B = 0$) based on a value measured by P-TIMS reported by Catanzaro et al. (1970). It is now backed by a range of more recently characterized boric acid reference materials (ERM AE120–122 also referred to as IRM 120–122) that span a $\delta^{11}B$ composition range from −20 to +40 ‰ (Vogl and Rosner 2011, Table 4.3). Secondary SRMs calibrated against NIST SRM 951, which have a similar composition to sample unknowns, can serve as equally, if not more valuable reference materials as they can be used to account for potential matrix related effects on measured boron isotope ratio and boron concentration analyses. Both seawater and carbonate standards are potentially useful secondary SRMs for boron isotope and concentration analysis. Seawater has a high boron abundance (432.6 $\mu mol\, kg^{-1}$ at salinity = 35; Lee et al. 2010), is isotopically homogeneous ($\delta^{11}B = +39.61 \pm 0.04$ ‰, Foster et al. 2010), and is readily obtained from several international suppliers (see Table 4.3). The only widely distributed calcium carbonate SRMs available for boron concentration and boron isotope analysis are the Geological Survey of Japan JCp-1 and JCt-1 SRMs. Unfortunately, these SRMs are no longer available and the international community is subsisting on reference materials, including remnants of JCp-1 and JCt-1, that are being distributed between a small number of laboratories. The lack of more widely available marine carbonate SRMs is a significant impediment to the analysis and reporting of high quality, accurate and precise, boron concentration, and boron isotope data. Ideally, these would comprise a suite of calcium carbonate SRMs with different $\delta^{11}B$ compositions and B concentrations.

The increasing use of microanalysis techniques for B concentration and isotope ratio analysis presents an even greater quality assurance challenge as microanalysis techniques require SRMs that are homogeneous at the length/volume scales of the technique being used (i.e. $<10^{-16}$ m^3 for SIMS up to 10^{-11} m^3 for LA-ICP-MS). NIST SRM 951 boric acid and otherwise homogeneous liquid reference materials (e.g. seawater) are inappropriate for

Table 4.3 Boron concentration and isotope ratio values of standard reference materials.

SRM or material	$\delta^{11}B$ ‰ (±2sd)	B concentration		Reference and comments
		$\mu g\,g^{-1}$	$mol\,mol^{-1}$ Ca	
Boric acid				
NIST 951	0			Catanzaro et al. (1970) $^{11}B/^{10}B = 4.04362 \pm 0.00137$
NIST 952	−986.84			Catanzaro et al. (1970) Highly enriched in ^{10}B $^{11}B/^{10}B = 0.053199 \pm 0.000032$;
IRMM-011	+0.16			Vogl and Rosner (2011)
Seawater				
OSIL IAPSO	+39.64 ± 0.42			Wang et al. (2010)
Seawater	+39.61 ± 0.04	4.68	2.4×10^{-2}	Foster et al. (2010), Lee et al. (2010) at salinity = 35
NASS-5 (NRC)	+39.89 ± 0.25			Louvat et al. (2011)
IAEA-B1	+39.20 ± 0.80			Foster et al. (2006) Not available
Boric acid solution				
ERM-AE 120	−20.2 ± 1.2			Vogl and Rosner (2011)
ERM-AE 121	19.9 ± 1.2			Vogl and Rosner (2011)
ERM-AE 122	39.7 ± 1.2			Vogl and Rosner (2011)
Marine carbonates				
GSJ JCt-1	+16.3 ± 1.6	20.5 ± 2.0	191.0 ± 18.6	Gutjahr et al. (2014), Hathorne et al. (2013)
GSJ JCp-1	+24.3 ± 0.34	49.3 ± 4.8	459.6 ± 45.4	Gutjahr et al. (2014), Hathorne et al. (2013)
IAEA-B-7	+9.7 ± 11.8			Gonfiantini et al. (2003) Natural limestone
Carrara		0.306	2.78×10^{-6}	Stoll et al. (2012) Natural limestone
Inorganic Carbonates				
OKA Carbonatite		0.676	6.15×10^{-6}	Stoll et al. (2012) High temperature natural calcite
Aluminosilicate glasses				
NIST 610/611	−0.78 – 0	350 ± 56	$1.59 \times 10^{-2} \pm 0.26 \times 10^{-2}$	Kasemann et al. (2001), Le Roux et al. (2004), Fietzke et al. (2010), Jochum et al. (2011)
NIST 612/613	−0.8 – 0.1	34.3 ± 1.7	$1.50 \times 10^{-3} \pm 0.08 \times 10^{-3}$	Le Roux et al. (2004), Fietzke et al. (2010), Jochum et al. (2011)
NIST 614/615	0.44 ± 1.40	1.49 ± 0.19	$6.50 \times 10^{-5} \pm 0.84 \times 10^{-5}$	Le Roux et al. (2004), Jochum et al. (2011)
NIST 616/617		0.94 ± 0.47	$4.13 \times 10^{-5} \pm 2.07 \times 10^{-5}$	Jochum et al. (2011)

IRMM = Institute for Reference Materials and Measurements; OSIL = Ocean Scientific International Ltd.; NRC = National Research Council of Canada; IAEA = International Atomic Energy Agency; ERM = European Reference Materials; GSJ = Geological Survey of Japan.

in situ analysis by SIMS or LA-ICP-MS. Moreover, it is problematic to find or synthesize suitable SRMs and it is challenging to evaluate their homogeneity over the range of length scales measurable by different microanalysis techniques. To date no systematic study has been undertaken to characterize B/Ca or boron isotope compositional variation in potential carbonate reference materials. Gabitov et al. (2013) assessed Mg/Ca, Sr/Ca, and Ba/Ca variability in natural calcites and aragonites using SIMS and NanoSIMS and found individual crystals of some materials to be sufficiently homogeneous for use as microanalysis reference materials. However, the heterogeneity and composition of each crystal needed to be assessed. Calcite from the Oka Carbonatite (OKA), one of the more widely available calcium carbonate materials, was reported to be the most homogeneous, although the extent to which this observation applies to its boron elemental and isotopic compositions is unknown. In lieu of no calcium carbonate materials being established as a suitable microanalysis SRM, many laboratories have resorted to using NIST SRM 610–617 soda-lime glasses as primary calibration materials and employing "in-house" calcium carbonate standards, such as natural calcite crystals or pressed powdered pellets made from precipitated or naturally occurring calcium carbonate minerals or biominerals (e.g. Fallon et al. 1999; Fietzke et al. 2010; Diez Fernández et al. 2015).

The use of the NIST SRM 610–617 glasses as reference materials for boron microanalysis in carbonates (and other materials) demands caution. These glasses were produced in the 1950s by the Corning glass company as SRMs for bulk sample analysis rather than microanalysis. Nonetheless they have proved useful as microanalysis reference materials because of the large number of contained trace elements, which have been doped at four nominal concentrations spanning four orders of magnitude (Erickson and Williams 1970): NIST SRM 610 and 611 at $500\,\mu g\,g^{-1}$; NIST SRM 612 and 613 at $50\,\mu g\,g^{-1}$; NIST SRM 614 and 615 at $1\,\mu g\,g^{-1}$; and NIST SRM 616 and 617 at $0.02\,\mu g\,g^{-1}$. Unfortunately, many of these doped elements, including boron are variably depleted and heterogeneously distributed within the NIST SRM 610–617 series glasses (Eggins and Shelley 2002). Moreover, boron is one of the most heterogeneously distributed and depleted of these elements, with concentrations being ~20% lower in both the bulk NIST SRM 610/611 bulk glass ($350 \pm 56\,ug\,g^{-1}$) and in the NIST SRM 612/613 bulk glass (34.3 ± 1.7) compared to the original doped trace element concentrations of $478\,\mu g\,g^{-1}$ and $40.75\,\mu g\,g^{-1}$ (Eggins and Shelley 2002). This level of depletion stems from the loss of volatile boron during glass manufacture, resulting in overall bulk glass depletion and the development of localized domains that are even more depleted in boron and other volatile elements than the bulk glass composition. In lieu of the likely progressive loss of volatile boron during manufacture, it is also possible that boron depletion varies over the 200 m length of glass rod produced for each of the NIST SRM 610–617 glass compositions. It has not yet been established whether and, if so, to what extent the boron isotope composition might vary within the NIST SRM 610–17 glasses.

4.4   Boron Concentration and Isotope Ratio Analysis

Boron concentrations can be measured by a wide variety of analytical techniques including nuclear, atomic absorption and emission spectroscopy, spectrophotometry, inductively coupled plasma-optical emission spectroscopy (ICP-OES) and inductively coupled plasma mass spectrometry (ICP-MS). Boron isotope ratios are likewise able to be measured by a variety of mass spectrometry techniques, most notably N-TIMS, P-TIMS, and MC-ICP-MS. Collectively these elemental and isotopic analysis methods offer a range of capabilities in terms of sensitivity, dynamic range, matrix tolerance, susceptibility to spectral interferences, ease, and cost of implementation, and different needs for pre-processing of samples. Many laboratories have now shifted to using mass spectrometry, particularly ICP-MS techniques which are sensitive, versatile, and capable of precise and accurate analysis of both boron concentrations and isotope ratios. Despite advances in analytical capability, a number of physical and chemical properties of boron make its concentration and isotope ratio analysis challenging. These include:

- the proportionally large mass difference between ^{10}B and ^{11}B, which can result in significant mass bias and make mass bias correction difficult, particularly in the presence of other matrix ions;
- the volatility of $B(OH)_3$ and other boron compounds, which can result in partial boron loss with associated isotopic fractionation from aqueous solutions, especially under acidic conditions;
- the presence of exotic boron in nature and in laboratory environments and reagents, which makes contamination problematic, particularly for analysis of small samples; and
- the presence of organic compounds in biogenic samples which can compromise boron yields and form species that create spectral interferences during analysis by mass spectrometry.

It is common for boron to be separated from samples using either ion-exchange chromatography or micro-distillation to avoid matrix-related mass bias effects, formation and potential loss of volatile species and to reduce spectral interferences (Aggarwal and Palmer 1995). However, both ion-exchange chromatography and micro-distillation can result in additional mass fractionation where separations are not quantitative, and can create other spectral interference and sample contamination issues.

4.4.1   Mass Bias Correction

The large (~10%) mass difference between ^{10}B and ^{11}B makes boron isotope ratio measurement susceptible to mass fractionation during both separation from the sample matrix and instrumental analysis. The absence of another stable or long-lived radioactive boron isotope precludes the use of a double-spike approach to correct for mass fractionation. As a consequence, instrumental mass bias changes during analysis must be monitored and corrected for, and the potential for isotopic fractionation during sample and standard processing avoided. In the case of MC-ICP-MS, a very large mass bias occurs against the light ^{10}B isotope due to "space charge" effects during ion beam extraction through the interface region of these instruments. This requires analysis of matrix separated (purified) boron at similar concentrations to standard solutions and also the use of "standard-sample-standard" bracketing to monitor and correct for mass bias drift (Foster 2008). In the case of TIMS, care is required to prepare and load samples onto filaments consistently, and to analyze samples at the same filament temperature using a reproducible filament heating regime. In the case of P-TIMS, the cesium metaborate ($Cs_2BO_2^+$) method which measures $^{11}B/^{10}B$ ratios at masses 309 and 308, greatly reduces the mass bias effects encountered with N-TIMS and other P-TIMS methods (Nakano and Nakamura 1998).

4.4.2 Boron Volatility

Boron can form a variety of volatile compounds, including boron halides and boric acid, during sample preparation involving aqueous solutions. Complete or partial loss of these compounds can result in both low recovery and isotopic fractionation of the residual sample, particularly where sample preparation involves heating and evaporation. Volatile boron compounds are also a source of the recalcitrant memory effects encountered during ICP-MS and ICP-OES analysis. The propensity for boron to form volatile compounds can be turned to advantage, however, to purify and concentrate boron by distillation (sublimation and condensation) of volatile boron compounds (Gaillardet et al. 2001).

The boron halides, BF_3 and BCl_3, are volatile gases at room temperature, with the boiling point (BP) of BF_3 being $-100.3\,°C$, and BCl_3 being $12.3\,°C$. Trimethyl borate (BP = $67.5\,°C$) and triethyl borate (BP = $118\,°C$) compounds are likely to be highly volatile when evaporating boron bearing ethanol and methanol solutions, as with the separation of boron as methyl borate from complex solutions (Spivack and Edmond 1987; Swihart 1996). In comparison, boric acid $B(OH)_3$ has a relatively low vapor pressure at room temperature (2.1×10^{-4} Pa at $25\,°C$) and even at typical evaporation temperatures (3.5×10^3 Pa at $100\,°C$; Pankajavalli et al. 2007), yet is still prone to loss from aqueous and acidic solutions. The volatile loss of boric acid occurs with significant isotope fractionation, due to the large positive isotopic fractionation (ca. $+27\,‰$) between boric acid and borate in aqueous solution (Figure 2.2) and to a large positive seawater-vapor fractionation ($\Delta_{seawater-vapor} = +25.5\,‰$; Chetelat et al. 2005; Rose-Koga et al. 2006). As a consequence, the composition of volatile boric acid emanating from the ocean surface has a boron isotopic composition of ~$+45\,‰$ at tropical surface seawater pH, which is consistent with the highest observed rainfall $\delta^{11}B$ values in coastal tropical sites at around +45 ‰. Lighter values ranging down to −12.8 ‰ are explained by rainout through Rayleigh distillation processes (Fogg and Duce 1985; Xiao et al. 1992; Eisenhut and Heumann 1997; Miyata et al. 2000; Chetelat et al. 2005). The significant amounts of boron present in the atmosphere and rainfall reflect contributions from both ocean surface derived volatile boric acid and sea salt, with increasingly large contributions of anthropogenic boron occurring in continental areas (for a review of the modern global boron cycle see Schlesinger and Vengosh 2016).

Procedures to isolate and purify boron can employ one or more evaporation steps (Swihart 1996). Mannitol has been widely used to complex boric acid in acidic solutions and minimize its loss and that of boron halide species during sample evaporation and processing for analysis (Ishikawa and Nakamura 1990; Nakamura et al. 1992). However, the mechanism by which mannitol reduces volatile loss of boric acid has been challenged by Gaillardet et al. (2001). These authors note the pKa of the mannitol-boron complexation reaction is around 7, making the amount of boric acid complexed with mannitol in acidic solutions minimal. They also report no difference between, or significant volatile boron loss from, 0.1 M HCl solutions evaporated with or without mannitol.

4.4.3   Environmental and Laboratory Sources of Boron Contamination

Boron concentration and boron isotope analysis usually entails extracting and analyzing small (often ng) amounts of sample boron in the presence of environmental or laboratory borne contaminants. This can be a challenge for the analysis of fossil foraminifer shells recovered from seafloor sediments, which contain fine clays that have adsorbed and concentrated boron from pore-waters and seawater. Isolating and purifying boron from foraminifer shells and other samples requires the use of low boron or boron "free" chemical reagents, careful handling of samples in the presence of laboratory borne contaminants, and avoiding cross contamination of samples and standards during instrumental analysis. The procedures for cleaning environmental contaminants from samples is covered in Section 4.5 while here we focus on laboratory and reagent sources of contamination.

The need to control boron contamination during sample preparation and analysis has long been appreciated when analyzing boron concentrations or boron isotope ratios. Kiss (1988) is notable for advocating the potential advantages of using *in situ* analysis techniques over methods that required digestion of samples using fluxes or acid reagents. Minimizing reagent and procedural blank levels has been an ongoing challenge for most analysis methods that separate and/or analyze boron in solution form. It is worth reiterating that these problems are exacerbated where small amounts of boron are available due to low boron abundances and/or sample size limitations.

Airborne contamination of samples is a significant risk due to the presence of volatile boron compounds and boron bearing particulates in the atmosphere, particularly in coastal areas due to proximity to marine sources, and also in urban and industrial locations due to anthropogenic sources. However, airborne boron contamination can be an even greater problem in laboratory environments due to the widespread use of boron compounds in general cleaning agents, in building construction, and in air filtration systems. Examples include the use of borax ($Na_2B_4O_7 \cdot 10H_2O$) and sodium perborate bleach ($NaBO_3 \cdot (1–3)H_2O$) in laundry products and as mild antiseptics, sodium borate pentahydrate ($Na_2B_4O_7 \cdot 5H_2O$) in the manufacture of fiberglass insulation, and the use of boric acid in the manufacture of textile fiberglass and cellulose insulation as a fire retardant. Filters used for treating air in buildings and laboratories can contain up to tens of percent boron by weight in the form of borax, boric acid, and borosilicate glass fibers, which appear to release boron over time (Rosner et al. 2005), possibly due to exposure to acid fumes. This makes otherwise nominally clean laboratory environments especially prone to boron contamination. It is recommended that laboratory clothing is used which has been cleaned with boron-free washing powders, and to avoid the transfer of boron from handling contact lens solutions, to avoid the use of glassware in preference for equipment made of Teflon and polyethylene, and to work in environments where air-handling systems are fitted with boron-free Teflon filters (Rosner et al. 2005).

Exposure to airborne contaminants in laboratory environments can occur during sample digestion, ion-exchange chromatography, and sample evaporation. Misra et al. (2014b) measured a boron blank accumulation of 1 ng per hour in a standard (non-PTFE HEPA) filtered laboratory environment and reports a light $\delta^{11}B$ composition (−5 ‰) from HEPA filter sourced boron. Such high rates of boron accumulation can be avoided by working in a laboratory fitted with PTFE HEPA air filters, else in a laboratory without a HEPA filtered air-supply. Foster et al. (2013) reported boron blank accumulation rate of only 4 pg hr^{-1} in a positively pressured PTFE HEPA filtered laboratory environment.

4.4.4 Chemical Reagent and Procedural Blanks

Chemical reagents and laboratory water used in sample processing also present potentially significant sources of boron contamination. It is important to characterize and report contaminant boron "blank" levels for individual reagents and/or complete procedural "blanks" used in any boron analysis method. Some reagents with low certified boron blank levels can be obtained from chemical suppliers, else reagents can be distilled using Teflon-ware to remove boron. Laboratory water can be a major contributor to reagent and procedural blanks, and the fitting of laboratory water purification systems with specialized resin cartridges to remove boron is required (e.g. Q-Gard˙ Boron Purification Cartridges for MilliQ™ water systems or equivalents). Misra et al. (2014b) have reported very low reagent blanks for ammonia, NaOH, and H_2O_2, and acid blanks at ~1 pg B ml^{-1}, and a total procedural blank of <0.01 ng B for a micro-sublimation (distillation) method to purify boron from digested samples (see Section 4.6.2). The latter represents a < 0.25% contribution to a minimum target sample yield of 4 ng of boron from benthic and planktic foraminifer species. By comparison, Lemarchand et al. (2002) reported a much higher blank contribution of 1.5 ng resulting from combined ion-exchange resin separation, micro-sublimation, and evaporation steps to a total procedural blank of 3.5 ± 1.5 ng. This indicated about 2 ng of boron was derived via carry over from the ion-exchange resin and/or airborne contamination, which represented almost 2% of the total 250 ng mass of boron analyzed. Whether or not a specific blank level is problematic depends both on its size relative to the amount of sample boron being processed and the boron isotope composition contrast between the procedural blank and the sample. Greater attention to characterizing boron blank levels and isotopic composition is clearly needed when smaller amounts of boron are being analyzed.

It is finally worth mentioning the practice of buffering live collected plankton samples in borax solution (e.g. Haynes 1981; Green 2001). A range of observations as to the effect of borax contamination on foraminiferal $\delta^{11}B$ and B/Ca has been made, ranging from minor (M. Henehan, pers. comm.) to crippling (M. Palmer, pers. comm.). Until a systematic assessment of this problem has been made, such samples should be treated with caution if used for B/Ca or boron isotope composition analysis.

4.5 Sample Preparation and Cleaning

Careful cleaning of samples is a general requirement prior to boron concentration and boron isotope analysis to remove contaminants derived from the sample's natural environment, during laboratory culture or during sample handling prior to laboratory processing. Any chemical cleaning thereafter should be conducted ideally in a boron-clean laboratory environment using low boron reagents to avoid any further sample contamination. A variety of different chemical cleaning procedures have been developed to suit different sample types and origins (Barker et al. 2003; Holcomb et al. 2015). Strong oxidizing agents are required to remove organic matter from most marine carbonate samples, and it may be necessary to apply more rigorous oxidative cleaning methods to remove larger amounts of organic matter associated with live-collected or cultured samples (Mashiotta et al. 1999; Pak et al. 2004). Reductive cleaning procedures are also used to solubilize and remove authigenic ferromanganese oxy-hydroxides from foraminifer samples that originate from deep sea sediments (Boyle 1983; Boyle and Keigwin 1985; Rosenthal et al. 1997; Martin and Lea 2002; Barker et al. 2003), however, evidence indicates reductive cleaning has a limited influence on boron elemental or isotopic compositions (Ni et al. 2007; Yu et al. 2007).

4.5.1 Foraminifera Shells

Procedures for extracting and selecting foraminifera from sediment cores entail first washing sediments in tap water or deionized water to remove most of the fine (<63 μm) fraction (clays, diatoms, coccolithophores, etc.). Washing is followed by sieving to select for different sediment size fractions and thereby foraminifer shell sizes. Individual foraminifers are then identified and sorted by species or species morphotypes by viewing at moderate magnification using a binocular microscope. Shells of the desired species are picked using a wetted natural sable ultrafine brush (ideally 000 or 0000 size), with care being needed to include only shells within a narrow size range (e.g. 400–450 μm), without any visible evidence of partial shell dissolution (e.g. holes, missing chambers) or overgrowths (e.g. black spots of manganese crusts), and of the same ontogenetic status. Both boron and boron isotope compositions can differ between similar species, with shell size within species, and with shell forms and the presence of crusts produced during gametogenesis (Hönisch and Hemming 2004; Ni et al. 2007; Henehan et al. 2013; Babila et al. 2014). Boron concentrations and isotope compositions may also differ between morphotypes and genotypes within morphospecies, due to differences in habitat preference (as in Mg/Ca ratios; Steinke et al. 2005) and/or possible biomineralization controls on shell chemistry (Sadekov et al. 2016). Careful attention to the selection of species is particularly critical when working with benthic foraminifera, as subtle differences in shell morphology and architecture occur between species and across species intergrades that have different habitat preferences within seafloor sediments (e.g. Rae et al. 2011). Reconstructions of bottom water carbonate chemistry invariably focus on epifaunal species such as *C. wuellerstorfi* and *Cibicidoides mundulus* where possible, as these epifaunal species are thought to record the water properties at the seafloor, unlike infaunal species, which are thought to record porewater properties (Katz et al. 2010).

The number of shells needed for analysis will depend on the average shell weight (size-class specific weight) and boron concentration of the target species, and of course the amount of boron required by the chosen analysis technique. An important and sometimes overlooked or difficult to meet consideration is the number of shells needed to be representative of a population. This stems from the inherent compositional differences between individual foraminifers that make up a sample population (Allen et al. 2011; Raitzsch et al. 2011). Limited information exists as to the variability of boron concentrations between individual shells within populations of particular foraminifer species, and there is no published data for boron isotope composition variability within or between individual foraminifers. Raitzsch et al. (2011) report on average a 43% range above and below the mean shell B/Ca composition for populations of the benthic foraminifer *Planulina wuellerstorfi* and noted this to also be similar to variation in Mg/Ca composition of the same shell populations. Studies of Mg/Ca variability within populations of planktic foraminifer species have indicated >20 shells are required to obtain composition estimates within 10% of the population mean with 95% confidence (e.g. Anand and Elderfield 2005; Sadekov et al. 2008) and, as a consequence, analyzing at least 20 shells is considered necessary to estimate seawater temperatures to better than $\pm 1\,°C$. Similar studies are needed to characterize within population compositional variation and required shells numbers to reconstruct specified uncertainties in seawater ΔCO_3^{2-} and pH when employing the B/Ca and $\delta^{11}B$ proxies. This may be more critical for B/Ca than B isotope analysis given the latter requires relatively large numbers of shells to obtain required amounts of boron for $\delta^{11}B$ analysis (i.e. ~1–3 ng using N-TIMS or MC-ICP-MS), which by necessity will average out within population variability.

Prior to analysis, foraminifer shells derived from seafloor sediments need to be cracked open and cleaned of detrital minerals and organic residues that accumulate within the shell interior, and of authigenic phases attached to internal and external shell surfaces. Shell cleaning procedures beginning at this step are documented in comprehensive studies conducted by Boyle and Keigwin (1985), Rosenthal et al. (1997) and Barker et al. (2003). In brief, following picking and washing, bulk shell samples are assembled on and gently crushed between two clean quartz glass plates or microscope slides (Barker et al. 2003). The crushed sample material is then placed in a small plastic or Teflon vial (e.g. 1 ml with snap lid) and cleaned of clay contaminants by ultrasonication in high purity water, sometimes followed by methanol (repeated up to three to five times). The sample is subsequently repeatedly rinsed and ultrasonicated in high purity water (typically five times), with the cloudy or discolored solution removed using a pipette after each ultrasonication until a clear solution is obtained. At this point samples are inspected at high magnification using a binocular microscope to assess cleanliness, looking for the presence of dark specs and particulates which might indicate the presence of clays or other residual terrigenous, authigenic, or diagenetic minerals.

Organic matter is considered a potential source of boron contamination in marine carbonates. A survey of boron contents in marine plankton by Yamamoto et al. (1973), reported dry weight concentrations from 18 to $104\,\mu g\,g^{-1}$ boron in phytoplankton and $33\,\mu g\,g^{-1}$ boron on average for zooplankton, however, evidence from subsequent studies on the boron content and isotopic composition of organic matter in marine environments is limited. Residual organic matter can be removed from cultured and live-collected samples from nature, as well as from seafloor sediment-derived foraminifer samples, by a variety of oxidative cleaning procedures. Many laboratories working with sediment-derived samples follow the method of Barker et al. (2003) which employs repeated immersion in a hot (60–80 °C) 1% H_2O_2 solution buffered with 0.1 M NH_4OH. This is undertaken in a water bath and followed by repeated rinsing (three to five times) with high purity water, then an optional methanol rinse, and again high purity water (Foster et al. 2013). The efficacy of the procedure is monitored following each step using a binocular microscope. A higher concentration of H_2O_2, typically a 50:50 mix of saturated H_2O_2 buffered with 0.001 M NaOH, is used to remove larger amounts of organic material from cultured and live-collected foraminifers (Pak et al. 2004). Organic material can also be removed by oxidation with dilute sodium hypochlorite (e.g. 3% NaClO solution; Holcomb et al. 2015) but because Na can affect plasma behavior during ICP-MS analysis, most laboratories avoid the use of sodium hypochlorite.

Reductive cleaning of foraminifer shells is used to remove ferromanganese oxy-hydroxides and possible manganese carbonate coatings and associated metals that can precipitate on foraminifer shells during seafloor diagenesis (Boyle 1981). Reductive cleaning can be applied either before or after oxidative cleaning, which involves heating samples in a small amount of dilute 1 M hydrous hydrazine solution in a 0.25 M citric acid and 16 M ammonia buffer (Boyle and Keigwin 1985). The method was developed specifically to obtain reliable Cd/Ca and Ba/Ca shell compositions (Boyle and Keigwin 1985). Reductive cleaning appears to have little or no effect on measured shell B/Ca (Yu and Elderfield 2007; Misra et al. 2014a) or $\delta^{11}B$ compositions (Misra et al. 2014b) but can cause significant additional dissolution of shells and bias measured Mg/Ca compositions to lower values by typically 10–15% compared to samples subjected only to oxidative cleaning (Barker et al. 2003).

After applying the oxidative or oxidative and reductive cleaning steps, it is usual for samples to be briefly rinsed once or twice in dilute 0.001 M HNO_3 to remove any residual adsorbed surface contaminants.

The efficacy of shell-cleaning procedures or specific steps can be assessed by analyzing elements that are indicative of the presence of authigenic, diagenetic or terrigenous contaminants. Aluminum and silicon are used to monitor for residual clays and other aluminosilicate phases, and manganese and iron for residual authigenic phases. Measured bulk shell Al/Ca ratio values <100 μmol mol^{-1} are taken to indicate the effectiveness of the cleaning protocol (Boyle 1981). Although this amount of Al is more than two orders of magnitude greater than in pure foraminiferal carbonate, it provides a satisfactory maximum acceptable level of detrital clay contamination. Clays can be a significant source of boron contamination, as they become enriched through boron adsorption onto mineral surfaces as well as diffusion of boron into clay sheet structures (Couch and Grim 1968; Deyhle and Kopf 2004).

4.5.2   Corals

The majority of coral samples subjected to B/Ca and B isotope analysis are either drilled from massive tropical corals (typically large *Porites* spp. coral heads or bommies), or are deep sea coral samples collected by dredge, benthic sled, or submersible. The tropical coral cores are typically rinsed in fresh water and sun-dried in the open air before being cut into slabs parallel to the growth axis. The coral slabs are then ultrasonically cleaned in ultrapure water, and rinsed multiple times before being dried in a clean laboratory environment and imaged to reveal growth banding. Growth increments are then milled to achieve the desired timescale resolution (see Hendy et al. 2002; Pelejero et al. 2005; Wei et al. 2009).

Aliquots of the milled coral powder are collected and transferred into acid-cleaned Teflon or other plastic vials for oxidative cleaning. This is to remove organic matter, and is usually accomplished with a saturated (30% v/v) H_2O_2 solution at elevated temperature (50–60 °C) in a water bath (Wei et al. 2009). The sample is centrifuged before removing the supernatant, then rinsed with high-purity water multiple (3–5) times prior to proceeding with sample digestion, boron separation, and purification for TIMS or MC-ICP-MS analysis (Section 4.6). For TIMS analyses, the processed boron sample is typically redissolved immediately prior to loading samples into the mass spectrometer (e.g. Wei et al. 2007, 2009). A recent study by Holcomb et al. (2015) investigated the effects of a range of commonly used cleaning protocols and reagents to coral and foraminiferal calcium carbonate, including rinsing with ultrapure H_2O, dilute bleach (3% NaOCl), saturated hydrogen peroxide solution (30% v/v), sodium hydroxide (0.006–0.1 M NaOH), and dilute nitric acid (0.05 M HNO_3). Only oxidizing and acid treatments that produced significant sample dissolution of compositionally heterogeneous samples were found to cause significant changes in measured $\delta^{11}B$ compositions.

4.6   Boron Separation and Purification

Boron is invariably separated from its sample matrix prior to determining B isotope ratios by P-TIMS or MC-ICP-MS, but not always in the case of N-TIMS analysis (Figure 4.4). The separation of boron from matrix ions can be achieved by ion chromatography using a boron specific resin or more generic cation/anion resins, by micro-sublimation (distillation) of volatile boric acid, or by a combination of these approaches (Aggarwal and Palmer 1995). The boron-specific ion-exchange resin Amberlite IRA-743 (also known as XE-243) is widely used for separating and purifying boron from marine biogenic carbonates (e.g. corals, bivalves, foraminifera; Vengosh et al. 1989, 1991; Hemming and Hanson 1992, 1994). More generic cation and anion exchange resins can also be used to extract and purify boron from complex matrixes, but require careful attention to the composition and ionic strength of the boron containing solution as these resins do not select specifically for boron (Xiao et al. 1988; Nakamura et al. 1992; Louvat et al. 2011; Ishikawa and Nagaishi 2011; McCulloch et al. 2014). It is more common for cation resins to be used in boron separation and purification procedures to remove Ca matrix ions (e.g. Louvat et al. 2011).

4.6.1 Ion-Exchange Chromatography

A range of procedures for separating boron from biogenic carbonates are based around the boron-specific ion-exchange resin Amberlite IRA-743. This resin has a strong affinity for boron and was developed in the 1960s for the purpose of extracting boron from water samples, and thereafter applied to complex matrices in the nuclear industry (Carlson and Paul 1968; Hill and Lash 1980) based on the capacity of its tertiary amine groups to convert borate to tetrafluoroborate ions. Kiss (1988) was the first to use the primary chelating resin sites (N-methylglucamine groups attached to polystyrene backbone) to adsorb borate ions, based on the strong binding of boron to the glucamine groups at neutral to alkaline pH via the exchange reaction shown in Figure 4.3. The partitioning of boron by IRA-743 resin increases with pH, with K_D values exceeding 1×10^5 at pH > 7 (Lemarchand et al. 2002). The resin's capacity to adsorb boron is reported to be ~5–6 mg B ml^{-1} for dilute solutions, increasing up to 10–15 mg B ml^{-1} for solutions with ionic strength similar to seawater (e.g. Kiss 1988; Leeman et al. 1991).

Quantitative adsorption of boron onto the IRA-743 resin can be achieved by adjusting and maintaining the dissolved sample at pH greater than five. This is facilitated by adding a base such as NH_4OH, NaOH, or Na acetate-acetic acid to elevate and buffer the sample pH (Lemarchand et al. 2002; Foster 2008). Care is required to avoid precipitating Ca and Mg hydroxides at higher pH when processing seawater or marine carbonate samples. Careful cleaning of the resin to remove Na^+, Cl^-, SO_4^{2-} and other ions left over from any previous use is also important, as residual matrix ions in the presence of eluted boron can induce significant matrix-dependent mass fractionation during subsequent analysis by mass spectrometry (P-TIMS or MC-ICP-MS; Wei

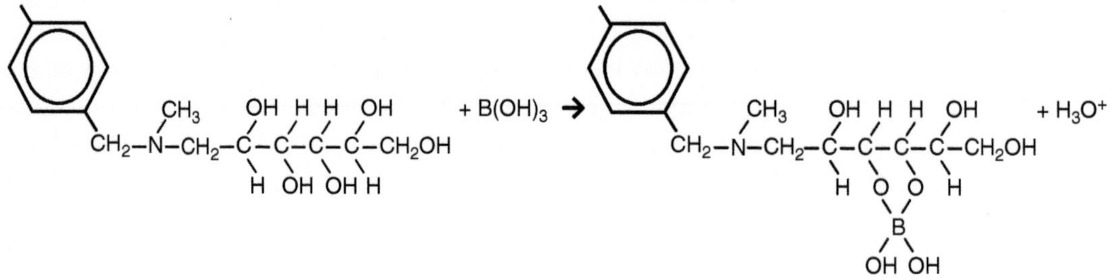

Figure 4.3 Boric acid binding reaction to the primary chelating N-methyl-glucamine groups of the boron specific ion-exchange resin, Amberlite IRA-743.

et al. 2009). Quantitative elution of boron from IRA-743 resin is achieved by passing moderately strong acid through the resin bed following elution of the sample matrix. Lemarchand et al. (2002) showed 0.1 M HCl to be most efficient with larger eluent volumes needed to achieve full boron recovery when using more concentrated acid. However, 0.5 M HNO_3 is more commonly used in MC-ICPMS analysis, as HNO_3 is regarded as a better carrier acid.

Generic cation resins, AG50W-x8 or similar, can also be employed to separate the dominant matrix ions prior to boron purification using IRA-743, provided separation is not carried out at high pH (>10) where boron can co-precipitate along with Ca and Mg hydroxides (Louvat et al. 2011; McCulloch et al. 2014; Figure 4.4). For example, Louvat et al. (2011) remove the Ca-matrix by passing calcium carbonate samples through AG50W-x8 cation resin. The collected boron is preconcentrated through a column containing 50 μl of Amberlite IRA-743 resin, and then purified in two steps in columns containing 10 and 3 μl of Amberlite IRA-743 resin (as in Lemarchand et al. 2002) using 0.05 M HNO_3. Boron free NaOH is used to maintain pH 7 to 9 within the column. An alternate strategy is to employ an anion resin following use of a cation resin. This approach facilitates elution of boron and removal of the dominant matrix ions from carbonate samples in a single procedure that uses both resins in series. McCulloch et al. (2014) describe a setup for purifying B from marine carbonate samples that employs 0.6 ml of AG50W-x8 resin packed in a 0.8 ml Bio-Rad Micro Bio-Spin polypropylene column that is positioned directly above and in line with a second slightly larger (1.2 ml)

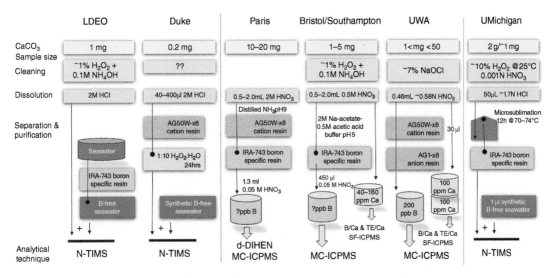

Figure 4.4 Schematic overview of different boron sample processing schemes used by various laboratories indicating the typical dissolved $CaCO_3$ sample quantity, cleaning reagents, sample digestion matrix, boron separation and purification procedure, and analysis technique employed. The LDEO, Duke University, Paris, and Southampton/Bristol schemes are based on procedures reported in Foster (2008), the UWA scheme on McCulloch et al. (2014), and the University of Michigan scheme on Liu et al. (2013). For further details on full cleaning steps, to remove clays for example, and the sample processing and analysis methods used by each laboratory readers should refer to the relevant studies.

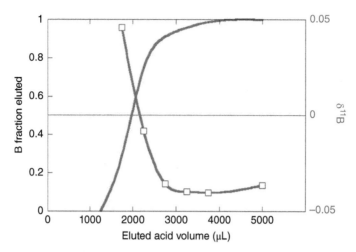

Figure 4.5 The relative fraction of boron eluted (gray "breakthrough curve") and variation of measured boron isotope ratio δ^{11}B composition in successive aliquots (blue squares) eluted through a 1 mL column of Amberlite IRA-743 resin shown as a function of the total eluted volume of 0.1 M HCl (Source: adapted from Lemarchand et al. 2002).

column packed with 1.0 ml of AG1-x8 resin. A 0.4 ml sample aliquot containing ~100 μg g^{-1} Ca in 0.1 M HNO$_3$ is loaded onto the first stage cation column in 2 ml of 0.0.75 M HNO$_3$, and allowed to flow directly onto the second stage anion column, and into a polypropylene collection vial. The resulting solution is suitable for boron isotope analysis by MC-ICP-MS without further processing or dilution. The method has been optimized for coral samples and might be adaptable to foraminifer samples, but is not suited for seawater analysis as it does not remove Cl$^-$ from the boron solution.

Significant boron isotope fractionation can occur during elution using IRA-743 or other ion-exchange resins if the elution does not achieve complete boron recovery. In the case of the Amberlite IRA-743 resin, Lemarchand et al. (2002) report large systematic changes in mass bias in different elution fractions, with the initial eluted boron being strongly enriched in ^{11}B and the tail being strongly enriched in ^{10}B (Figure 4.5). If complete boron recovery is not achieved, by missing either the start or end of the elution, the eluted boron sample will be enriched or depleted in ^{11}B. Similar biases have been reported when using generic cation and anion exchange resins (Conrard et al. 1972), emphasizing the importance of achieving quantitative recovery of boron from samples when using any ion-exchange method for separating and purifying boron.

4.6.2   Micro-Distillation (Micro-Sublimation)

Micro-distillation (also known as micro-sublimation) provides a relatively straightforward method for separating and purifying boron but, like ion-exchange

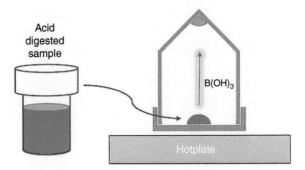

Acid
digested
sample

B(OH)₃

Hotplate

Figure 4.6 Schematic diagram of boron micro-sublimation setup showing acid sample digestion in a Teflon beaker (on left), with transfer of solution micro-aliquot to the inverted lid of conical Teflon beaker (on right), which is assembled upside-down on a hotplate. Source: Diagram adapted after Wang et al. (2010).

methods, requires care to achieve high yields and avoid loss of volatile boron species (Gaillardet et al. 2001; Wang et al. 2010; Misra et al. 2014b). The volatility of boric acid is purposely used to achieve high purity separation of boron from its dissolved matrix in a simple single-step procedure. The method is reported to give high yields and low blank levels due to reduced handling of samples ($\leq 0.05 \pm 0.01$ ng B; Misra et al. 2014b). It requires loading small volumes ($<50\,\mu l$) of dissolved calcium carbonate sample along with high purity concentrated HCl into a simple micro-distillation vessel, specifically the upturned lid of a small (5 ml) conical Teflon° beaker. The beaker is then assembled upside-down on a temperature-controlled hotplate (Figure 4.6). Boron is transferred as B(OH)$_3$ via the vapor phase from the lid to condense at the cooler conical end of the up-turned beaker. Complete recovery of boron requires significant distillation time (>18 hours). Recent studies have used temperatures in the range $\geq 95\,°C$ (Misra et al. 2014b) rather than the 60–65 °C originally reported by Gaillardet et al. (2001). Misra et al. (2014b) report incomplete (95%) transfer after five hours and the need to keep the sample load below $50\,\mu l$ when using small 5 ml Teflon Savillex conical beakers, in order to avoid generating a large droplet at the conical end that may fall back onto the cap. Raitzsch et al. (2018) report $99.8 \pm 0.9\%$ (1SD) recovery using a distillation time of 24 hours at 100 °C following the method of Misra et al. (2014b). Wang et al. (2010) have found no significant change in measured isotopic compositions in the presence of matrix ion concentrations up to $5000\,mg\,l^{-1}$ Na and $8000\,mg\,l^{-1}$ Ca by adjusting the analyte to pH <3 and the temperature to 98 °C. Following the distillation period, samples are cooled for ~15 minutes, the lid is unscrewed, the beaker is returned to an upright position and a small amount (0.5 ml) of dilute 0.3 M HF is added (Misra et al. 2014b). The beaker is then capped with a replacement acid-cleaned cap in preparation for boron isotope analysis. The sample residue on the original cap may be taken up in dilute acid for sample trace element analysis and boron yield assessment by ICP-MS or ICP-OES (Misra et al. 2014b).

4.6.3   Removal of Organic Matter

Boron separation and purification procedures can result in soluble organic matter derived from the sample or breakdown products from ion-exchange resins being transferred to the purified boron sample. This organic matter can be a source of isobaric interferences during subsequent boron isotope analysis, with CNO^- ions interfering with $^{10}B^{16}O_2^-$ at mass 42 during N-TIMS analysis (Hemming and Hanson 1994), and $^{133}Cs_2^{12}C^{14}N^{16}O^+$ ions interfering with $^{133}Cs_2^{12}B^{16}O_2^+$ at mass 308 during P-TIMS analysis (Wei et al. 2009). Different studies have addressed this problem by treating residual organic matter with UV light or strong oxidants such as hydrogen peroxide (e.g. Hemming and Hanson 1994; Foster 2008; Wei et al. 2009). The problem is claimed to be avoided where micro-distillation is used to extract boron from organic rich samples (Gaillardet et al. 2001; Section 4.6.2), however, Van Hoecke et al. (2014) report microsublimation (cf. ion-exchange chromatography) to be "plagued by isotope fractionation" with large deviations (~8 ‰) from reference values occurring when measuring spinach.

4.7   Instrumental Techniques

4.7.1   Thermal Ionization Mass Spectrometry (TIMS)

Thermal ionization mass spectrometers are relatively simple instruments that produce very stable ion emissions. TIMS analysis involves placing a small amount of sample containing boron onto a metal filament which is then heated to high temperature under high vacuum by passing a current through the filament. Positive or negative ionized boron species desorb from the hot metal filament and are collected and analyzed in techniques referred to as positive-ion TIMS (P-TIMS) and negative-ion TIMS (N-TIMS). In respective cases positive or negative ions are focused using electrostatic fields through an entrance slit into a magnetic field to separate ions based on their mass to charge ratio. The very low energy spread of ions generated by the thermal ionization source makes the use of an electrostatic sector to filter ion energies unnecessary (cf. MC-ICP-MS, SIMS). The efficiency with which an element is ionized, the proportion of ionic to neutral species (n_+/n_0), can be described by the Saha-Langmuir equation;

$$\frac{n_+}{n_0} = exp\left(\frac{(W - IP)}{kT}\right) \tag{4.3}$$

where W is the work function of the filament surface, IP is the ionization energy of the element, k is Boltzmann's constant, and T is temperature in °K.

TIMS methods are best suited to analysis of ions or charged molecular species that have low ionization energies. In the case of boron, maximum achievable filament temperatures around 2500 °C are unable to produce

B^+ ions efficiently, because of boron's relatively high 1st ionization potential (i.e. ~8.3 eV). As such, P-TIMS and N-TIMS techniques have been developed to measure positive alkali-borate molecular boron species ($M_2BO_2^+$) and negative borate species (BO_2^-) respectively.

4.7.2   Positive-ion TIMS (P-TIMS)

Most P-TIMS methods generate and measure either sodium borate ions ($Na_2^{10}BO_2^+$ and $Na_2^{11}BO_2^+$) at 89 and 88 atomic mass units (amu) (Catanzaro et al. 1970; Sah and Brown 1998; Rao et al. 2008), or cesium borate ions ($Cs_2^{10}BO_2^+$ and $Cs_2^{11}BO_2^+$) at 309 and 308 amu (Ramakumar et al. 1985; Spivack and Edmond 1986). These methods involve loading the dissolved sample along with Na_2CO_3, NaOH, or CsOH onto the metal filament in proportions which give B/Na or B/Cs molar ratios that generate the most efficient molecular ion yields (Rao et al. 2008, 2009, 2012). Graphite is added (as a "promoter") to enhance ion yield and thereby measured signal intensities to achieve the best possible measurement precision (Xiao et al. 1991; Rao et al. 2009). Because Na and Cs are monoisotopic these borate compounds produce simple mass spectra which require only a simple correction for the small contribution of ^{17}O to the higher mass molecule (i.e. $^{11}B/^{10}B = {}^{309}M/{}^{308}M - 0.00078$; Spivack and Edmond 1986). The smaller relative mass difference between $Cs_2^{11}BO_2^+$ and $Cs_2^{10}BO_2^+$ (0.32%) compared to that between $Na_2^{11}BO_2^+$ and $Na_2^{10}BO_2^+$ (1.14%) is beneficial in delivering a smaller mass-dependent fractionation and reducing analytical variability. Nonetheless, some laboratories persist with analyzing boron isotopes using $Na_2BO_2^+$ methods (e.g. Rao et al. 2008, 2009, 2012).

P-TIMS analysis can be compromised by isobaric interferences, which are reported to bias measured $^{11}B/^{10}B$ isotope ratios to lower values when using the $Cs_2BO_2^+$ method (Hemming and Hanson 1992; Aggarwal and Palmer 1995). This was thought to be due to a Cs_2CNO^+ interference at 308 amu resulting from HNO_3 use in sample preparation, but has since been attributed to the presence of organic CN-based compounds (Wei et al. 2004; Wu et al. 2012). Formation of these compounds can be limited by oxidizing organic-matter using either H_2O_2 or $HClO_4$. Alternatively, samples can be loaded along with dilute (1 v/v %) H_3PO_4, which depresses $Cs_2BO_2^+$ ionization until higher (unspecified) filament temperatures are reached and allows Cs_2CNO^+ to be "burnt off" at lower temperature (Wei et al. 2004). It is also claimed that H_3PO_4 may prevent the formation of Cs_2CNO^+ in the presence of nitrate ions (Wei et al. 2004). Some studies report monitoring Cs_2Cl^+ formation at 303 amu as a proxy for the production of the Cs_2CNO^+ interference, and only integrating $^{11}B/^{10}B$ measurements after the $Cs_2Cl^+/Cs_2BO_2^+$ ratio (301 amu/309 amu) falls below a minimum value (~$10^{-3} - 10^{-4}$; see figure 3 in Wei et al. 2009). These approaches and other method refinements have been reported to facilitate analysis of $^{11}B/^{10}B$ compositions of natural coral samples, including the GSJ JCp-1 coral standard with a reproducibility of ±0.08 ‰ (2σ; Ishikawa and Nagaishi 2011), and the Davies Reef and NEP B in-house coral standards to a reproducibility of ±0.11 ‰ (2σ; Wei et al. 2009) and ± 0.44 ‰ (2σ; D'Olivo et al. 2015).

A drawback of the P-TIMS technique is that it produces low ion yields and requires larger amounts of sample boron for analysis than N-TIMS, typically 100 to 1000 ng (cf. ~1 ng for N-TIMS; Section 4.7.3). This makes P-TIMS impractical for the analysis of typical planktic and benthic foraminifer samples, which are often limited to milligram or lesser amounts of shell calcite and only 10s of ng of available boron. The sample size requirement of P-TIMS is generally not an issue for other marine carbonates, such that P-TIMS has been adopted by laboratories specializing in $\delta^{11}B$ analysis of corals (e.g. Wei et al. 2009; Ishikawa and Nagaishi 2011; Xiao et al. 2013; D'Olivo et al. 2015).

4.7.3   Negative-Ion TIMS (N-TIMS)

The N-TIMS method measures $^{11}B/^{10}B$ isotope ratios by monitoring negatively charged $^{11}BO_2^-$ and $^{10}BO_2^-$ ions at 43 and 42 amu. The yield of negative charged boron ions is two to three orders of magnitude higher than the P-TIMS method (Zeininger and Heumann 1983; Duchateau and de Bièvre 1983; Aggarwal and Palmer 1995). This has made N-TIMS a clear method of choice for analyzing the $^{11}B/^{10}B$ isotope compositions of foraminiferal and other samples with limited available amounts of boron, until the recent development of MC-ICP-MS. Laboratories employing the N-TIMS method reported the ability to analyze 1–5 ng of boron routinely (Hemming and Hanson 1992; Pelejero et al. 2005; Farmer et al. 2016).

Another advantage of the N-TIMS method is its straightforward implementation, which in its simplest form entails direct loading of acid digested carbonate samples along with either natural or synthetic boron-free seawater onto a single rhenium metal filament. There is no need for prior boron separation and purification from carbonate or seawater samples, although some laboratories choose to separate boron from Ca and other matrix cations (Duke University laboratory method reported in Foster et al. 2013). The salts in seawater enhance the ionization efficiency by reducing the electronic work function of the filament (Hemming and Hanson 1992, 1994; Eq. 4.3). Boron-free natural seawater can be prepared using the boron specific Amberlite IRA 743 ion-exchange resin (Barth 1997; Figure 4.4) else a synthetic seawater analogue prepared by mixing high purity Ca, Mg, Na, and K solutions in 5% HCl (Dwyer and Vengosh 2008). The latter approach eliminates the potential for occurrence of CNO^- molecular interferences due to small amounts of resin bleeding during seawater purification (Hemming and Hanson 1994; Foster et al. 2006).

The analysis of small amounts of boron by N-TIMS necessitates more care in the handling and processing of samples to avoid contamination from laboratory boron sources (Rosner et al. 2005; Section 4.4.3) and to also avoid generating interfering organic species (Hemming and Hanson 1994). The potential presence of a significant $^{12}C^{14}N^{16}O^-$ interference at 42 amu can be determined by monitoring the production of $^{12}C^{14}N^-$ at 26 amu (Hemming and Hanson 1994). This is usually undertaken using an electron multiplier and rejecting analyses or parts of an analysis where the measured $^{12}C^{14}N^-$ signal exceeds a threshold value, usually ~2000 cps (cf. typical count rates of a few tens or hundreds cps; Hemming and Hanson 1994; Foster et al. 2013).

N-TIMS analyses are carried out by raising the filament current slowly over a period of ~25 minutes until reaching the analysis temperature of 970–980 °C, which is monitored using an external filament pyrometer (Pelejero et al. 2005; Foster et al. 2013). The analysis is commenced once the $^{11}BO_2^-$ signal intensity reaches ~100 mV, and is continued for a period of ~25–45 minutes over which time the signal intensity ideally rises to between several hundred mV and a few V (as measured by amplifiers fitted with 10^{11} Ω resistors). Coral samples have been noted to ionize more efficiently than foraminifer samples, and to reach higher signal intensities. As noted above some N-TIMS methods employ ion chromatography to remove Ca and other matrix ions from samples prior to loading and analysis, so that pure boron sample solutions can be generated and analyzed that closely match dissolved NIST SRM 951 boric acid. Samples processed in this way are reported to pre-heat and ionize at a lower temperature (~900 °C cf. 970–980 °C), with reduced organic molecule interference (<500 cps for CN^-), and exhibit lower mass dependent fractionation (0.2 to 0.5 ‰; see Duke University laboratory method in Foster et al. 2013).

Total evaporation N-TIMS (TE-N-TIMS) has been promoted as a means of accommodating instrumental mass fractionation effects and maximizing the measured sample signal to minimize the amount of boron required for analysis (Foster et al. 2006). The TE-N-TIMS method entails analyzing the loaded sample until it is completely exhausted from the filament (Kanno 1971). Samples comprising ≤1 ng boron are ideally analyzed so the sample measurement can be completed in an acceptable time (several hours). Unfortunately, the TE-N-TIMS technique is reported to produce biased results (by ~2–6 ‰) relative to both MC-ICP-MS and traditional N-TIMS approaches (Ni et al. 2010), while also delivering much larger analytical uncertainties (often ≥0.7 ‰, 2σ; Ni et al. 2007). An internal mass fractionation correction schema, based on the oxygen isotope composition of ReO_4^- ions generated during TE-N-TIMS at low filament temperatures, has been proposed by Aggarwal et al. (2009b) but has not been widely adopted. The greater reliability and simplicity of the N-TIMS method (as opposed to TE-N-TIMS) has prevailed (Pelejero et al. 2005; Hönisch et al. 2009; Farmer et al. 2015).

The precision of individual NTIMS analyses is typically <0.1 ‰ (2se), but this is usually significantly better than the external reproducibility of repeat analyses. The relatively large atomic mass difference between the measured $^{11}BO_2^-$ and $^{10}BO_2^-$ ions can result in significant mass dependent isotope fractionation. This is corrected for by reference to analysis of the primary isotopic reference material (NIST SRM 951 boric acid) at identical conditions. However, time dependent fractionation during analysis is an inherent issue with TIMS that cannot be corrected for because boron only has two isotopes. This is addressed by adopting strict analysis protocols that involve consistent slow ramping of filament temperature accompanied by repeated ion beam focusing, and analysis at the lowest possible filament temperature. Most N-TIMS analysis protocols usually also involve measuring separate aliquots of the same prepared sample solution in triplicate. Individual analyses are rejected if the mass fractionation of the 43/42 mass ratio measured within an analysis exceeds a threshold value (e.g. 1 ‰; Sanyal et al. 1996; Hönisch and Hemming 2004). In this event, or if the reproducibility of repeated $\delta^{11}B$ measurements for a sample exceeds a quality control threshold value (e.g. $\geq \pm 0.35$ ‰ or outlier test based on reproducibility calculated as $2\sigma/\sqrt{n}$), more samples can be run in effort to improve the precision (Hönisch et al. 2004; Pelejero et al. 2005). However, where care is taken with sample and filament preparation, and consistency of the analysis protocol maintained, the need to reject spurious analyses is rare. For instance, Bartoli et al. (2011) report 219 successful analyses of which only three outliers were excluded based on Chauvenet's criterion (i.e. they deviated by more than 2sd from the population mean).

Readers are cautioned that significant biases may persist between published N-TIMS data from different laboratories. Pearson and Palmer (2000), for example, report values for planktic foraminifera that are significantly higher than reported by subsequent studies of similar samples. In this context, it is notable that Pearson and Palmer (2000) employed a measurement protocol that waits until observed mass fractionation stabilizes. Similarly, Sanyal et al. (1995, 2001) also waited ~30 minutes before commencing data acquisition and conducted analyses at a higher temperature (reported in Hönisch and Hemming 2005). As the fractionation behavior of carbonate samples and the NBS 951 standard may differ over time, these wait times result in measured unknowns being biased compared to analyses commenced immediately upon reaching the analysis temperature and a signal intensity of 100 mV. By way of example, Hönisch and Hemming (2005) obtained ~1 ‰ lower $\delta^{11}B$ values than Sanyal et al. (1995, 2001) for for the same coretop *Globigerinoides sacculifer* samples when measured on the same N-TIMS instrument at the same temperature as NBS 951 analyses but without the 30 minutes wait time. Inter-laboratory biases have also been reported where different mass spectrometers have been used (e.g. Finnigan MAT 262, NBS 6″, and Thermo Finnigan Triton). This may be due to differences in the behavior and extraction efficiency of their ion sources and/or in the actual analysis temperatures when using different instruments, noting the pyrometers used to monitor filament temperatures may not all be calibrated equivalently.

4.7.4 Inductively Coupled Plasma Mass Spectrometry (ICP-MS)

4.7.4.1 Boron Concentration and B/Ca Molar Ratio Analysis

Since development of the first commercial instruments in the early 1980s (Houk et al. 1980), ICP-MS has become the method of choice for routine analysis of boron concentration and isotope ratio in a wide variety of Earth and environmental materials including marine carbonates. The widespread adoption of ICP-MS is underpinned by the capacity of an Ar-inductively coupled plasma (Ar-ICP) to atomize liquid and solid aerosols and to efficiently ionize most elements. The Ar-ICP employs collisional energy transfer between Ar gas and electrons accelerated in a radiofrequency magnetic field (26 or 42 MHz) to generate a high temperature (~7000–10 000 K) plasma in which atom and ion energies are buffered near the 1st excitation state of Ar (~10.6 eV). Under these conditions elements with ionization potentials significantly lower than ~7.5 eV are fully ionized (Turner and Montaser 1998), and those with higher ionization potentials up to 10.6 eV are partially ionized. Boron, which has a first ionization potential of 8.3 eV, is partly ionized, typically with about 50% efficiency subject to operating conditions of the Ar-ICP. Only molecular species with very high bond strengths and a few doubly-charged ions are stable under these plasma conditions, such that sampling of the Ar-ICP results in spectra that are relatively free of interferences. However, ion sampling from the Ar-ICP is relatively inefficient, particularly at low mass where only one in ten thousand, or fewer, boron ions generated within the Ar-ICP is extracted through the interface region into the mass spectrometer (Figure 4.7). Higher accelerating voltages applied to the ion extraction region immediately downstream from the interface enable more efficient ion sampling and account for the higher sensitivity of sector field (SF) than quadrupole (Q)-ICP-MS instruments. Despite the relative inefficiency of ion extraction, ICP-MS provides high sensitivity, a large dynamic range (~6 orders of magnitude) and rapidity and ease of use, resulting in exceptional capabilities for accurate and precise analysis of boron and other elements. Its many advantages aside, ICP-MS is a drift prone technique that requires attention to minimizing and, where necessary, correcting for varying instrument mass fractionation bias and varying bias in the efficiency of ionization of different elements. For these reasons, standard-sample-standard bracketing and matrix matching are important in attaining accurate results.

Inductively coupled plasma optical emission spectrometry (ICP-OES, also known as ICP-AES) is a precursor technique to ICP-MS that retains some advantages over ICP-MS for element ratio analysis. Although ICP-OES lacks sensitivity for analysis of boron and most other elements compared to ICP-MS and has a more restricted linear dynamic range (ca. 2–3 orders of magnitude; Olesik 1991), its simultaneous analysis capabilities facilitate more precise trace metal/Ca ratio measurements than ICP-MS (i.e. <0.2–0.5% vs ~1–2% RSD). Accordingly, ICP-OES is sometimes used in

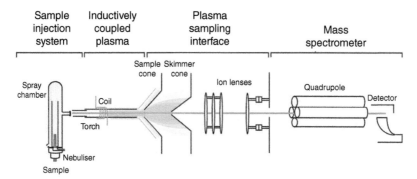

Figure 4.7 Schematic diagram of a quadrupole ICP-MS instrument showing the sample introduction system on far left (comprising nebulizer, spray chamber, and torch), the Argon-inductively couple plasma generated by the radiofrequency coil surrounding the torch, the plasma sampling interface (comprising sample and skimmer cones) followed by the ion extraction and ion-focusing optics, and the quadrupole mass filter and ion detector on the far right.

combination with ICP-MS to generate high quality data sets for a range of trace elements including boron in marine carbonates. Recently, Kaczmarek et al. (2015) used an optic fiber to view the Ar-ICP source of a multi-collector (MC)-ICP-MS instrument and enable simultaneous analysis of boron concentrations along with boron isotope ratios in marine carbonates sampled by laser ablation. They report the ability to measure B concentrations and isotope ratios to a typical relative precision of ±3% (RSD) and 0.53 ‰ (RSE).

4.7.4.2 Solution-ICP-MS Analysis of Boron and Other Trace Elements in Marine Carbonates

Numerous laboratories have developed and reported protocols for analysis of boron concentrations of marine carbonates using single-collector quadrupole- and sector field-ICP-MS instruments (e.g. Rosenthal et al. 1999; Yu et al. 2005; Marchitto 2006; Babila et al. 2014; Misra et al. 2014a). Typically, boron concentrations are measured along with Ca and as part of analysis routines that measure multiple elements of interest in marine carbonates (e.g. Li, B, Mg, Ca, Mn, Fe, Zn, Sr, Cd, Ba, and U; Yu et al. 2005; Babila et al. 2014). A comparison of B/Ca measurement methods by Q-ICP-MS and SF-ICP-MS is provided in Table 4.4. Because both these instrument types are only able to measure a single mass at any time, they rely on the ability to switch (peak-hop) rapidly between assigned element isotope masses during the analysis. Quadrupole-inductively coupled plasma mass spectrometers (Q-ICP-MS) are able to do this more nimbly than sector field instruments, but SF-ICP-MS is more sensitive than Q-ICP-MS when operated at low mass resolution (M/ΔM = 300). Sector field-ICP-MS instruments can also be operated at higher mass resolution (M/ΔM = 3000 and 7500) giving the ability to resolve interfering species from analyte masses, however these capabilities are generally not required for boron concentration or isotope ratio analysis.

Table 4.4 Comparison of ICP-MS methods used for boron concentration analysis.

	Q-ICP-MS	SF-ICP-MS (normal sensitivity interface)	SF-ICP-MS (high sensitivity interface)
Primary reference	Yu et al. (2005)	Babila et al. (2014) after Rosenthal et al. (1999)	Misra et al. (2014a)
Measured isotope and abundance (%)	$^{11}B - 80\%$ $^{46}Ca - 0.004\%$	$^{11}B - 80\%$ $^{43}Ca - 0.31\%$	$^{11}B - 80\%$ $^{43}Ca - 0.31\%$ $^{44}Ca - 2.1\%$ $^{46}Ca - 0.004\%$
Sensitivity (cps/ppb)	10×10^3	–	2.5×10^6
B blank level (% sample intensity)	5	2	<1
Reproducibility (RSD %)	4.2	3.3	2
Minimum sample size (μg $CaCO_3$)	60	10	5
Detection Limit ($\mu mol\,mol^{-1}$)	15	–	2
Diluted Ca matrix concentration (ppm)	100	60–160	5–10
Spray chamber	Cyclonic – Quartz	Cyclonic – Quartz	Scott type (single pass) – Teflon®
Nebulizer	Micro-concentric - 60 $\mu l\,min^{-1}$ (Glass Expansion, Micromist™)	Teflon micro-concentric – 100 $\mu l\,min^{-1}$ (T-1, CETAC)	ESI™ – 50 $\mu l\,min^{-1}$
Measures to reduce boron memory effects	Ammonia gas added to spray chamber	Ammonia gas added to spray chamber	Mixed acid matrix of 0.1 M HNO_3 + 0.3 M HF
Comments including elements co-analyzed with B	Fast sequential peak hopping Li, Mg, Al, Si, Ca, Mn, Fe, Zn, Sr, Y, Cd, Ba, LREE, Pb, U	Fast sequential peak hopping Li, Mg, Al, Si, Ca, Mn, Fe, Zn, Sr, Y, Cd, Ba, LREE, Pb, U	Fast sequential peak hopping Li, Mg, Al, Si, Ca, Mn, Fe, Zn, Sr, Y, Cd, Ba, LREE, Pb, U

Boron is one of the more difficult elements to analyze by ICP-MS. This is largely due to the self-repulsion of positively charged ions that reduces ion transmission efficiency, particularly of low mass ions, during extraction into the ICP-MS (Douglas and Tanner 1998). This so-called space-charge effect strongly discriminates against the lighter $^{10}B^+$ than $^{11}B^+$ ion, resulting in large positive mass bias of measured $^{11}B/^{10}B$ isotope ratios relative to other techniques (e.g. $\Delta M/M$ between masses 10 and 11 is significantly larger than that between 308 and 309 measured for $Cs_2BO_2^+$ by PTIMS, ~10% vs. 0.3%). Greater ion densities and heavier matrix ions result in larger space charge effects (ion beam self-repulsion) such that mass bias can change with the nature and amount of matrix and analyte ions generated from different samples and standards in an analytical run. Overall sector-field-ICP-MS instruments are less prone to mass bias drift than Q-ICP-MS instruments, due to the higher accelerating voltages used to extract ion beams from the interface region. The Ar-ICP operating conditions and the extent to which different elements are ionized within the plasma are subject to the amount of introduced matrix material. Higher matrix loads tend to decrease the ionization efficiency of boron and other elements that have higher ionization potentials. The much higher first ionization potential of B compared to Ca (8.30 vs. 6.11 eV) and its lower mass, can combine to decrease measured B/Ca ratios in the presence of higher matrix contents. Internal standard elements and enriched isotopes can be added to samples and standards to monitor and correct for space charge and ionization efficiency changes encountered during ICP-MS analysis (Eggins et al. 1997). However, many marine carbonate analysts choose to use large and closely matched matrix dilutions (e.g. 100 µg of dissolved $CaCO_3$ per gram of solution) to minimize matrix effects across all samples and standards (Rosenthal et al. 1999; Yu et al. 2005; Misra et al. 2014a).

The boron isotopes are largely free from isobaric interferences in ICP-MS analysis, except for multiply-charged plasma and matrix ions, specifically $^{40}Ar^{4+}$ (59.81 eV, 9.99622 amu), $^{20}Ne^{2+}$ (40.96 eV, 9.99622 amu), and possible overlap of $^{40}Ca^{4+}$ (67.27 eV, 9.99065 amu) on $^{10}B^+$ (10.01294 amu), and $^{44}Ca^{4+}$ (67.27 eV; 10.98887 amu) along with $^{10}BH^+$ (11.020877 amu) on $^{11}B^+$ (11.01294 amu) (see Figure 4.8). These multiply-charged plasma and sample matrix ions are prominent during SF-ICP-MS and MC-ICP-MS analysis but virtually non-existent with Q-ICP-MS analysis. The formation of these multiple charged ions requires very high ionization energies that cannot be generated in the Ar-ICP but rather, are generated by high energy collisions

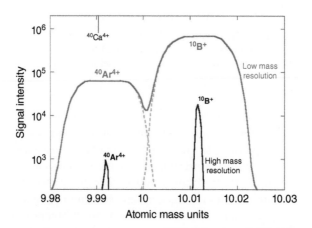

Figure 4.8 Mass spectrum measured in the vicinity of $^{10}B^+$ measured at low mass resolution (M/ΔM < 400, solid blue curve) and at high mass resolution (M/ΔM > 7000, solid black peaks). The shoulder of the $^{40}Ar^{4+}$ peak partially overlaps with the shoulder of the $^{10}B^+$ peak when measured at low resolution (interpolated dotted blue lines). Both peaks are full resolved when measured at high mass resolution, but the measured signal intensity is reduced by more than an order of magnitude. The approximate position of the potential interfering species $^{40}Ca^{4+}$ is indicated and is absent with the measured high-resolution mass spectra (Source: adapted from Fietzke et al. 2010).

under high accelerating voltages (typically 2000 V) within the interface region of sector field single- and multiple-collector ICP-MS instruments (cf. 0–200 V used in Q-ICP-MS instruments). Fortunately, the mass/charge ratios of $^{40}Ar^{4+}$, $^{40}Ca^{4+}$, and $^{44}Ca^{4+}$ can be fully resolved from $^{10}B^+$ and $^{11}B^+$ at low mass resolution (M/ΔM = 300) of the sector field instruments given well optimized peak shapes and mass resolution (Figure 4.8).

Boron concentrations in marine carbonates are generally reported as B/Ca molar ratios and so entail analysis of both B and Ca. The most abundant Ca isotope (^{40}Ca = 96.94%) cannot be resolved from the abundant interfering plasma ion $^{40}Ar^+$, requiring analysis of one of the minor calcium isotopes, usually $^{43}Ca^+$ (0.163%) or $^{44}Ca^+$ (2.02%) although some laboratories measure $^{46}Ca^+$ (0.004%), and $^{48}Ca^+$ (0.186%) is also a possibility. These minor Ca isotopes all have significant isobaric interference with molecular species deriving from the plasma gas and acidic or aqueous media, as well as with some doubly-charged sample matrix species (e.g. $^{40}ArH_2^+$, $^{16}N_3^+$, and $^{84}Sr^{2+}$ on $^{42}Ca^+$; $^{86}Sr^{2+}$ on $^{43}Ca^+$, and $^{14}N_2O^+$, $^{12}CO_2^+$, $^{88}Sr^{2+}$ on $^{44}Ca^+$). In general, contributing plasma gas interferences can be corrected for by subtracting "blank" spectra of the aqueous or acid media, or can be fully resolved using the high mass resolution capabilities of sector-field ICP-MS.

Perhaps the most problematic issue for boron concentration and boron isotope analysis is accumulated memory within, and slow washout of boron from, the sample introduction system following prior analysis of other standards and samples (e.g. Al-Ammar et al. 1999, 2000). The memory effect can be severe where boron occurs as boric acid in droplets on the walls of the spray chamber under acidic and neutral conditions (Al-Ammar et al. 2000) and is exacerbated when using sample introduction systems constructed from inert plastics, which do not wet and drain as readily as surfaces of quartz glass spray chambers. This problem can be counteracted by using surfactants to increase the wettability of spray chamber internal surfaces, or by using higher solution aspiration rates and/or spray chamber designs that provide more effective "wash down" of the spray chamber walls. For example, a dual port spray chamber (Hirata 2000) allows a second "boron-free" solution to be aspirated into the spray chamber to enhance the "wash-down" of the chamber walls and reduce washout times. This can be particularly effective when using low flow nebulizers (20–100 µl min^{-1}) in combination with larger volume spray chambers to measure small samples.

The volatility of boron can be reduced by cooling the spray chamber or by adding ammonia gas or dissolved ammonia solution to raise the pH within the spray chamber. Cooling the spray chamber from room temperature to 0 °C reduces the vapor pressure of boric acid by an order of magnitude (Brandani et al. 1988) but this approach has not been adopted in preference to adding ammonia gas, which is used to raise the pH of the spray chamber environment above the pKa of boric acid (9.14) and quantitatively convert boric acid to the non-volatile borate species (Al-Ammar et al. 2000). Ammonia gas has been employed to reduce boron memory problems with both boron elemental and isotopic analysis by ICP-MS (Al-Ammar et al.

1999, 2000; Foster 2008). It requires accurate control over the input of ammonia at small flow rates (a few cm^3/min) using a low-flow mass flow controller that is chemically resistant to ammonia. Ammonia also brings risk to the operator and laboratory due to the toxicity of ammonia gas, which is dangerous to life and health where concentrations exceed 300 ppmv in air or extended exposure (>15 minutes) occurs at concentrations >25 ppmv (ATSDR 2004).

An alternative strategy used for reducing boron memory and washout problems is to employ hydrofluoric acid (HF), which has a strong affinity for boron (Makishima et al. 1997; Misra et al. 2014b). In an HF matrix boric acid reacts via BF_3 to form tetrafluoroboric acid (H_3OBF_4; pKa = −0.4), which dissociates to its conjugate base BF_4^-. Both BF_3 and H_3OBF_4 are highly soluble in water, and H_3OBF_4 has a boiling point of 130 °C such that in the presence of HF matrix it should have low volatility under standard laboratory conditions.

The range of strategies outlined above are used to minimize boron memory problems when making simultaneous measurements of a wide range of elements in addition to boron. The reported accuracy and precision of B/Ca determinations as part of multiple element routines is generally in the range ~3–5% (2σ, Rae et al. 2011; Babila et al. 2014), with reported reproducibility for B/Ca measurements using SF-ICP-MS being <4% (2σ) when analyzing $CaCO_3$ samples as small as 5 μg (Misra et al. 2014a). Similar levels of precision are also obtained using high sensitivity Q-ICP-MS instruments albeit on larger samples of foraminiferal calcite (Yu et al. 2013).

4.7.5   MC-ICP-MS

Multi-collector inductively coupled plasma mass spectrometry (MC-ICP-MS) is increasingly being employed for boron isotope analysis (e.g. Guerrot et al. 2011; Louvat et al. 2011; Rae et al. 2011; McCulloch et al. 2014), due to its ability to measure small samples with high accuracy and precision, and relative ease of use and rapidity of analysis (Foster 2008). Because both boron isotopes are measured simultaneously (as opposed to single-collector ICP-MS), MC-ICP-MS affords more precise measurement of isotope ratios when sampling the noisy Ar-ICP ion source and because it also increases (up to double) the measurement efficiency.

Notwithstanding these advantages, MC-ICP-MS is subject to the same suite of analytical challenges encountered with single-collector ICP-MS, namely: (i) memory problems arising from the volatility of boric acid; (ii) a large instrument induced mass bias (10% per u) which is susceptible to change with matrix composition and to drift with time during an analytical run; (iii) potential interference at low mass resolving power (M/ΔM = 300); and (iv) relatively poor transmission efficiency and ionization of boron. Accordingly, many of the same approaches have been adopted to address these problems as with boron analysis by single-collector ICP-MS.

Memory problems have been reduced by introducing ammonia gas to the spray chamber (Foster 2008; Farmer et al. 2016), employing direct injection nebulizers (DIHENs) to circumvent the use of spray chambers (Louvat et al. 2011), and adding HF to complex boron in solution as non-volatile BF_3 and tetrafluoroboric acid species (Misra et al. 2014b). Studies using these approaches report similar abilities to "wash-out" boron signals to <2% of peak sample or standard intensities within 100–120 seconds.

Isobaric interferences can be resolved by using a mass resolving power (M/ΔM) up to and exceeding 7000. However, because MC-ICP-MS instruments lack collector slits, analyte masses are measured as shoulders on the sides of the overlapping peaks. In the case of boron, the main interfering species (e.g. $^{40}Ar^{4+}$ and $^{20}Ar^{2+}$ on $^{10}B^+$) are sufficiently different in mass that they are fully resolved at M/ΔM = 300, given optimized peak shapes and low production of $^{40}Ar^{4+}$ relative to $^{10}B^+$. The potential for $^{10}BH^+$ interference on $^{11}B^+$ has been shown to be insignificant by Foster (2008).

The very large mass bias (~10–15% at 10–11 amu) encountered with MC-ICP-MS compared to N-TIMS and P-TIMS can vary over the course of a MC-ICP-MS analytical run. This necessitates correction using a standard-sample-standard bracketing to correct for mass bias, rapid analysis of individual solutions combined with fast washout of any memory so the drift between repeat standard analyses is minimized, and preparation of samples and standards to have matching boron concentrations and minimal matrix contents (Foster 2008; Louvat et al. 2011; Holcomb et al. 2014).

4.7.6   ICP-MS and MC-ICP-MS Sample Introduction

4.7.6.1   Conventional High Efficiency Nebulizers (HENs) and Spray Chambers

MC-ICP-MS analysis procedures are notable for employing different ways of introducing dissolved standards and samples into the Ar-ICP. This includes the use of different types of nebulizers and spray chambers, and the addition (or omission) of ammonia gas to the spray chamber. These variations reflect underlying efforts to improve the accuracy and precision of boron isotope analysis by maximizing method sensitivity and minimizing memory effects, as well to reduce sample size requirements and speed-up analysis so that more samples can be measured with improved drift correction.

Most established procedures use HENs, almost exclusively low-flow (40–60 µl min^{-1}), micro-concentric, pneumatic nebulizers constructed of chemically resistant Teflon (e.g. Foster 2008; Wang et al. 2010; McCulloch et al. 2014). Lower and higher flow HENs could be used as alternatives, but lower flow rate nebulizers are more susceptible to blockage and require longer total analysis times, and higher flow rate nebulizers use more sample. Micro-concentric nebulizers produce very fine aerosols within a narrow size range (typically 1–10 µm). These small droplets pass more efficiently

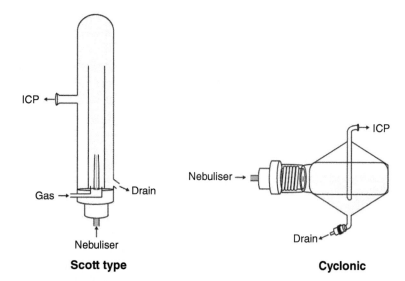

Figure 4.9 Schematic
diagrams comparing the Scott
and cyclonic spray chamber
designs which are commonly
used for solution sample
introduction in inductively
coupled plasma-mass
spectrometry.

through spray chambers, which are designed to remove larger droplets through collision with the spray chamber walls as larger droplets result in noisier ion signals. The spray chambers are designed to be wetted by and to drain away the aerosol fraction that collects on the chamber walls.

Most MC-ICPMS procedures employ Scott-type double-pass spray chambers made of acid resistant Teflon™ (Foster 2008; Wang et al. 2010; Rae et al. 2011; Farmer et al. 2016; Figure 4.9). These spray chambers produce very stable signals, however, they accumulate droplets and suffer memory problems because of their relatively large surface area which is difficult to maintain in a fully wetted and well drained state. Consequently, most laboratories have resorted to adding small flows of ammonia gas (typically $<5\,cm^3\,min^{-1}$) sufficient to elevate the pH within spray chambers so that memory effects are reduced to an acceptable level (<2–3% of peak sample signal intensity after ~120 seconds) and to maintain sample throughput (Rae et al. 2011; Foster et al. 2013; Farmer et al. 2016). In a departure from this approach, McCulloch et al. (2014) employ a cyclonic quartz spray chamber (Figure 4.9) which produces faster washout due its smaller volume and improved surface wetting and draining (cf. Teflon). They report not needing to introduce ammonia gas to reduce boron memory effects to levels commensurate with or better than other procedures.

4.7.6.2 Direct Injection High Efficiency Nebulizers (DIHENs)

Direct injection nebulizers (DINs) aspirate solutions as fine aerosols directly into the Ar-ICP, circumventing the need for a spray chamber and avoiding associated boron memory problems. Much faster wash-out of boron is reported, requiring only 100 seconds to achieve ^{11}B signals <1 ‰ of prior peak intensities (Louvat et al. 2011). These nebulizers also provide advantages in terms of aerosol transfer to the Ar-ICP, delivering 2 to 5x greater improvement in analysis sensitivity compared to methods using conventional spray chambers (Smith et al. 1991; Louvat et al. 2011, 2014). Nonetheless, this increase is smaller than expected, as DIHENs generate some larger droplets (up to 30 µm diameter) which are difficult to fully evaporate, atomize, and ionize in the Ar-ICP (Minnich et al. 2001). The difficulty DIHENs have with generating a narrow size distribution of small droplets can also make igniting and maintaining a stable Ar-ICP more problematic than conventional nebulizer-spray chamber configurations. The d-DIHEN (demountable version of the DIHEN) used by Louvat et al. (2011, 2014), and first described by Westphal et al. (2004), permits the aerosol size distribution characteristics to be adjusted using a needle and capillary that can be changed separately. Boron isotope data reported from the Institut de Physique du Globe de Paris (Louvat et al. 2014) indicate broadly comparable capabilities and results to those obtained using more conventional MC-ICPMS methods (see Foster et al. 2013).

4.8 Laser Ablation-Inductively Coupled Plasma-Mass Spectrometry (LA-ICP-MS)

Laser ablation-(MC)-ICP-MS provides a highly flexible, sensitive, and rapid method for making spatially resolved boron concentration and boron isotope measurements on calcium carbonate and a wide range of other sample materials. An advantage of laser ablation sampling, as with direct injection nebulization, is sample boron is transferred directly into the Ar-ICP in particulate form and is not subject to volatility and memory issues encountered with conventional solution nebulization. However, the drawback of this technique is that boron is not separated from the sample matrix prior to analysis. This creates the potential for variable sample matrix-dependent plasma loading and associated mass bias effects. These problems increase with higher laser ablation sampling rates (i.e. with laser ablation spot size, pulse energy, and/or pulse repetition rate), which can be required to generate signal intensities sufficient to make precise isotope ratio measurements on marine carbonates due to their low boron contents.

Laser ablation sampling of calcium carbonate can produce a narrow size distribution of fine particulates that are readily digested, atomized, and ionized by the Ar-ICP. This requires use of high powered lasers that have either very short wavelengths (λ <200 nm), as calcite is transparent at longer wavelengths, or have ultrashort (femtosecond) pulse lengths that facilitate multi-photon absorption (Park and Haglund 1997). Controlled laser sampling of calcium carbonate was first achieved with the use of nanosecond ArF ($\lambda = 193$ nm) excimer lasers in the mid-late 1990s (Sinclair et al. 1998), around the same time Jeffries et al. (1998) demonstrated improved laser ablation of calcite using $\lambda = 213$ nm versus $\lambda = 266$ nm laser light.

Laser ablation ICP-MS was first used to quantify the B/Ca compositions of marine carbonates by Sinclair et al. (1998) and Fallon et al. (1999), who reported large cyclic seasonal variations in tropical corals containing 50–60 $\mu g\,g^{-1}$ boron. These measurements involved translating a large rectangular laser "spot" (500 x 50 μm), aligned parallel to growth banding, along the direction of coral growth. The broad ablation area helped smooth out fine-scale compositional heterogeneity within the corals and produced a large sample yield that compensated for the low sensitivity of the Q-ICP-MS instrument used. These studies were able to quantify B/Ca and other trace element variations at approximately fortnightly to monthly growth resolution, and to generate continuous records spanning several centuries by splicing together analyzed sections from long coral cores (e.g. Alibert and Kinsley 2008).

The LA-ICP-MS technique has proved useful for addressing questions surrounding the nature and extent of B/Ca variability at small length scales in a range of marine carbonates, including B/Ca variability within individual shells and between different shells comprising foraminifer populations grown in culture or sampled from nature. Raitzsch et al. (2011) reported B/Ca to vary by a factor of two within individual shells of the benthic foraminifer *P. wuellerstorfi*, and found significant differences between chambers in the same shell. Raitzsch et al. (2011) were also able to demonstrate, despite significant B/Ca variability within and between shells, that mean shell population compositions varied with seawater $\Delta[CO_3^{2-}]$ in a way consistent with the behavior of bulk shell compositions measured using solution ICP-MS by Yu and Elderfield (2007). Allen et al. (2011) subsequently employed a high resolution (sub-micron) depth profiling method (Eggins et al. 2004) to measure B/Ca banding in the shell walls of the laboratory cultured planktic foraminifera, *Orbulina universa*. This cyclic B/Ca banding has since been shown to be antiphased with observed diurnal Mg/Ca banding in the same *O. universa* shells, and consistent with diurnal changes in carbonate system chemistry within the foraminiferal microenvironment regulated by respiration and symbiont photosynthesis (Holland et al. 2017).

Laser ablation sampling can be combined with MC-ICP-MS for precise analysis of boron isotope ratios in marine carbonates. By comparison the precision achieved using single-collector, quadrupole-, and sector field-ICP-MS instruments is inherently limited by the "noisy" and often rapidly changing signals that are generated by laser sampling. The first boron isotope measurements made using laser ablation-MC-ICP-MS were obtained on natural silicate glasses by Le Roux et al. (2004) who reported achieving precision of ~ ± 1.0 ‰ (2se) albeit using extreme laser ablation conditions (200 μm diameter spot, 20 Hz laser pulse rate and a pulse energy of 5 mJ per laser pulse to sample > 60 μg of silicate glasses containing between 0.39 and 30.2 $\mu g\,g^{-1}$ boron over a 360 seconds analysis period). The higher boron contents of some marine carbonates have facilitated reports of similar precision/reproducibility (i.e. ~ ± 1.0 ‰ 2se) at more moderate laser sampling conditions (0.2 ng total boron, Fietzke et al. 2010; Kaczmarek et al. 2015).

The analysis of boron isotopes by LA-MC-ICP-MS in marine carbonates with relatively low boron concentrations results in small $^{10}B^+$ and $^{11}B^+$ signal intensities that are subject to proportionally higher effects of Johnson noise when using amplifiers fitted with standard 10^{11} or 10^{12} Ω resistors. This problem has been addressed recently with newly developed low-noise 10^{13} Ω amplifiers that produce higher measured signal intensities. However, these amplifiers have much slower response times than 10^{11} Ω amplifiers, and if the amplifier response times are not perfectly matched, measured isotope ratio values will vary with the rate and sign of signal intensity change (Iizuka et al. 2011). This is problematic when measuring transient signals generated by laser ablation sampling, as measured B isotope ratios can be artificially biased to higher/lower values depending on whether the ^{11}B and ^{10}B signal intensities are increasing or decreasing with time. An alternative approach is to measure small $^{10}B^+$ and $^{11}B^+$ signal intensities with ion counters (e.g. Kaczmarek et al. 2015), which requires ion count rates to be maintained below thresholds where the counting system dead-time correction becomes non-linear (i.e. $<1 \times 10^6$ counts per second). This can require long analysis times to achieve desirable high levels of precision due to counting statistic limitations (i.e. needing to count 10^7 to 10^8 $^{10}B^+$ ions to achieve a $\delta^{11}B$ precision in the range 0.2–0.5 ‰ 2se; where the relative se (‰) = $1000 \times \sqrt{((1/\Sigma N_{10}) + (1/\Sigma N_{11}))}$, here ΣN_{10} and ΣN_{11} are the total number of counted $^{10}B^+$ and $^{11}B^+$ ions).

As with solution-based analysis, standard-sample-standard bracketing is used for instrument calibration, and to monitor and correct for mass bias drift when measuring B/Ca or B isotopes by LA-ICP-MS. The outstanding issue for analysis of marine carbonates is the lack of suitable matrix-matched standards, as most natural calcium carbonates are heterogeneous at the length scales of laser ablation sampling (tens to hundreds of microns) and so are unsuitable as microanalysis standards. Fallon et al. (1999) addressed this problem by producing a pressed pellet of a well character-ized coral powder sample mixed with fine Teflon powder. The inherent compositional heterogeneity of the coral powder was overcome by scanning and integrating the calibration measurement over a very large area of the pellet. The same approach cannot be applied to static laser ablation sam-pling using typical spot sizes in the range 10–100 μm, as this overlaps the size range of compositionally distinct particles that make up the pressed powdered coral pellet. Accordingly, most analysts have resorted to using the NIST SRM 610–617 soda-lime glasses for calibration of LA-ICP-MS analysis (Eggins et al. 2003; Hathorne et al. 2003). This works well for the analysis of alkaline earths (Mg, Ca, Sr, and Ba) but may incur matrix and plasma loading-related element biases in the case of B/Ca ratio analysis due to the very different atomization and ionization behaviors of B and Ca in the Ar-ICP.

Fietzke and Frische (2016) have shown matrix-dependent mass biases can be mitigated by establishing "robust," hot Ar-ICP, operating conditions and maintaining these by avoiding overloading the plasma using large laser spot sizes and repetition rates. It remains to be established just how accurately B/Ca and other trace elements can be determined in calcium carbonate samples by calibration against the NIST SRM 610–617 soda-lime glasses. In the case of boron isotopes, Fietzke et al. (2010) have demonstrated accurate and precise analysis of calcium carbonate materials when calibrated against the NIST SRM 610 glass using a robust, hot plasma approach. Ultimately, addressing the effects of matrix-dependent bias on the accuracy of boron concentration and isotope ratio analysis of marine carbonates will require the community of analysts to have suitable, matrix-matched SRMs. Accordingly, efforts to find suitable homogeneous natural calcium carbonates and to synthesize homogeneous calcium carbonate standards deserve to be given greater priority (e.g. development of the MAC series standards distributed by Stephen Wilson of the USGS, see Jochum et al. 2012). Although a number of studies report using homogeneous "in-house" calcium carbonate materials to verify the quality of LA-ICP-MS analyses (Hathorne et al. 2008) these materials are not widely available. Promise may lie in wet-milling to produce much finer, nano-particulate ($<1.5\,\mu m$ grain size) pressed powder pellets (Garbe-Schönberg and Müller 2014).

4.9 Secondary Ion Mass Spectrometry (SIMS)

Secondary ion mass spectrometry (SIMS) is a surface analysis technique that has occasionally been used for *in situ* microanalysis of boron concentrations and boron isotope ratios in marine carbonates. The technique can measure secondary ions as positive species ($^{11}B^+$ and $^{10}B^+$) or negative species ($^{11}B^-$ and $^{10}B^-$) by sputtering ions from the surface of a sample under high vacuum using an incident (primary) beam of $^{16}O^-$ ions or $^{133}Cs^+$ ions. Sputtered ions and charged species are collected and passed through electrostatic and magnetic sectors of the mass spectrometer for measurement by either single or multiple detectors (ion counters or Faraday cups). To date boron concentrations and isotopic compositions have been measured in corals, bivalves, foraminifers, and reference glasses with both smaller single-collector instruments (e.g. Cameca ims 4f; Kasemann et al. 2009) and larger multiple-collector instruments (e.g. Cameca IMS 1270 and IMS 1280; Rollion-Bard et al. 2003, 2011b; Blamart et al. 2007; Kasemann et al. 2009; Marschall and Monteleone 2015). The larger geometry SIMS instruments provide higher ion transmission efficiency, stability, and sensitivity, which facilitate more precise B isotope ratio determination in samples with lower boron contents or when using smaller incident beam sizes. SIMS analysis requires a clean, very flat, and highly polished sample surface (Kita et al. 2009). A shortcoming of the technique lies with its ion yield dependence on the sample chemical matrix and mineral structure orientation. This necessitates calibration using matrix-matched and potentially equivalently oriented reference materials that are also homogeneous in isotopic composition. Analysis also requires first cleaning the target site by sputtering the sample surface for about one minute prior to analysis using the incident ion beam (Kasemann et al. 2009).

Boron concentration and B/Ca molar ratio analysis in marine carbonates requires sequential analysis of at least one B and one Ca isotope. Invariably the most abundant boron isotope ^{11}B is selected to achieve the highest possible sensitivity and a lower abundance calcium isotope (e.g. ^{44}Ca) may need to be selected to avoid exceeding the ion counter dead time threshold. Kasemann et al. (2009) used a small radius, single collector, Cameca ims 4f SIMS instrument to measure ^{26}Mg$^+$, ^{27}Al$^+$, ^{30}Si$^+$, ^{55}Mn$^+$, and ^{88}Sr$^+$ in addition to ^{11}B$^+$ and ^{44}Ca$^+$ over 15 measurement cycles requiring a total analysis time of ~10 minutes (including a pre-analysis sputter period to clean the sample surface). The measured count rates for each element were normalized to Ca and calibrated against the NIST SRM 610 glass and several "in-house" calcium carbonate standards (NCC – Norman Cross Calcite, Swart 1990, and OKA – Oka Carbonatite calcite from Quebec, Canada, Allison et al. 2007). In a departure from other laboratories, McCoy et al. (2011) report measuring secondary ^{11}B$^-$ and ^{40}Ca$^-$ ions by rastering an incident ^{133}Cs$^+$ beam over a 30 x 30 μm area, from which a central region was selected using a 15 x 15 μm field aperture to avoid "edge effects." The Cameca IMS 1280 instrument used was operated with a mass resolution (M/ΔM) of 3100 to separate ^{40}Ca$^-$ from ^{24}Mg^{16}O$^-$. Analyses were conducted over a period of 600 seconds following a 120 seconds pre-analysis sputter cleaning period to remove surface contamination that was reported to affect B (total analysis time ~12 minutes). McCoy et al. (2011) report sensitivities of 250 cps for ^{11}B$^-$ and 50 000 cps for ^{40}Ca$^-$ when using 5–6 nA primary ^{133}Cs$^+$ ion beam. Measured ^{11}B$^-$/^{40}Ca$^-$ ratios of polished sections of *Mytilus californianus* shells were calibrated against a suite of calcium carbonate standards, and were measured with a typical internal precision of ±1.2% (2se) for B/Ca compositions in the range 20 to 120 μmol mol^{-1}.

The first reported analyses of ^{11}B/^{10}B isotope ratios by SIMS employed a small radius Cameca IMS 3f instrument to measure ^{11}B$^+$/^{10}B$^+$ ratios to a precision ±1.5 ‰ (2se) in alumino-silicate materials containing only 0.5 μg g^{-1} boron (Chaussidon et al. 1997). Rollion-Bard et al. (2003, 2011b) subsequently used a large radius Cameca IMS 1270 instrument to measure boron isotopes in a massive *Porites* coral using a mass resolution of (M/ΔM) 2000 to resolve the potentially significant isobaric interference of ^{10}BH$^+$ on ^{11}B$^+$. The ^{11}B and ^{10}B isotopes were measured sequentially using a single electron multiplier which yielded intensities of 1x10^4 cps for ^{10}B$^+$ and of 4x10^4 cps for ^{11}B$^+$. This produced an internal precision of ±1.2 ‰ (2se) after several minutes analysis, and an external reproducibility for carbonate standards of ±1.8 ‰ (2se). Kasemann et al. (2009) analyzed boron isotope ratios of marine carbonates using both a small radius Cameca IMS 4f instrument and a large radius Cameca IMS 1270 instrument. Secondary ^{10}B$^+$ and ^{11}B$^+$ ions were measured using a single electron multiplier in both cases, by cycling measurements between the two isotopes (5 seconds for ^{10}B$^+$ and 3 seconds for ^{11}B$^+$), for 200 cycles using the Cameca IMS 4f (total analysis time of ~30 minutes) and for 60 cycles using the Cameca IMS 1270 (total analysis time of ~10 minutes).

Samples with high boron concentrations (corals, bivalves, and reference glasses) were measured with a mass resolution (M/ΔM) of ~1600 to resolve potential interference of $^{10}B^1H^+$ on $^{11}B^+$ in samples and standards, and interferences of $^9Be^1H^+$ and $^{30}Si^{3+}$ on $^{10}B^+$ in silicate glass standards. Due to lower boron concentrations in foraminifer shells the analysis of boron isotopes required a lower mass resolution (M/ΔM ~500) to be used to increase ion transmission through the instrument. The size of the $^{10}B^1H^+$ interference on $^{11}B^+$ was found to be only ~$\pm$0.3 ‰ and contributed an insignificant level of uncertainty (~0.1 ‰) to the measured $\delta^{11}B$ composition of unknowns.

The accuracy and precision of boron isotope analysis of marine carbonates by SIMS is poorer than other techniques (Figure 4.2). Reported analysis precision of $\pm$1 ‰ is only marginally useful for palaeo-proxy applications which demand $\pm$0.2–0.4 ‰ but can be useful for characterizing compositional variation within proxy archives at the spatial resolution of SIMS analysis (10–30 μm). For example, Rollion-Bard et al. (2003) found large variability in the $\delta^{11}B$ composition in different structural elements of a *Porites* coral. This was interpreted to reflect pH variation between 7.1 and 9.0 at the site of calcification, and to be consistent with $\delta^{18}O$ variability measured by SIMS in the same coral due to pH induced changes in the dissolved carbonate species at the site of calcification as a result of slow kinetic equilibration with water. Similarly, Blamart et al. (2007) used a large Cameca IMS 1270 instrument to determine boron isotope ratios in the aragonitic, deep-sea coral *Lophelia pertusa*. They also report a large 10‰ range in $\delta^{11}B$ isotopic composition in different skeletal parts of the coral and were unable to account for this by variation in deep ocean pH. It has been subsequently shown by microsampling fibrous aragonite components of the deep-sea coral *Desmophyllum dianthus*, that much more restricted $\delta^{11}B$ isotopic composition ranges (<2 ‰) can be obtained from analyses of specific microstructures in individual corals (i.e. fibrous aragonite versus calcification centres; Stewart et al. 2016). Rollion-Bard and Erez (2010) have also reported a large range in $\delta^{11}B$ compositions within the calcite shells of the benthic foraminifer *Amphistegina lobifera* which had been cultured at different seawater pH. They concluded these foraminifers were able to significantly elevate their calcification site pH relative to ambient seawater (Section 2.4.4 and Figure 2.18). Collectively these studies have made significant contributions to our understanding of the extent to which biology controls the conditions of calcification in these organisms, and modifies these conditions from that of the external seawater environment.

4.10 Other Techniques

A range of other techniques can be used to determine the distribution of boron and boron isotopes at sub-micron scales in marine biogenic carbonates and to determine the co-ordination state of boron in marine biogenic carbonates.

4.10.1   Methods for Determining Boron Co-Ordination in Marine Carbonates

Understanding the co-ordination state of boron in marine carbonates is of significant interest given the basis for the B/Ca and boron isotope–seawater pH proxies has been assumed to be the selective uptake and incorporation of borate ions from seawater into precipitated biogenic carbonate. A range of techniques can be used to detect and quantify the co-ordination state of boron in marine carbonates and other materials including [11]B solid-state nuclear magnetic resonance (NMR), electron energy loss spectroscopy (EELS) and scanning transmission X-ray microscopy (STXM). However, these techniques have very different capabilities in terms of the amount and concentration of boron required for analysis. Moreover, they are specialized and non-routine, such that only the basic capabilities of these techniques are reported here, with the reader being referred to more detailed accounts in the literature where these techniques have been applied.

4.10.1.1   [11]B Solid-State Nuclear Magnetic Resonance (NMR)

Solid-state NMR is a bulk analysis technique that measures nuclear spin interactions which reflect the arrangement of proximal atomic nuclei in solid phases. The technique can be applied to marine carbonates containing only tens of $\mu g\,g^{-1}$ levels of boron but entails analysis of large amounts (100–200 mg) of powdered sample (Klochko et al. 2009; Rollion-Bard et al. 2011a). Studies to date suggest tetrahedral boron groups to be the more predominant form of boron co-ordination in biogenic aragonite, with more variable proportions of tetrahedral and trigonal co-ordination in calcite.

Sen et al. (1994) reported the presence of only BO_4 groups in a sample of *Montastrea spp.* coral. Rollion-Bard et al. (2011a) also report a high proportion (82 ± 5% 2se) of tetrahedrally co-ordinated boron in aragonite fibers of the deep-sea coral *L. pertusa*, with the balance comprising trigonal boron groups. A significantly smaller proportion (52 ± 11% 2se) of tetrahedrally co-ordinated boron was found in the centers of calcification of the same coral, and when combined these results are broadly consistent with the 64% BO_4 groups reported for *Diploria strigose* and *Porites* spp. corals by Klochko et al. (2009).

In the case of biogenic calcite, reported BO_4 group substitutions range from 20% in *Gonolithion* coralline algae (Sen et al. 1994) up to 70% in the red coralline alga *Lithothamnion glaciale* (Cusack et al. 2015). Klochko et al. (2009) report roughly equal proportions of tetrahedral (54%) and trigonal boron groups (46%) in shells of the foraminifer *Assilina ammonoides*. This NMR result is notable for being inconsistent with the *in situ* results obtained by Branson et al. (2015) for the benthic foraminifer *Amphistegina lessonii* where the maximum level of tetrahedrally co-ordinated boron was found to be <10.5% (95% confidence; see Section 4.10.1.2 below).

Theoretical modeling of NMR spectra by Balan et al. (2016) indicates boron substitutes for CO_3^{2-} ions in calcium carbonates, and that the substituted boron species depends on the $CaCO_3$ mineral polymorph. In aragonite borate $(B(OH)_4^-)$ species dominate, whereas in calcite deprotonated trigonal $BO_2(OH)^{2-}$ species dominate over borate$^-$ groups. Balan et al. (2016) note the additional incorporation of boric acid $(B(OH)_3)$ molecules and BO_3^{3-} groups in biogenic carbonate samples, and suggest the incorporation of trigonal species in calcite occurs by deprotonation of adsorbed borate species on the growing mineral surface.

4.10.1.2   Scanning Transmission X-Ray Microscopy (STXM)

STXM is capable of determining both the co-ordination and the distribution of boron and other elements in calcium carbonate samples at very fine scale (e.g. <50 nm; see Bluhm et al. (2006) for overview of technique capabilities and Branson et al. (2015) for application to B in a marine carbonate). The technique employs the absorption characteristics of X-rays at and exceeding the ionization energy of boron, which produces a near-edge X-ray absorption fine structure (NEXAFS) spectra that reflects the co-ordination environment of boron in the sample. Branson et al. (2015) analyzed boron K-edge spectra at the Advanced Light Source synchrotron by transmitting a focused soft X-ray beam through an ultra-thin (100–700 nm) section, cut from a *A. lessonii* shell by focused ion beam (FIB) milling. The characteristic 194 eV peak of trigonally co-ordinated boron was found in all parts of the sample whereas evidence for tetrahedrally co-ordinated boron was unable to be found, suggesting the complete absence of tetrahedral boron in foraminiferal calcite. Branson et al. (2015) further used the two-dimensional mapping capabilities of the technique to determine the relative distributions of boron and magnesium at better than 50 nm resolution within shell wall cross sections.

4.10.1.3   Electron Energy Loss Spectroscopy (EELS)

Electron energy loss spectroscopy (EELS) measures the energy loss spectrum of an incident beam of electrons with a narrow range of kinetic energy within a sample. Rollion-Bard et al. (2011a) employed the technique using a transmission electron microscope fitted with an electron energy loss spectrometer to measure energy loss spectra in the vicinity of the boron K-edge in ion beam thinned foils cut from the deep sea coral *L. pertusa*. The low boron concentrations in the coral required use of a high electron beam current which resulted in degraded spectral resolution. This precluded quantitative analysis but indicated the qualitative presence of BO_3 groups in both fibers and calcification centres of *L. pertusa*.

4.10.2   Methods for Two- and Three-dimensional Mapping of Boron in Marine Carbonates

Only a handful of mass spectrometry techniques are capable of two- and three-dimensional analysis of boron concentrations or boron isotope distributions at sub-micron resolution. These techniques are in general precluded from producing useful high precision boron isotope analyses of marine carbonates. This reflects fundamental counting statistic limitations that stem from a combination of low ion yield efficiencies and the small number of boron atoms (i.e. $\leq 10^6$) contained in sub-micron volumes. Nonetheless these methods have yet undemonstrated potential to map the nano-scale distribution of trace and minor amounts of boron in marine carbonates.

4.10.2.1 NanoSIMS and TOF-SIMS

NanoSIMS is a high spatial resolution surface mapping technique that, like SIMS, measures boron and boron isotopes as secondary ions generated by focusing a sub-micron incident ion beam on flat polished sample surfaces. Unfortunately, the low concentrations of boron in marine carbonates and very low boron ion yields obtained have not been demonstrated to be practical for mapping the distribution of boron at sub-micron scale. TOF-SIMS (Time-of-Flight-SIMS) shares many similarities to SIMS and NanoSIMS but employs a pulsed primary ion beam and a time-of-flight mass spectrometer to achieve mass separation that provides mass unit resolution (as compared to NanoSIMS mass resolution of $M/\Delta M \geq 6000$). TOF-SIMS affords similar two-dimensional mapping capabilities and lateral spatial resolution to NanoSIMS (~100 nm), and requires similar analysis times (typically hours) to map equivalent size areas (>10 000 μm^2). For a comparison of the capabilities of the techniques applied to boron analysis in glass samples readers are referred to Wang et al. (2016) and, to compare examples of NanoSIMS and TOF-SIMS applied to analysis of element and isotope distributions in marine calcites, are referred to recent articles by Fehrenbacher et al. (2017) and Branson et al. (2016).

4.10.2.2 Laser Resonance TOF-SNMS

Vering et al. (2006) employed resonant laser secondary neutral, time of flight mass spectrometry (laser resonance TOF-SNMS) to make *in situ* measurements of boron isotope ratios in foraminifer shells. The technique employed an electron gun and primary Ga$^+$ beam to sputter boron neutral atoms from samples, and two tunable dye lasers and a Nd-YAG laser were used to ionize extracted boron neutrals via a 3-step resonant ionization process. Results indicate a reproducibility of only 5% (2σ) that was attributed to a combination of underlying counting statistic limitations and the presence of contaminants on the analyzed sample surface.

4.10.2.3 Atom-Probe Tomography (APT)

Atom-probe tomography has the potential to measure the three-dimensional arrangement of individual atoms within tiny needle-shaped sample volumes (hundreds of nm long by tens of nm in diameter) that have been cut from samples by focused ion-beam milling. The technique is ideally suited to analysis of conductive materials and is capable of very high detection efficiencies (40–80%), however the low concentration of boron in natural marine carbonates preclude useful detection of boron concentration distributions. The insulating properties of calcium carbonates also make analysis challenging. For an overview of capabilities of the technique readers are referred to Devaraj et al. (2017), and for insight into capabilities applied to the measurement of Mg and other element distributions in foraminiferal calcite, see Branson et al. (2016).

4.11   Outlook and Future Directions

In this chapter we have reviewed the state-of-the-art, as it exists, for the analysis of boron concentration and boron isotope ratio compositions of marine carbonates. The past two decades have witnessed rapid progress in instrumental analysis capabilities with the advent of high sensitivity ICPMS and MC-ICPMS based techniques, ongoing improvements to existing TIMS based techniques, and refinement of methods for separating boron from seawater and calcium carbonate samples. It is now feasible to generate boron concentration and isotope ratio data in sufficient quantity and with sufficient accuracy and precision to reconstruct carbonate system parameters in past oceans with useful fidelity. Nonetheless, the field remains notable for the diversity of instrumental techniques used and different approaches to sample preparation, the persistent challenges posed by sample contamination and volatile boron loss, and the lack of appropriate calcium carbonate SRMs needed for data quality control across the growing global community of analysts.

For the $\delta^{11}B$-pH proxy to reach its full potential, accurate, and precise determination of the isotopic composition of the small amounts of boron hosted in marine carbonates would ideally become more routine and standardized. Automated systems that can perform inline or offline matrix removal and preconcentration of elements using ion-exchange resins are becoming more common as a front-end to ICPMS analysis and could realize significant improvements over multi-step manual procedures currently employed to prepare samples for ICPMS analysis in particular. In addition to offering greater reproducibility to existing procedures, automated sample processing would also reduce the potential for exposure to laboratory environment contamination. In a similar way, automated sample heating systems may provide greater reproducibility over boron extraction and purification by micro-sublimation, as could automated robotic systems for preparing, loading, and preheating/drying samples on filaments for TIMS analysis.

The lack of readily-available, well-characterized SRMs for boron concentration and boron isotope ratios in calcium carbonate matrixes remains an impediment to verifying the quality of data produced. This problem will become increasingly acute given the increasing number of laboratories that are reporting and applying boron isotope data for the purpose of reconstructing past seawater carbonate system parameters, especially given the dwindling reserves of standards such as JCp-1 and JCt-1

(Okai et al. 2002). Development of a series of appropriate standards for both bulk analysis and *in situ* microanalysis is critically needed, and this would constitute a significant contribution to the field of geochemical paleoceanography.

New $10^{13}\Omega$ resistors developed for both TIMS and MC-ICPMS deliver improved response times and lower noise characteristics, enabling more precise analysis of isotope ratios at low ion beam intensities (Breton et al. 2015). This has the potential to make boron isotope ratio measurements with equivalent precision on samples that are an order of magnitude smaller than is currently feasible using $10^{11}\Omega$ resistors. Even with such advances the intrinsic compositional variability within samples that has been revealed by microanalysis techniques in recent years indicates sample heterogeneity may ultimately limit the achievable accuracy and precision of both boron concentration and isotope ratio analysis.

Significant advances are being made in the area of *in situ* microanalysis and nano-scale analysis, and also with new TOF-MS systems. Due to underlying counting statistic limitations resulting from the low boron contents of marine carbonates, useful isotopic ratio variations may not be achievable at sub-micron scale resolution, but significant opportunities do exist for mapping out two- and three-dimensional variations in the distribution of boron concentrations in biominerals. Current advances occurring with laser ablation TOF-ICP-MS capabilities could make rapid boron and other element mapping in marine carbonates feasible at or near micron scale in the not-so-distant future. Laser ablation MC-ICPMS techniques are already capable of making high precision (sub per mil) isotope ratio composition measurements at the scale of tens of micrometers, and being extended to rapid two-dimensional mapping of large areas or line scan imaging over cm to dm and even larger linear length scales (e.g. Fietzke et al. 2015).

Finally, STXM methods based on synchrotron or alternate soft X-ray sources are poised to provide further clarity about how boron is co-ordinated within and incorporated into different calcium carbonate polymorphs and biominerals. Ongoing effort using these approaches is likely to contribute to an improved mechanistic understanding of boron incorporation into different marine biocarbonates, and will help to resolve the origin of observed species and taxon specific differences in boron content and $\delta^{11}B$ compositions. Ultimately this should greatly improve our fundamental understanding of how the boron proxies work and lead to more robust interpretations of proxy results.

References

Aggarwal, J.K. and Palmer, M.R. (1995). Boron isotope analysis. A review. *Analyst* *120*: 1301–1307.

Aggarwal, J., Böhm, F., Foster, G. et al. (2009a). How well do non-traditional stable isotope results compare between different laboratories: results from the interlaboratory comparison of boron isotope measurements. *Journal of Analytical Atomic Spectrometry 24*: 825–831.

Aggarwal, S.K., Wang, B.-S., You, C.-F., and Chung, C.-H. (2009b). Fractionation correction methodology for precise and accurate isotopic analysis of boron by negative thermal ionization mass spectrometry based on BO_2^- ions and using the $^{18}O/^{16}O$ ratio from ReO_4^- for internal normalization. *Analytical Chemistry 81*: 7420–7427.

Al-Ammar, A., Gupta, R.K., and Barnes, R.M. (1999). Elimination of boron memory effect in inductively coupled plasma-mass spectrometry by addition of ammonia. *Spectrochimica Acta Part B: Atomic Spectroscopy 54*: 1077–1084.

Al-Ammar, A.S., Gupta, R.K., and Barnes, R.M. (2000). Elimination of boron memory effect in inductively coupled plasma-mass spectrometry by ammonia gas injection into the spray chamber during analysis. *Spectrochimica Acta Part B: Atomic Spectroscopy 55*: 629–635.

Alibert, C. and Kinsley, L. (2008). A 170-year Sr/Ca and Ba/Ca coral record from the western Pacific warm pool: 1. What can we learn from an unusual coral record? *Journal of Geophysical Research. Oceans 113*: c4008. doi: 10.1029/2006JC003979.

Allen, K.A. and Hönisch, B. (2012). The planktic foraminiferal B/Ca proxy for seawater carbonate chemistry: a critical evaluation. *Earth and Planetary Science Letters 345–348*: 203–211.

Allen, K.A., Hönisch, B., Eggins, S.M. et al. (2011). Controls on boron incorporation in cultured tests of the planktic foraminifer *Orbulina universa*. *Earth and Planetary Science Letters 309*: 291–301.

Allison, N., Finch, A.A., Webster, J.M., and Clague, D.A. (2007). Palaeoenvironmental records from fossil corals: the effects of submarine diagenesis on temperature and climate estimates. *Geochimica et Cosmochimica Acta 71*: 4693–4703.

Anagnostou, E., Huang, K.F., You, C.F. et al. (2012). Evaluation of boron isotope ratio as a pH proxy in the deep sea coral *Desmophyllum dianthus*: evidence of physiological pH adjustment. *Earth and Planetary Science Letters 349–350*: 251–260.

Anand, P. and Elderfield, H. (2005). Variability of Mg/Ca and Sr/Ca between and within the planktonic foraminifers *Globigerina bulloides* and *Globorotalia truncatulinoides*. *Geochemistry, Geophysics, Geosystems 6*: doi: 10.1029/2004GC000811.

ATSDR, U., 2004. Toxicological profile for ammonia. US Department of Health and Human Services, Agency for Toxic Substances and Disease Registry.

Babila, T.L., Rosenthal, Y., and Conte, M.H. (2014). Evaluation of the biogeochemical controls on B/Ca of *Globigerinoides ruber* white from the oceanic flux program, Bermuda. *Earth and Planetary Science Letters 404*: 67–76.

Balan, E., Pietrucci, F., Gervais, C. et al. (2016). First-principles study of boron speciation in calcite and aragonite. *Geochimica et Cosmochimica Acta 193*: 119–131.

Barker, S., Greaves, M., and Elderfield, H. (2003). A study of cleaning procedures used for foraminiferal Mg/Ca paleothermometry. *Geochemistry, Geophysics, Geosystems 4*: doi: 10.1029/2003GC000559.

Barth, S. (1997). Boron isotopic analysis of natural fresh and saline waters by negative thermal ionization mass spectrometry. *Chemical Geology 143*: 255–261.

Bartoli, G., Hönisch, B., and Zeebe, R.E. (2011). Atmospheric CO_2 decline during the Pliocene intensification of Northern Hemisphere glaciations. *Paleoceanography 26*: doi: 10.1029/2010PA002055.

Blamart, D., Rollion-Bard, C., Meibom, A. et al. (2007). Correlation of boron isotopic composition with ultrastructure in the deep-sea coral Lophelia pertusa: implications for biomineralization and paleo-pH. *Geochemistry, Geophysics, Geosystems 8*: doi: 10.1029/2007GC001686.

Bluhm, H., Andersson, K., Araki, T. et al. (2006). Soft X-ray microscopy and spectroscopy at the molecular environmental science beamline at the advanced light source. *Journal of Electron Spectroscopy and Related Phenomena 150*: 86–1004.

Boyle, E.A. (1981). Cadmium, zinc, copper, and barium in foraminifera tests. *Earth and Planetary Science Letters 53*: 11–35.

Boyle, E.A. (1983). Manganese carbonate overgrowths on foraminifera tests. *Geochimica et Cosmochimica Acta 47*: 1815–1819.

Boyle, E.A. and Keigwin, L.D. (1985). Comparison of Atlantic and Pacific paleochemical records for the last 215,000 years: changes in deep ocean circulation and chemical inventories. *Earth and Planetary Science Letters 76*: 135–150.

Brandani, V., Del Re, G., and Di Giacomo, G. (1988). Thermodynamics of aqueous solutions of boric acid. *Journal of Solution Chemistry 17*: 429–434.

Branson, O., Kaczmarek, K., Redfern, A.T. et al. (2015). The coordination and distribution of B in foraminiferal calcite. *Earth and Planetary Science Letters 416*: 67–72.

Branson, O., Bonnin, E.A., Perea, D.E. et al. (2016). Nanometer-scale chemistry of a calcite biomineralization template: implications for skeletal composition and nucleation. *Proceedings of the National Academy of Science 113*: 12934–12939.

Breton, T., Lloyd, N.S., Trinquier, A. et al. (2015). Improving precision and signal/noise ratios for MC-ICP-MS. *Procedia Earth and Planetary Science 13*: 240–243.

Carlson, R. and Paul, J. (1968). Potentiometric determination of boron as tetrafluoroborate. *Analytical Chemistry 40*: 1292–1295.

Catanzaro, E.J., Champion, C.E., Garner, E.L. et al. (1970). *Boric Acid: Isotopic and Assay Standard Reference Materials*, National Bureau of Standards. Institute for Materials Research.

Chaussidon, M., Robert, F., Mangin, D. et al. (1997). Analytical procedures for the measurement of boron isotope compositions by ion microprobe in meteorites and mantle rocks. *Geostandards Newsletter 21*: 7–17.

Chetelat, B., Gaillardet, J., Freydier, R., and Négrel, P. (2005). Boron isotopes in precipitation: experimental constraints and field evidence from French Guiana. *Earth and Planetary Science Letters 235*: 16–30.

Conrard, P., Caude, M., and Rosset, R. (1972). Separation of close species by displacement development on ion exchangers. I. A theoretical equation of steady-state permutation front. Experimental verification in the case of the separation of boron isotopes. *Separation Science 7*: 465–486.

Couch, E.L. and Grim, R.E. (1968). Boron fixation by illites. *Clays and Clay Minerals 16*: 249–256.

Cusack, M., Kamenos, N.A., Rollion-Bard, C., and Tricot, G. (2015). Red coralline algae assessed as marine pH proxies using ^{11}B MAS NMR. *Scientific Reports 8175*. doi: 10.1038/srep08175.

Devaraj, A., Perea, D.E., Liu, J. et al. (2017). Three-dimensional nanoscale characterisation of materials by atom probe tomography. *International Materials Reviews* doi: 10.1080/09506608.2016.1270728.

Deyhle, A. and Kopf, A. (2004). Possible influence of clay contamination on B isotope geochemistry of carbonaceous samples. *Applied Geochemistry 19*: 737–745.

Diez Fernández, S., Ruiz Encinar, J., Sanz-Medel, A. et al. (2015). Determination of low B/Ca ratios in carbonates using ICP-QQQ. *Geochemistry, Geophysics, Geosystems 16*: doi: 10.1002/2015GC005817.

D'Olivo, J., McCulloch, M.T., Eggins, S., and Trotter, J. (2015). Coral records of reef-water pH across the central great barrier reef, Australia: assessing the influence of river runoff on inshore reefs. *Biogeosciences 12*: 1223.

Douglas, D. and Tanner, S. (1998). Fundamental considerations in ICPMS. In: *Inductively Coupled Plasma Mass Spectrometry* (ed. A. Montaser), 615–679. New York: Wiley-VCH.

Duchateau, N.L. and de Bièvre, P. (1983). Boron isotopic measurements by thermal ionization mass spectrometry using the negative BO_2^- ion. *International Journal of Mass Spectrometry and Ion Processes 54*: 289–297.

Dwyer, G. and Vengosh, A., 2008. Alternative filament loading solution for accurate analysis of boron isotopes by negative thermal ionization mass spectrometry, AGU Fall Meeting Abstracts.

Eggins, S.M. and Shelley, J.M.G. (2002). Compositional heterogeneity in NIST SRM 610-617 glasses. *Geostandards Newsletter 26*: 269–286.

Eggins, S.M., Woodhead, J.D., Kinsley, L.P.J. et al. (1997). A simple method for the precise determination of ≥40 trace elements in geological samples by ICPMS using enriched isotope internal standardisation. *Chemical Geology 134*: 311–326.

Eggins, S., De Deckker, P., and Marshall, J. (2003). Mg/Ca variation in planktonic foraminifera tests: implications for reconstructing palaeo-seawater temperature and habitat migration. *Earth and Planetary Science Letters 212*: 291–306.

Eggins, S.M., Sadekov, A., and De Deckker, P. (2004). Modulation and daily banding of Mg/Ca in *Orbulina universa* tests by symbiont photosynthesis and respiration: a complication for seawater thermometry? *Earth and Planetary Science Letters 225*: 411–419.

Eisenhut, S. and Heumann, K. (1997). Identification of ground water contaminations by landfills using precise boron isotope ratio measurements with negative thermal ionization mass spectrometry. *Fresenius' Journal of Analytical Chemistry 359*: 375–377.

Erickson, A. and Williams, J., 1970. A series of trace element standards in a glass matrix for the National Bureau of Standards. Corning Glass Works Report, L-1169.

Fallon, S.J., McCulloch, M.T., van Woesik, R., and Sinclair, D.J. (1999). Corals at their latitudinal limits: laser ablation trace element systematics in Porites from Shirigai Bay, Japan. *Earth and Planetary Science Letters 172*: 221–238.

Farmer, J.R., Hönisch, B., Robinson, L.F., and Hill, T.M. (2015). Effects of seawater-pH and biomineralization on the boron isotopic composition of deep-sea bamboo corals. *Geochimica et Cosmochimica Acta 155*: 86–106.

Farmer, J.R., Hönisch, B., and Uchikawa, J. (2016). Single laboratory comparison of MC-ICP-MS and N-TIMS boron isotope analyses in marine carbonates. *Chemical Geology 447*: 173–182.

Fehrenbacher, J., Russell, A.S., Davis, C.V. et al. (2017). Link between light-triggered Mg-banding and chamber formation in the planktic foraminifera *Neogloboquadrina dutertrei*. *Nature Communications* doi: 10.1038/ncomms15441.

Fietzke, J. and Frische, M. (2016). Experimental evaluation of elemental behavior during LA-ICP-MS: influences of plasma conditions and limits of plasma robustness. *Journal of Analytical Atomic Spectrometry 31*: 234–244.

Fietzke, J., Heinemann, A., Taubner, I. et al. (2010). Boron isotope ratio determination in carbonates via LA-MC-ICP-MS using soda-lime glass standards as reference material. *Journal of Analytical Atomic Spectrometry 25*: 1953–1957.

Fietzke, J., Ragazzola, F., Halfar, J. et al. (2015). Century-scale trends and seasonality in pH and temperature for shallow zones of the Bering Sea. *Proceedings of the National Academy of Sciences 112*: 2960–2965.

Fogg, T.R. and Duce, R.A. (1985). Boron in the troposphere: distribution and fluxes. *Journal of Geophysical Research. Atmospheres 90*: 3781–3796.

Foster, G.L. (2008). Seawater pH, pCO$_2$ and [CO$_3{}^{2-}$] variations in the Caribbean Sea over the last 130 kyr: a boron isotope and B/Ca study of planktic foraminifera. *Earth and Planetary Science Letters 271*: 254–266.

Foster, G.L., Ni, Y., Haley, B., and Elliott, T. (2006). Accurate and precise isotopic measurement of sub-nanogram sized samples of foraminiferal hosted boron by total evaporation NTIMS. *Chemical Geology 230*: 161–174.

Foster, G.L., Pogge von Strandmann, P.A.E., and Rae, J.W.B. (2010). Boron and magnesium isotopic composition of seawater. *Geochemistry, Geophysics, Geosystems 11*: doi: 10.1029/2010GC003201.

Foster, G.L., Hönisch, B., Paris, G. et al. (2013). Interlaboratory comparison of boron isotope analyses of boric acid, seawater and marine CaCO$_3$ by MC-ICPMS and NTIMS. *Chemical Geology 358*: 1–14.

Gabitov, R.I., Gagnon, A.C., Guan, Y. et al. (2013). Accurate Mg/Ca, Sr/Ca, and Ba/Ca ratio measurements in carbonates by SIMS and NanoSIMS and an assessment of heterogeneity in common calcium carbonate standards. *Chemical Geology 356*: 94–108.

Gaillardet, J.M. and Allègre, C.J. (1995). Boron isotopic compositions of corals: seawater or diagenesis record? *Earth and Planetary Science Letters 136*: 665–676.

Gaillardet, J., Lemarchand, D., Göpel, C., and Manhès, G. (2001). Evaporation and sublimation of boric acid: application for boron purification from organic rich solutions. *Geostandards and Geoanalytical Research 25*: 67–75.

Garbe-Schönberg, D. and Müller, S. (2014). Nano-particulate pressed powder tablets for LA-ICP-MS. *Journal of Analytical Atomic Spectrometry 29*: 990–1000.

Gonfiantini, R., Tonarini, S., Gröning, M. et al. (2003). Intercomparison of boron isotope and concentration measurements. Part II: evaluation of results. *Geostandards Newsletter 27*: 41–57.

Green, O.R. (2001). Collecting techniques for microfossil and live foraminifera. In: *A Manual of Practical Laboratory and Field Techniques in Palaeobiology*, 27–42. Springer.

Guerrot, C., Millot, R., Robert, M., and Négrel, P. (2011). Accurate and high-precision determination of boron isotopic ratios at low concentration by MC-ICP-MS (Neptune). *Geostandards and Geoanalytical Research 35*: 275–284.

Gutjahr, M., Bordier, L., Douville, E. et al. 2014. Boron Isotope Intercomparison Project (BIIP): Development of a new carbonate standard for stable isotopic analyses, EGU General Assembly Conference Abstracts, p. 5028.

Hathorne, E.C., Alard, O., James, R.H., and Rogers, N.W. (2003). Determination of intratest variability of trace elements in foraminifera by laser ablation inductively coupled plasma-mass spectrometry. *Geochemistry, Geophysics, Geosystems 4*: doi: 10.1029/2003GC000539.

Hathorne, E.C., James, R.H., Savage, P., and Alard, O. (2008). Physical and chemical characteristics of particles produced by laser ablation of biogenic calcium carbonate. *Journal of Analytical Atomic Spectrometry 23*: 240–243.

Hathorne, E.C., Gagnon, A., Felis, T. et al. (2013). Interlaboratory study for coral Sr/Ca and other element/Ca ratio measurements. *Geochemistry, Geophysics, Geosystems 14*: 3730–3750. doi: 10.1002/ggge.20230.

Haynes, J.R. (1981). *Foraminifera*. Wiley Online Library.

He, M.-y., Xiao, Y.-k., Jin, Z.-d. et al. (2013). Accurate and precise determination of boron isotopic ratios at low concentration by positive thermal ionization mass spectrometry using static multicollection of Cs$_2$BO$_2{}^+$ ions. *Analytical Chemistry 85*: 6248–6253.

Heinemann, A., Fietzke, J., Melzner, F. et al. (2012). Conditions of Mytilus edulis extracellular body fluids and shell composition in a pH-treatment experiment: acid-base status, trace elements and δ^{11}B. *Geochemistry, Geophysics, Geosystems 13*: doi: 10.1029/2011GC003790.

Hemming, N.G. and Hanson, G.N. (1992). Boron isotopic composition and concentration in modern marine carbonates. *Geochimica et Cosmochimica Acta 56*: 537–543.

Hemming, N. and Hanson, G. (1994). A procedure for the isotopic analysis of boron by negative thermal ionization mass spectrometry. *Chemical Geology 114*: 147–156.

Hendy, E.J., Gagan, M.K., Alibert, C.A. et al. (2002). Abrupt decrease in tropical Pacific Sea surface salinity at end of little ice age. *Science 295*: 1511–1514.

Henehan, M.J., Rae, J.W.B., Foster, G.L. et al. (2013). Calibration of the boron isotope proxy in the planktonic foraminifera *Globigerinoides ruber* for use in palaeo-CO_2 reconstruction. *Earth and Planetary Science Letters 364*: 111–122.

Henehan, M.J., Foster, G.L., Rae, J.W.B. et al. (2015). Evaluating the utility of B/Ca ratios in planktic foraminifera as a proxy for the carbonate system: a case study of Globigerinoides ruber. *Geochemistry, Geophysics, Geosystems 16*: 1052–1069.

Henehan, M.J., Foster, G.L., Bostock, H.C. et al. (2016). A new boron isotope-pH calibration for *Orbulina universa*, with implications for understanding and accounting for 'vital effects'. *Earth and Planetary Science Letters 454*: 282–292.

Hill, C. and Lash, R. (1980). Ion chromatographic determination of boron as tetra-fluoroborate. *Analytical Chemistry 52*: 24–27.

Hirata, T. (2000). Development of a flushing spracy chamber for inductively couples plasma-mass spectrometry. *Journal of Analytical Atomic Spectroscopy 15*: 1447–1450.

Holcomb, M., Rankenburg, K., and McCulloch, M. 2014. High-precision MC-ICP-MS measurements of δ^{11}B: matrix effects in direct injection and spray chamber sample introduction systems. In K. Grice (ed.) Principles and practice of analytical techniques in geosciences, Royal Society of Chemistry, pp. 251–270.

Holcomb, M., DeCarlo, T.M., Schoepf, V. et al. (2015). Cleaning and pre-treatment procedures for biogenic and synthetic calcium carbonate powders for determina-tion of elemental and boron isotopic compositions. *Chemical Geology 398*: 11–21.

Holland, K., Eggins, S.M., Hönisch, B. et al. (2017). Calcification rate and shell chem-istry response of the planktic foraminifer *Orbulina universa* to changes in micro-environment seawater carbonate chemistry. *Earth and Planetary Science Letters 464*: 124–134.

Hönisch, B., Bijma, J., Russell, A.D. et al. (2003). The influence of symbiont photo-synthesis on the boron isotopic composition of foraminifera shells. *Marine Micropaleontology 49*: 87–96.

Hönisch, B. and Hemming, N.G. (2004). Ground-truthing the boron isotope-paleo-pH proxy in planktonic foraminifera shells: partial dissolution and shell size effects. *Paleoceanography 19*: doi: 10.1029/2004PA001026.

Hönisch, B. and Hemming, N.G. (2005). Surface Ocean pH response to variations in pCO_2 through two full glacial cycles. *Earth and Planetary Science Letters 236*: 305–314.

Hönisch, B., Hemming, N.G., Grottoli, A.G. et al. (2004). Assessing scleractinian corals as recorders for paleo-pH: empirical calibration and vital effects. *Geochimica et Cosmochimica Acta 68*: 3675–3685.

Hönisch, B., Bickert, T., and Hemming, N.G. (2008). Modern and Pleistocene boron isotope composition of the benthic foraminifer *Cibicidoides wuellerstorfi*. *Earth and Planetary Science Letters 272*: 309–318.

Hönisch, B., Hemming, N.G., Archer, D. et al. (2009). Atmospheric carbon dioxide concentration across the mid-Pleistocene transition. *Science 324*: 1551–1554.

Houk, R.S., Fassel, V.A., Flesch, G.D. et al. (1980). Inductively coupled argon plasma as an ion source for mass spectrometric determination of trace elements. *Analytical Chemistry 52*: 2283–2289.

Iizuka, T., Eggins, S.M., McCulloch, M.T. et al. (2011). Precise and accurate determination of $^{147}Sm/^{144}Nd$ and $^{143}Nd/^{144}Nd$ in monazite using laser ablation-MC-ICPMS. *Chemical Geology 282*: 45–57.

Ishikawa, T. and Nagaishi, K. (2011). High-precision isotopic analysis of boron by positive thermal ionization mass spectrometry with sample preheating. *Journal of Analytical Atomic Spectrometry 26*: 359–365.

Ishikawa, T. and Nakamura, E. (1990). Suppression of boron volatilization from a hydrofluoric acid solution using a boron-mannitol complex. *Analytical Chemistry 62*: 2612–2616.

Jeffries, T.E., Jackson, S.E., and Longerich, H.P. (1998). Application of a frequency quintupled Nd: YAG source ($\lambda = 213$ nm) for laser ablation inductively coupled plasma mass spectrometric analysis of minerals. *Journal of Analytical Atomic Spectrometry 13*: 935–940.

Joachimski, M.M., Simon, L., Van Geldern, R., and Lécuyer, C. (2005). Boron isotope geochemistry of Paleozoic brachiopod calcite: implications for a secular change in the boron isotope geochemistry of seawater over the Phanerozoic. *Geochimica et Cosmochimica Acta 69*: 4035–4044.

Jochum, K.P., Weis, U., Stoll, B. et al. (2011). Determination of reference values for NIST SRM 610–617 glasses following ISO guidelines. *Geostandards and Geoanalytical Research 35*: 397–429.

Jochum, K.P., Scholz, D., Stoll, B. et al. (2012). Accurate trace element analysis of speleothems and biogenic calcium carbonates by LA-ICP-MS. *Chemical Geology 318*: 31–44.

Kaczmarek, K., Horn, I., Nehrke, G., and Bijma, J. (2015). Simultaneous determination of $\delta^{11}B$ and B/Ca ratio in marine biogenic carbonates at nanogram level. *Chemical Geology 392*: 32–42.

Kanno, H. (1971). Isotopic fractionation in a thermal ion source. *Bulletin of the Chemical Society of Japan 44*: 1808–1812.

Kasemann, S., Meixner, A., Rocholl, A. et al. (2001). Boron and oxygen isotope composition of certified reference materials NIST SRM 610/612 and reference materials JB-2 and JR-2. *Geostandards and Geoanalytical Research 25*: 405–416.

Kasemann, S.A., Schmidt, D.N., Bijma, J., and Foster, G.L. (2009). In situ boron isotope analysis in marine carbonates and its application for foraminifera and palaeo-pH. *Chemical Geology 260*: 138–147.

Katz, M.E., Cramer, B.S., Franzese, A. et al. (2010). Traditional and emerging proxies in foraminifera. *The Journal of Foraminiferal Research 40*: 165–192.

Kiss, E. (1988). Ion-exchange separation and spectrophotometric determination of boron in geological materials. *Analytica Chimica Acta 211*: 243–256.

Kita, N.T., Ushikubo, T., Fu, B., and Valley, J.W. (2009). High precision SIMS oxygen isotope analysis and the effect of sample topography. *Chemical Geology 264*: 43–57.

Klochko, K., Cody, G.D., Tossell, J.A. et al. (2009). Re-evaluating boron speciation in biogenic calcite and aragonite using ^{11}B MAS NMR. *Geochimica et Cosmochimica Acta 73*: 1890–1900.

Krief, S., Hendy, E.J., Fine, M. et al. (2010). Physiological and isotopic responses of scleractinian corals to ocean acidification. *Geochimica et Cosmochimica Acta 74*: 4988–5001.

Le Roux, P., Shirey, S., Benton, L. et al. (2004). In situ, multiple-multiplier, laser ablation ICP-MS measurement of boron isotopic composition (δ^{11}B) at the nanogram level. *Chemical Geology 203*: 123–138.

Lécuyer, C., Grandjean, P., Reynard, B. et al. (2002). ^{11}B/^{10}B analysis of geological materials by ICP–MS plasma 54: application to the boron fractionation between brachiopod calcite and seawater. *Chemical Geology 186*: 45–55.

Lee, K., Kim, T.-W., Byrne, R.H. et al. (2010). The universal ratio of boron to chlorinity for the North Pacific and North Atlantic oceans. *Geochimica et Cosmochimica Acta 74*: 1801–1811.

Leeman, W.P., Vocke, R.D., Beary, E.S., and Paulsen, P.J. (1991). Precise boron isotopic analysis of aqueous samples: ion exchange extraction and mass spectrometry. *Geochimica et Cosmochimica Acta 55*: 3901–3907.

Lemarchand, D., Gaillardet, J., Göpel, C., and Manhès, G. (2002). An optimized procedure for boron separation and mass spectrometry analysis for river samples. *Chemical Geology 182*: 323–334.

Liu, Y.-W., Aciego, S.M., Wanamaker, A.D. Jr., and Sell, N.K. (2013). A high-throughput system for boron microsublimation and isotope analysis by total evaporation thermal ionization mass spectrometry. *Rapid Communications in Mass Spectrometry 27*: 1705–1714.

Louvat, P., Bouchez, J., and Paris, G. (2011). MC-ICP-MS isotope measurements with direct injection nebulisation (d-DIHEN): optimisation and application to boron in seawater and carbonate samples. *Geostandards and Geoanalytical Research 35*: 75–88.

Louvat, P., Moureau, J., Paris, G. et al. (2014). A fully automated direct injection nebulizer (d-DIHEN) for MC-ICP-MS isotope analysis: application to boron isotope ratio measurements. *Journal of Analytical Atomic Spectrometry 29*: 1698–1707.

Makishima, A., Nakamura, E., and Nakano, T. (1997). Determination of boron in silicate samples by direct aspiration of sample HF solutions into ICPMS. *Analytical Chemistry 69*: 3754–3759.

Marchitto, T.M. (2006). Precise multielemental ratios in small foraminiferal samples determined by sector field ICP-MS. *Geochemistry, Geophysics, Geosystems 7*: doi: 10.1029/2005GC001018.

Marschall, H.R. and Monteleone, B.D. (2015). Boron isotope analysis of silicate glass with very low boron concentrations by secondary ion mass spectrometry. *Geostandards and Geoanalytical Research 39*: 31–46.

Martin, P.A. and Lea, D.W. (2002). A simple evaluation of cleaning procedures on fossil benthic foraminiferal Mg/Ca. *Geochemistry, Geophysics, Geosystems 3*: 1–8.

Mashiotta, T.A., Lea, D.W., and Spero, H.J. (1999). Glacial–interglacial changes in Subantarctic Sea surface temperature and δ^{18}O-water using foraminiferal Mg. *Earth and Planetary Science Letters 170*: 417–432.

McCoy, S., Robinson, L.F., Pfister, C.A. et al. (2011). Exploring B/Ca as a pH proxy in bivalves: relationships between *Mytilus californianus* B/Ca and environmental data from the Northeast Pacific. *Biogeosciences 8*: 2567–2579.

McCulloch, M.T., Holcomb, M., Rankenburg, K., and Trotter, J.A. (2014). Rapid, high-precision measurements of boron isotopic compositions in marine carbonates. *Rapid Communications in Mass Spectrometry 28*: 2704–2712.

Minnich, M.G., McLean, J.A., and Montaser, A. (2001). Spatial aerosol characteristics of a direct injection high efficiency nebulizer via optical patternation. *Spectrochimica Acta Part B: Atomic Spectroscopy 56*: 1113–1126.

Misra, S., Greaves, M., Owen, R. et al. (2014a). Determination of B/Ca of natural carbonates by HR-ICP-MS. *Geochemistry, Geophysics, Geosystems 15*: 1617–1628.

Misra, S., Owen, R., Kerr, J. et al. (2014b). Determination of $\delta^{11}B$ by HR-ICP-MS from mass limited samples: application to natural carbonates and water samples. *Geochimica et Cosmochimica Acta 140*: 531–552.

Miyata, Y., Tokieda, T., Amakawa, H. et al. (2000). Boron isotope variations in the atmosphere. *Tellus B 52*: 1057–1065.

Montagna, P., McCulloch, M., Mazzoli, C. et al. (2007). The non-tropical coral *Cladocora caespitosa* as the new climate archive for the Mediterranean: high-resolution (~weekly) trace element systematics. *Quaternary Science Reviews 26*: 441–462.

Nakamura, E., Ishikawa, T., Birck, J.-L., and Allègre, C.J. (1992). Precise boron isotopic analysis of natural rock samples using a boron-mannitol complex. *Chemical Geology: Isotope Geoscience 94*: 193–204.

Nakano, T. and Nakamura, E. (1998). Static multicollection of $Cs_2BO_2^+$ ions for precise boron isotope analysis with positive thermal ionization mass spectrometry. *International Journal of Mass Spectrometry 176*: 13–21.

Ni, Y., Foster, G.L., Bailey, T. et al. (2007). A core top assessment of proxies for the ocean carbonate system in surface-dwelling foraminifers. *Paleoceanography 22*: doi: 10.1029/2006PA001337.

Ni, Y., Foster, G.L., and Elliott, T. (2010). The accuracy of $\delta^{11}B$ measurements of foraminifers. *Chemical Geology 274*: 187–195.

Okai, T., Suzuki, A., Kawahata, H. et al. (2002). Preparation of a new geological survey of Japan geochemical reference material: coral JCp-1. *Geostandards Newsletter: The Journal of Geostandards and Geoanalysis 26*: 95–99.

Olesik, J.W. (1991). Elemental analysis using ICP-OES and ICP-MS. *Analytical Chemistry 63*: 12A–21A.

Pak, D.K., Lea, D.W., and Kennett, J.P. (2004). Seasonal and interannual variation in Santa Barbara Basin water temperatures observed in sediment trap foraminiferal Mg/Ca. *Geochemistry, Geophysics, Geosystems 5*: doi: 10.1029/2004GC000760.

Pankajavalli, R., Anthonysamy, S., Ananthasivan, K., and Rao, P.V. (2007). Vapour pressure and standard enthalpy of sublimation of H_3BO_3. *Journal of Nuclear Materials 362*: 128–131.

Park, H.K. and Haglund, R.F. Jr. (1997). Laser ablation and desorption from calcite from ultraviolet to mid-infrared wavelengths. *Applied Physics A 64*: 431–438.

Pearson, P.N. and Palmer, M.R. (2000). Atmospheric carbon dioxide over the past 60 million years. *Nature 406*: 695–699.

Pelejero, C., Calvo, E., McCulloch, M.T. et al. (2005). Preindustrial to modern interdecadal variability in coral reef pH. *Science 309*: 2204–2207.

Penman, D.E., Hönisch, B., Rasbury, E.T. et al. (2013). Boron, carbon, and oxygen isotopic composition of brachiopod shells: intra-shell variability, controls, and potential as a paleo-pH recorder. *Chemical Geology 340*: 32–39.

Rae, J.W.B., Foster, G.L., Schmidt, D.N., and Elliott, T. (2011). Boron isotopes and B/Ca in benthic foraminifera: proxies for the deep ocean carbonate system. *Earth and Planetary Science Letters 302*: 403–413.

Raitzsch, M., Hathorne, E.C., Kuhnert, H. et al. (2011). Modern and late Pleistocene B/Ca ratios of the benthic foraminifer *Planulina wuellerstorfi* determined with laser ablation ICP-MS. *Geology 39*: 1039–1042.

Raitzsch, M., Bijma, J., Benthien, A. et al. (2018). Boron isotope-based seasonal paleo-pH reconstruction for the Southeast Atlantic – a multispecies approach using habitat preference of planktonic foraminifera. *Earth and Planetary Science Letters 487*: 138–150.

Ramakumar, K., Parab, A., Khodade, P. et al. (1985). Determination of isotopic composition of boron. *Journal of Radioanalytical and Nuclear Chemistry 94*: 53–61.

Rao, R.M., Parab, A.R., Sasibhushan, K., and Aggarwal, S.K. (2008). Studies on the isotopic analysis of boron by thermal ionisation mass spectrometry using NaCl for the formation of $Na_2BO_2^+$ species. *International Journal of Mass Spectrometry 273*: 105–110.

Rao, R.M., Parab, A.R., Sasibhushan, K., and Aggarwal, S.K. (2009). A robust methodology for high precision isotopic analysis of boron by thermal ionization mass spectrometry using $Na_2BO_2^+$ ion. *International Journal of Mass Spectrometry 285*: 120–125.

Rao, R.M., Parab, A.R., and Aggarwal, S.K. (2012). The preparation and use of synthetic isotope mixtures for testing the accuracy of the PTIMS method for $^{10}B/^{11}B$ isotope ratio determination using boron mannitol complex and NaCl for the formation of $Na_2BO_2^+$. *Analytical Methods 4*: 3593–3599.

Rollion-Bard, C. and Erez, J. (2010). Intra-shell boron isotope ratios in the symbiont-bearing benthic foraminiferan *Amphistegina lobifera*: implications for $\delta^{11}B$ vital effects and paleo-pH reconstructions. *Geochimica et Cosmochimica Acta 74*: 1530–1536.

Rollion-Bard, C., Chaussidon, M., and France-Lanord, C. (2003). pH control on oxygen isotopic composition of symbiotic corals. *Earth and Planetary Science Letters 215*: 275–288.

Rollion-Bard, C., Blamart, D., Trebosc, J. et al. (2011a). Boron isotopes as pH proxy: a new look at boron speciation in deep-sea corals using ^{11}B MAS NMR and EELS. *Geochimica et Cosmochimica Acta 75*: 1003–1012.

Rollion-Bard, C., Chaussidon, M., and France-Lanord, C. (2011b). Biological control of internal pH in scleractinian corals: implications on paleo-pH and paleo-temperature reconstructions. *Comptes Rendus Geoscience 343*: 397–405.

Rose-Koga, E., Sheppard, S., Chaussidon, M., and Carignan, J. (2006). Boron isotopic composition of atmospheric precipitations and liquid–vapour fractionations. *Geochimica et Cosmochimica Acta 70*: 1603–1615.

Rosenthal, Y., Boyle, E.A., and Slowey, N. (1997). Temperature control on the incorporation of magnesium, strontium, fluorine, and cadmium into benthic foraminiferal shells from Little Bahama Bank: prospects for thermocline paleoceanography. *Geochimica et Cosmochimica Acta 61*: 3633–3643.

Rosenthal, Y., Field, M.P., and Sherrell, R.M. (1999). Precise determination of element/calcium ratios in calcareous samples using sector field inductively coupled plasma mass spectrometry. *Analytical Chemistry 71*: 3248–3253.

Rosner, M., Romer, R.L., and Meixner, A. (2005). Air handling in clean laboratory environments: the reason for anomalously high boron background levels. *Analytical and Bioanalytical Chemistry 382*: 120–124.

Sadekov, A., Eggins, S.M., De Deckker, P., and Kroon, D. (2008). Uncertainties in seawater thermometry deriving from intratest and intertest Mg/Ca variability in *Globigerinoides ruber*. *Paleoceanography 23*: doi: 10.1029/2007PA001452.

Sadekov, A.Y., Darling, K.F., Ishimura, T. et al. (2016). Geochemical imprints of genotypic variants of *Globigerina bulloides* in the Arabian Sea. *Paleoceanography 31*: 1440–1452.

Sah, R.N. and Brown, P.H. (1998). Isotope ratio determination in boron analysis. *Biological Trace Element Research 66*: 39–53.

Sanyal, A., Hemming, N., Hanson, G.N., and Broecker, W.S. (1995). Evidence for a higher pH in the glacial ocean from boron isotopes in foraminifera. *Nature 373*: 234–236.

Sanyal, A., Hemming, N., Broecker, W. et al. (1996). Oceanic pH control on the boron isotopic composition of foraminifera: evidence from culture experiments. *Paleoceanography 11*: 513–517.

Sanyal, A., Bijma, J., Spero, H.J., and Lea, D.W. (2001). Empirical relationship between pH and the boron isotopic composition of *G. sacculifer*: implications for the broon isotope paleo-pH proxy. *Paleoceanography 16*: 515–519.

Schlesinger, W.H. and Vengosh, A. (2016). Global boron cycle in the Anthropocene. *Global Biogeochemical Cycles 30*: 219–230.

Sen, S., Stebbins, J.F., Hemming, N.G., and Ghosh, B. (1994). Coordination environments of B impurities in calcite and aragonite polymorphs; a ^{11}B MAS NMR study. *American Mineralogist 79*: 819–825.

Sinclair, D.J., Kinsley, L.P., and McCulloch, M.T. (1998). High resolution analysis of trace elements in corals by laser ablation ICP-MS. *Geochimica et Cosmochimica Acta 62*: 1889–1901.

Smith, F.G., Wiederin, D.R., Houk, R.S. et al. (1991). Measurement of boron concentration and isotope ratios in biological samples by inductivey coupled plasma mass spectrometry with direct injection nebulization. *Analytica Chimica Acta 248*: 229–234.

Spivack, A.J. and Edmond, J.M. (1986). Determination of boron isotope ratios by thermal ionization mass spectrometry of the dicesium metaborate cation. *Analytical Chemistry 58*: 31–35.

Spivack, A. and Edmond, J. (1987). Boron isotope exchange between seawater and the oceanic crust. *Geochimica et Cosmochimica Acta 51*: 1033–1043.

Steinke, S., Chiu, H.-Y., Yu, P.-S. et al. (2005). Mg/Ca ratios of two *Globigerinoides ruber* (white) morphotypes: implications for reconstructing past tropical/subtropical surface water conditions. *Geochemistry, Geophysics, Geosystems 6*: doi: 10.1029/2005GC000926.

Stewart, J.A., Anagnostou, E., and Foster, G.L. (2016). An improved boron isotope pH proxy calibration for the deep-sea coral *Desmophyllum dianthus* through subsampling of fibrous aragonite. *Chemical Geology 447*: 148–160.

Stoll, H., Langer, G., Shimizu, N., and Kanamaru, K. (2012). B/Ca in coccoliths and relationship to calcification vesicle pH and dissolved inorganic carbon concentrations. *Geochimica et Cosmochimica Acta 80*: 143–157.

Swart, P.K. (1990). Calibration of the ion microprobe for the quantitative determination of strontium, iron, manganese, and magnesium in carbonate minerals. *Analytical Chemistry 62*: 722–728.

Swihart, G.H. (1996). Instrumental techniques for boron isotope analysis. *Reviews in Mineralogy and Geochemistry 33*: 845–862.

Tonarini, S., Pennisi, M., Adorni-Braccesi, A. et al. (2003). Intercomparison of boron isotope and concentration measurements. Part I: selection, preparation and homogeneity tests of the intercomparison materials. *Geostandards Newsletter 27*: 21–39.

Trotter, J., Montagna, P., McCulloch, M. et al. (2011). Quantifying the pH 'vital effect' in the temperate zooxanthellate coral *Cladocora caespitosa*: validation of the boron seawater pH proxy. *Earth and Planetary Science Letters 303*: 163–173.

Turner, I. and Montaser, A. (1998). Plasma generation in ICPMS. In: *Inductively Coupled Plasma Mass Spectrometry* (ed. A. Montaser), 265–334. New York: Wiley–VCH.

Van Hoecke, K., Devulder, V., Claeys, P. et al. (2014). Comparison of microsublimation and ion exchange chromatography for boron isolation preceding its isotopic analysis via multi-collector ICP-MS. *Journal of Analytical Atomic Spectrometry 29*: 1819–1826.

Vengosh, A., Chivas, A.R., and McCulloch, M.T. (1989). Direct determination of boron and chlorine isotopic compositions in geological materials by negative thermal-ionization mass spectrometry. *Chemical Geology: Isotope Geoscience 79*: 333–343.

Vengosh, A., Kolodny, Y., Starinsky, A. et al. (1991). Coprecipitation and isotopic fractionation of boron in modern biogenic carbonates. *Geochimica et Cosmochimica Acta 55*: 2901–2910.

Vering, G., Crone, C., Kathers, P. et al. (2006). Resonant laser-SNMS of boron for analysis of paleoceanographic samples. *Applied Surface Science 252*: 7163–7166.

Vogl, J. and Rosner, M. (2011). Production and certification of a unique set of isotope and delta reference materials for boron isotope determination in geochemical, environmental and industrial materials. *Geostandards and Geoanalytical Research 36*: 161–175.

Wang, B.-S., You, C.-F., Huang, K.-F. et al. (2010). Direct separation of boron from Na- and Ca-rich matrices by sublimation for stable isotope measurement by MC-ICP-MS. *Talanta 82*: 1378–1384.

Wang, Z., Liu, J., Zhou, Y. et al. (2016). Nanoscale imaging of Li and B in nucler waste glass, a comparison of TOF-SIMS, NanoSIMS, and APT. *Surface and Interface Analysis 48*: 1392–1401.

Wei, H., Xiao, Y., Sun, A. et al. (2004). Effective elimination of isobaric ions interference and precise thermal ionization mass spectrometer analysis for boron isotope. *International Journal of Mass Spectrometry 235*: 187–195.

Wei, G., Deng, W., Yu, K. et al. (2007). Sea surface temperature records in the northern South China Sea from mid-Holocene coral Sr/Ca ratios. *Paleoceanography 22*: doi: 10.1029/2006PA001270.

Wei, G., McCulloch, M.T., Mortimer, G. et al. (2009). Evidence for ocean acidification in the Great Barrier Reef of Australia. *Geochimica et Cosmochimica Acta 73*: 2332–2346.

Westphal, C.S., Kahen, K., Rutkowski, W.F. et al. (2004). Demountable direct injection high efficiency nebulizer for inductively coupled plasma mass spectrometry. *Spectrochimica Acta Part B: Atomic Spectroscopy 59*: 353–368.

Wu, H.-P., Jiang, S.-Y., Wei, H.-Z., and Yan, X. (2012). An experimental study of organic matters that cause isobaric ions interference for boron isotopic measurement by thermal ionization mass spectrometry. *International Journal of Mass Spectrometry 328*: 67–77.

Xiao, Y.-K., Beary, E., and Fassett, J. (1988). An improved method for the high-precision isotopic measurement of boron by thermal ionization mass spectrometry. *International Journal of Mass Spectrometry and Ion Processes 85*: 203–213.

Xiao, Y.-K., Jin, L., and Qi, H.-P. (1991). Investigation of thermal ion emission characteristics of graphite. *International Journal of Mass Spectrometry and Ion Processes 107*: 205–213.

Xiao, Y., Sun, D., Wang, Y. et al. (1992). Boron isotopic compositions of brine, sediments, and source water in Da Qaidam Lake, Qinghai, China. *Geochimica et Cosmochimica Acta 56*: 1561–1568.

Xiao, J., Xiao, Y., Jin, Z. et al. (2013). Boron isotopic compositions in growing corals from the South China Sea. *Journal of Asian Earth Sciences 62*: 561–567.

Yamamoto, T., Yamaoka, T., Fojita, T., and Isoda, C. (1973). Boron content in marine plankton. *Records Oceanography Works in Japan 12*: 13–21.

Yu, J. and Elderfield, H. (2007). Benthic foraminiferal B/Ca ratios reflect deep water carbonate saturation state. *Earth and Planetary Science Letters 258*: 73–86.

Yu, J., Day, J., Greaves, M., and Elderfield, H. (2005). Determination of multiple element/calcium ratios in foraminiferal calcite by quadrupole ICP-MS. *Geochemistry, Geophysics, Geosystems 6*: doi: 10.1029/2005GC000964.

Yu, J., Elderfield, H., Greaves, M., and Day, J. (2007). Preferential dissolution of benthic foraminiferal calcite during laboratory reductive cleaning. *Geochemistry, Geophysics, Geosystems 8* (6): doi: 10.1029/2006GC001571.

Yu, J., Foster, G.L., Elderfield, H. et al. (2010). An evaluation of benthic foraminiferal B/Ca and δ^{11}B for deep ocean carbonate ion and pH reconstructions. *Earth and Planetary Science Letters 293*: 114–120.

Yu, J., Thornalley, D.J.R., Rae, J.W.B., and McCave, N.I. (2013). Calibration and application of B/Ca, Cd/Ca, and δ^{11}B in *Neogloboquadrina pachyderma* (sinistral) to constrain CO_2 uptake in the subpolar North Atlantic during the last deglaciation. *Paleoceanography 28*: 237–252.

Zeebe, R.E. (2005). Stable boron isotope fractionation between dissolved $B(OH)_3$ and $B(OH)_4^-$. *Geochimica et Cosmochimica Acta 69*: 2753–2766.

Zeininger, H. and Heumann, K.G. (1983). Boron isotope ratio measurement by negative thermal ionization mass spectrometry. *International Journal of Mass Spectrometry and Ion Physics 48*: 377–380.

Zhang, S., Henehan, M.J., Hull, P.M. et al. (2017). Investigating controls on boron isotope ratios in shallow marine carbonates. *Earth and Planetary Science Letters 458*: 380–393.

Index

Please note: Page numbers in italics refer to figures, page numbers in bold refer to tables.

Boron Proxies in Paleoceanography and Paleoclimatology, First Edition. Bärbel Hönisch, Stephen M. Eggins, Laura L. Haynes, Katherine A. Allen, Katherine D. Holland, and Katja Lorbacher. © 2019 John Wiley & Sons Ltd. Published 2019 by John Wiley & Sons Ltd. Companion website: www.wiley.com/go/Hönisch/Boron_Paleoceanography